Combinatorial Geometry and Its Algorithmic Applications

The Alcalá Lectures

Mathematical
Surveys
and
Monographs

Volume 152

Combinatorial Geometry and Its Algorithmic Applications

The Alcalá Lectures

János Pach
Micha Sharir

American Mathematical Society
Providence, Rhode Island

EDITORIAL COMMITTEE

Jerry L. Bona Michael G. Eastwood
Ralph L. Cohen J. T. Stafford, Chair

Benjamin Sudakov

2000 Mathematics Subject Classification. Primary 05C35, 05C62, 52C10, 52C30, 52C35, 52C45, 68Q25, 68R05, 68W05, 68W20.

For additional information and updates on this book, visit
www.ams.org/bookpages/surv-152

Library of Congress Cataloging-in-Publication Data
Pach, Janos.
  Combinatorial geometry and its algorithmic applications : The Alcala lectures / Janos Pach, Micha Sharir.
    p. cm. — (Mathematical surveys and monographs ; v. 152)
  Includes bibliographical references and index.
  ISBN 978-0-8218-4691-9 (alk. paper)
  1. Combinatorial geometry. 2. Algorithms. I. Sharir, Micha. II. Title.

QA167.p332 2009
516′.13–dc22
                                                                        2008038876

Copying and reprinting. Individual readers of this publication, and nonprofit libraries acting for them, are permitted to make fair use of the material, such as to copy a chapter for use in teaching or research. Permission is granted to quote brief passages from this publication in reviews, provided the customary acknowledgment of the source is given.

Republication, systematic copying, or multiple reproduction of any material in this publication is permitted only under license from the American Mathematical Society. Requests for such permission should be addressed to the Acquisitions Department, American Mathematical Society, 201 Charles Street, Providence, Rhode Island 02904-2294, USA. Requests can also be made by e-mail to reprint-permission@ams.org.

© 2009 by the American Mathematical Society. All rights reserved.
The American Mathematical Society retains all rights
except those granted to the United States Government.
Printed in the United States of America.

∞ The paper used in this book is acid-free and falls within the guidelines
established to ensure permanence and durability.
Visit the AMS home page at http://www.ams.org/

10 9 8 7 6 5 4 3 2 1   14 13 12 11 10 09

Contents

Preface and Apology	vii

Chapter 1. Sylvester–Gallai Problem:
The Beginnings of Combinatorial Geometry — 1
1. James Joseph Sylvester and the Beginnings — 1
2. Connecting Lines and Directions — 3
3. Directions in Space vs. Points in the Plane — 6
4. Proof of the Generalized Ungar Theorem — 7
5. Colored Versions of the Sylvester–Gallai Theorem — 10

Chapter 2. Arrangements of Surfaces:
Evolution of the Basic Theory — 13
1. Introduction — 13
2. Preliminaries — 16
3. Lower Envelopes — 20
4. Single Cells — 27
5. Zones — 29
6. Levels — 32
7. Many Cells and Related Problems — 37
8. Generalized Voronoi Diagrams — 40
9. Union of Geometric Objects — 42
10. Decomposition of Arrangements — 49
11. Representation of Arrangements — 54
12. Computing Arrangements — 56
13. Computing Substructures in Arrangements — 58
14. Applications — 63
15. Conclusions — 70

Chapter 3. Davenport–Schinzel Sequences:
The Inverse Ackermann Function in Geometry — 73
1. Introduction — 73
2. Davenport–Schinzel Sequences and Lower Envelopes — 74
3. Simple Bounds and Variants — 79
4. Sharp Upper Bounds on $\lambda_s(n)$ — 81
5. Lower Bounds on $\lambda_s(n)$ — 86
6. Davenport–Schinzel Sequences and Arrangements — 89

Chapter 4. Incidences and Their Relatives:
From Szemerédi and Trotter to Cutting Lenses — 99
1. Introduction — 99
2. Lower Bounds — 102

3.	Upper Bounds for Incidences via the Partition Technique	104
4.	Incidences via Crossing Numbers—Székely's Method	106
5.	Improvements by Cutting into Pseudo-segments	109
6.	Incidences in Higher Dimensions	112
7.	Applications	114

Chapter 5. Crossing Numbers of Graphs:
           Graph Drawing and its Applications   119
1. Crossings—the Brick Factory Problem   119
2. Thrackles—Conway's Conjecture   120
3. Different Crossing Numbers?   122
4. Straight-Line Drawings   125
5. Angular Resolution and Slopes   126
6. An Application in Computer Graphics   127
7. An Unorthodox Proof of the Crossing Lemma   129

Chapter 6. Extremal Combinatorics:
           Repeated Patterns and Pattern Recognition 133
1. Models and Problems   133
2. A Simple Sample Problem: Equivalence under Translation 135
3. Equivalence under Congruence in the Plane   137
4. Equivalence under Congruence in Higher Dimensions 139
5. Equivalence under Similarity   141
6. Equivalence under Homothety or Affine Transformations 143
7. Other Equivalence Relations for Triangles in the Plane 144

Chapter 7. Lines in Space:
           From Ray Shooting to Geometric Transversals 147
1. Introduction   147
2. Geometric Preliminaries   149
3. The Orientation of a Line Relative to n Given Lines 152
4. Cycles and Depth Order   158
5. Ray Shooting and Other Visibility Problems   163
6. Transversal Theory   167
7. Open Problems   170

Chapter 8. Geometric Coloring Problems:
           Sphere Packings and Frequency Allocation 173
1. Multiple Packings and Coverings   173
2. Cover-Decomposable Families and Hypergraph Colorings 175
3. Frequency Allocation and Conflict-Free Coloring   178
4. Online Conflict-Free Coloring   181

Chapter 9. From Sam Loyd to László Fejes Tóth:
           The 15 Puzzle and Motion Planning   183
1. Sam Loyd and the Fifteen Puzzle   183
2. Unlabeled Coins in Graphs and Grids   185
3. László Fejes Tóth and Sliding Coins   187
4. Pushing Squares Around   194

Bibliography   197

Index   227

Preface and Apology

These lecture notes are a compilation of surveys of the topics that were presented in a series of talks at Alcalá, Spain, August 31 – September 5, 2006, by János Pach and Micha Sharir. To a large extent, these surveys are adapted from earlier papers written by the speakers and their collaborators. In their present form, the notes aptly describe both the history and the state of the art of these topics.

The notes are arranged in an order that roughly parallels the order of the talks. Chapter 1 describes the beginnings of combinatorial geometry: Starting with Sylvester's problem on the existence of "ordinary lines," we introduce a number of exciting problems on incidences between points and lines in the plane and in space. This chapter uses the material in Pach, Pinchasi, and Sharir [**588, 589**]. In Chapter 2 we survey many aspects of the theory of *arrangements of surfaces* in higher dimensions. It is adapted from Agarwal and Sharir [**51**]. Readers that have some familiarity with the basic theory of arrangements can start their reading on this topic from this chapter, while those that are complete novices may find it useful to look first at Chapter 3, which studies arrangements of *curves* in the plane, with special emphasis on *Davenport–Schinzel sequences* and the major role they play in the theory of arrangements. This chapter is adapted from Agarwal and Sharir [**52**].

Chapter 4 covers the topic of incidences between points and curves and its many relatives, where a surge of activity has taken place in the past decade. It is adapted from two similar surveys by Pach and Sharir [**600, 601**]. The study of combinatorial and topological properties of planar arrangements of curves has become a separate discipline in discrete and computational geometry, under the name of *Graph Drawing*. Some basic aspects of this emerging discipline are discussed in Chapter 5, which is based on the survey by Pach [**583**]. Some classical questions of Erdős on repeated interpoint distances in a finite point set can be reformulated as problems on the maximum number of incidences between points and circles, spheres, etc. In fact, these questions motivated and strongly influenced the early development of the theory of incidences a quarter of a century ago and they led to the discovery of powerful new combinatorial and topological tools. Many of Erdős's questions can be naturally generalized to problems on larger repeated subpatterns in finite point sets. Based on Brass and Pach [**175**], in Chapter 6 we outline some of the most challenging open problems of this kind, whose solution would have interesting consequences in *pattern matching* and recognition.

Chapter 7 treats the special topic of lines in three-dimensional space, which is a nice application (or showpiece, if you will) of the general theory of arrangements on one hand, and shows up in a variety of only loosely related topics, ranging from *ray shooting* and *hidden surface removal* in computer graphics to geometric *transversal theory*. This chapter partially builds upon a somewhat old paper by Chazelle, Edelsbrunner, Guibas, Sharir, and Stolfi [**220**], but its second half is new, and presents (some of) the recent developments. Some combinatorial properties of arrangements of spheres, boxes, etc., are discussed in Chapter 8. They raise difficult questions on the *chromatic numbers* and other similar parameters of certain geometrically defined graphs and hypergraphs, with possible applications to *frequency allocation* in cellular telephone networks. Here we borrowed some material from Pach, Tardos, and Tóth [**608**].

An old and rich area of applications of Davenport–Schinzel sequences and the theory of geometric arrangements is *motion planning*. Starting with Sam Loyd's coin puzzles, in Chapter 9 we discuss a number of problems that can be regarded as discrete variants of the "piano movers' problem" on graphs and grids. Some of the results have applications to the reconfiguration of metamorphic *robotic systems*. This chapter is based on recent joint papers with Bereg, Călinescu, and Dumitrescu [**138, 190, 280**].

While we have made our best attempts to make these notes comprehensive, they are not at the level of a polished monograph, nor do they provide full coverage of all the relevant recent results and developments. It is our hope that they be a useful source of reference to the rich and extensive theory presented at the Alcalá series of talks.

Apart from those friends and coauthors whose names were mentioned above, we freely borrowed from joint work with D. Pálvölgyi and R. Wenger. We are extremely grateful to all of them for the enjoyable and fruitful collaboration and for their kind permission to reproduce their ideas and "plagiarize" their words. Our thanks are also due to Sergei Gelfand, Gabriel Nivasch, and Deniz Sarioz for their invaluable help in finalizing the manuscript.

János Pach (New York and Budapest)
Micha Sharir (Tel Aviv and New York)

CHAPTER 1

Sylvester–Gallai Problem: The Beginnings of Combinatorial Geometry

1. James Joseph Sylvester and the Beginnings

James Joseph Sylvester (1814–1897) was one of the most colorful figures of nineteenth century British mathematics. He started his studies at the University of London at the age of 14, where he was a student of the logician, Augustus de Morgan. In spite of his brilliant achievements in Cambridge, he was not granted a degree there until 1872, because as a Jew, he declined to take the Thirty-Nine Articles of the Church of England. In 1838, however, he became Professor of Natural Philosophy at University College London, but he could not obtain a teaching position until 1855. In 1841, he was awarded a BA and an MA by Trinity College Dublin. While working in London as an actuary, together with his lifelong friend, Arthur Cayley, he made important contributions to matrix theory and invariant theory. In 1877, Sylvester became the inaugural professor of mathematics at the new Johns Hopkins University, and one year later he founded the American Journal of Mathematics. He did pioneering work in combinatorics, in number theory, and in the theory of partitions.

"The early study of Euclid made me a hater of geometry," said Sylvester [**521**]. It is somewhat ironic that most likely the hatred would not be mutual! The following innocent looking question of Sylvester was first proposed as a problem in the *Educational Times* [**708**]. Euclid would have probably loved this question, because to formulate it we need only three notions: points, lines, and incidences, the three basic elements of his geometry! *Is it true that any finite set of points in the Euclidean plane, not all on a line, has two elements whose connecting line does not pass through a third?* Such a connecting line is called an *ordinary* line.

In the same year, the journal published an incorrect solution by Woodall and another argument, which was characterized as "equally incomplete, but may be worth notice." In the early 1930s, the question was rediscovered by Erdős, and shortly thereafter, an affirmative answer was given by T. Grünwald (alias Gallai). In 1943, Erdős [**325**] posed the problem in the *American Mathematical Monthly*. In the following year, it was solved by Steinberg [**337**] and others. However, the oldest published proof is due to Melchior [**547**], who established the dual statement, as a corollary to a more general inequality: any finite family of lines in the plane, not all of which pass through the same point, determines a *simple* intersection point, i.e., a point that belongs to precisely two lines. Both the primal and dual forms of the result have become known as the *Sylvester–Gallai theorem*. Many alternative proofs and generalizations have been found by de Bruijn and Erdős [**184**], Coxeter [**253**], Motzkin [**557**], Lang [**505**], Williams [**749**], Lin [**515**], Edelstein *et al.* [**312**, **311**], Borwein [**164**, **165**], Giering [**378**], Herzog and Kelly [**441**], Kupitz [**502**], Watson

[**741**], and Lenchner [**510**]. A survey of these results was given by P. Borwein and W. Moser [**166**]. The history of this problem is somewhat complicated; much of the confusion can be explained by interruptions to academic life caused by World War II. In fact, it is not even clear whether originally Sylvester had any solution in mind.

THEOREM 1.1 (Sylvester–Gallai theorem). *Any finite noncollinear set of points in the plane has two elements whose connecting line does not pass through a third.*

The following "book proof" – using Erdős's terminology – was found by Kelly. Take a set P of n points in the plane, not all collinear. Consider all pairs (ℓ, p), where ℓ is a line passing through at least two elements of P, and $p \in P$ does not lie on ℓ. Among all these pairs, pick one, for which the distance between p and ℓ is minimal. Denoting such a pair by (ℓ, p), with a slight abuse of notation, we claim that ℓ cannot pass through more than two points. Assume for contradiction that it does. Then at least two points $q, r \in P \cap \ell$ lie on the same side of the foot p' of the perpendicular from p to ℓ. Assume without loss of generality that r is closer to p' than q is. Clearly, the distance between $r \in P$ and the line $\ell' := qp$ is smaller than the distance between p and ℓ, contradicting the minimality of (ℓ, p). □

The next natural task is to find the minimum number $ol(n)$ of ordinary lines (passing through precisely two points) determined by n noncollinear points in the plane.

CONJECTURE 1.2 (Dirac [**268**], Motzkin). *For every $n \neq 7, 13$, the number of ordinary lines determined by n noncollinear points in the plane is at least $\lceil \frac{1}{2}n \rceil$.*

Kelly and W. Moser [**474**] proved that $ol(n) \geq \frac{3}{7}n$, with equality for $n = 7$. The dual statement also holds for pseudo-line arrangements [**475**]. The best known lower bound, $ol(n) \geq \frac{6}{13}n$, was found by Csima and Sawyer [**255, 256**]. Equality is attained for the configuration of 13 points depicted in Figure 1. The following exact values were determined by Crowe and McKee [**254**], and by Brakke [**169**]:

n	3	4	5	6	7	8	9	10	11	12	13	14	15	16	17	18	19	21	22
$ol(n)$	3	3	4	3	3	4	6	5	6	6	6	7	?	8	?	9	?–?		11

Observe that the function $ol(n)$ is not monotonically increasing. There are two configurations with 7 and 13 points with exceptionally few ordinary lines. Apart from these values, the general upper bound is $ol(n) \leq \frac{n}{2}$ if n is even and $ol(n) \leq 3\lfloor \frac{n}{4} \rfloor$ if n is odd.

For even n, the construction consists of a regular $\frac{n}{2}$-gon, which determines $\frac{n}{2}$ directions, and the $\frac{n}{2}$ projective points corresponding to these directions. If $n \equiv 1 \pmod 4$, the best known example can be obtained by adding the center of the polygon to the construction for $n - 1$. If $n \equiv 3 \pmod 4$, one of the "points at infinity" has to be deleted from the construction for $n+1$. The substantial difference between the cases n even and n odd is quite unusual. It is especially strange in view of the fact that there are several other configurations of both parities that determine very few different directions and may serve as bases for constructions of point sets with few ordinary lines.

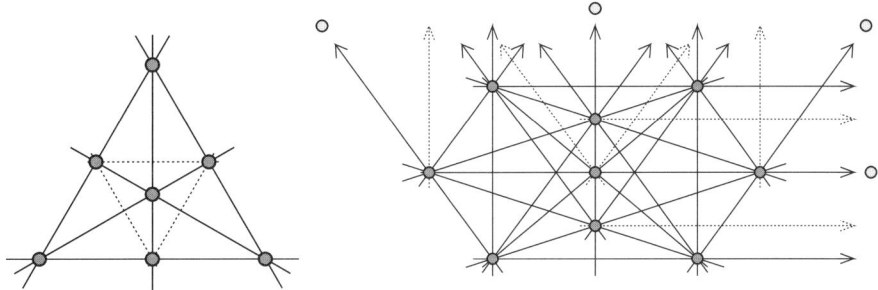

FIGURE 1. Exceptional sets with 7 and 13 points and few ordinary lines.

PROBLEM 1.3. *For odd values of n, find sets of n points in the plane that determine only $\frac{1}{2}n + O(1)$ ordinary lines.*

A ghost reference related to the Dirac–Motzkin conjecture is the dissertation of S. Hansen from 1981. It claims to prove Conjecture 1.2, but one of its key lemmas turned out to be false [**255**].

2. Connecting Lines and Directions

Erdős pointed out the following immediate consequence of the Gallai–Sylvester theorem on ordinary lines, i.e., lines passing through precisely two elements of the set.

THEOREM 2.1 (Erdős theorem). *Any set of n noncollinear points in the plane determines at least n different connecting lines. Equality is attained if and only if all but one of the points are collinear.*

PROOF. To establish the first statement, it is enough to notice that if we delete one of the points that lie on an ordinary line, then the number of connecting lines induced by the remaining point set will decrease by at least one. Hence, the result follows by induction, unless the remaining point set is collinear. However, in the latter case, our original set determines precisely n connecting lines.

The easy proof of the second statement is left to the reader. □

Erdős's theorem was the starting point of many important investigations that led to the birth of the theory of *finite projective planes* and, more generally, to the theory of *block designs* [**738**].

Here we recall a far-reaching combinatorial generalization of the theorem, together with its "book proof," due to Motzkin and Conway.

THEOREM 2.2. *Let P be a set of $n \geq 3$ elements and let $\ell_1, \ldots, \ell_m$ be proper subsets of P such that any pair of elements of P is contained in precisely one set ℓ_i. Then we have $m \geq n$.*

PROOF. Suppose $m < n$. For any $p \in P$, let $d(p)$ denote the number of ℓ_i's containing p.

If $p \notin \ell_i$, then $d(p) \geq |\ell_i|$, which, in turn, implies that $m|\ell_i| < nd(p)$. From here, we get
$$m(n - |\ell_i|) > n(m - d(p)).$$

Thus, we have

$$1 = \sum_{p \in P} \sum_{\ell_i \not\ni p} \frac{1}{n(m-d(p))} > \sum_{\ell_i} \sum_{p \notin \ell_i} \frac{1}{m(n-|\ell_i|)} = 1.$$

This contradiction completes the proof of the theorem. □

In the same spirit, Scott [**665**] posed two similar questions in 1970: Is it true that the minimum number of different directions assumed by the connecting lines of

(1) $n \geq 4$ noncollinear points in the plane is $2\lfloor n/2 \rfloor$?

(2) $n \geq 6$ noncoplanar points in 3-space is $2n-3$ if n is even and $2n-2$ if n is odd?

Twelve years later, the first question was answered in the affirmative by Ungar.

THEOREM 2.3 (Ungar's theorem [**728**]). *Any set of $n \geq 4$ points in the plane, not all on a line, determine at least $2\lfloor n/2 \rfloor$ different directions.*

Ungar's proof is a real gem, a brilliant application of the method of *allowable sequences* invented by Goodman and Pollack [**382, 385**]. Moreover, it solves the problem in an elegant combinatorial setting, for "pseudolines", as was suggested independently by Goodman and Pollack and by Cordovil [**252**]. For even n, Ungar's theorem generalizes Erdős's above mentioned result. However, in contrast to Erdős's result, here there is an overwhelming diversity of extremal configurations, for which equality is attained. Four infinite families and more than one hundred sporadic configurations were catalogued by Jamison and Hill [**453**] (see also [**452**] for an excellent survey).

Progress on the second question of Scott has been much slower. As Jamison [**452**] noticed, unless we impose some further restriction on the point set, for odd n, the number of directions determined by n points in 3-space can be as small as $2n-5$. Indeed, equality is attained, e.g., for the n-element set obtained from the vertex set of a regular $(n-3)$-gon P_{n-3} (or from any other centrally symmetric extremal configuration for the planar problem) by adding its center c and two other points whose midpoint is c and whose connecting line is orthogonal to the plane of P_{n-3}; see Figure 2. Blokhuis and Seress [**154**] introduced a natural condition excluding the above configurations: they assumed that no three points are collinear. Under this assumption, they proved that every noncoplanar set of n points in 3-space determines at least $1.75n - 2$ different directions.

Following Pach *et al.* [**588, 589**], one can answer Scott's second question in the affirmative, using the same assumption as Blokhuis and Seress, and in almost full generality.

THEOREM 2.4. *Every set of $n \geq 6$ points in $\mathbb{R}^3$, not all of which are on a plane and no three are on a line, determine at least $n + 2\lceil n/2 \rceil - 3$ different directions. This bound is sharp.*

THEOREM 2.5. *Every set of $n \geq 6$ points in $\mathbb{R}^3$, not all of which are on a plane, determine at least $2n-5$ different directions if n is odd, and at least $2n-7$ different directions if n is even. This bound is sharp for every odd $n \geq 7$.*

In the next two sections, we outline a proof of Theorem 2.4.

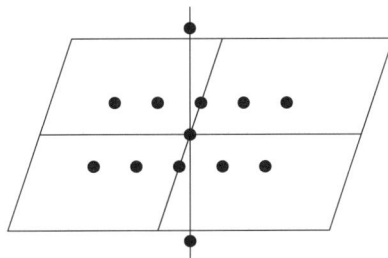

FIGURE 2. n noncoplanar points in space with minimum number of directions.

According to a beautiful result of Motzkin [**558**], Rabin, and Chakerian [**198**] (see also [**60**]), any set of n noncollinear points in the plane, colored with two colors *red* and *green*, determines a monochromatic line. Motzkin and Grünbaum [**395**] initiated the investigation of *biased* colorings, i.e., colorings without monochromatic red lines. Their motivation was to justify the intuitive feeling that if there are many red points in such a coloring and not all of them are collinear, then the number of green points must also be rather large. Denoting the sets of red and green points by R and G, respectively, it is a challenging unsolved question to decide whether the "surplus" $|R| - |G|$ of the coloring can be arbitrarily large. We do not know any example where this quantity exceeds 6 [**396**].

The problem of biased colorings was rediscovered by Erdős and Purdy [**335**], who formulated it as follows:

PROBLEM 2.6 (Grünbaum, Erdős–Purdy). *Let $m(n)$ denote the smallest number m such that for every set P of n noncollinear points in the plane, the smallest set disjoint from P that meets all lines spanned by P has at least m elements. Is it true that $n - m(n) \to \infty$?*

An $\Omega(n)$ lower bound on $m(n)$ follows from the "weak Dirac conjecture" proved by Szemerédi and Trotter [**712**] and Beck [**136**], according to which there is a point that lies on $\Omega(n)$ different connecting lines. Each of these connecting lines has to be represented by a different point.

CONJECTURE 2.7 (Dirac's conjecture [**268**]). *There exists a constant c such that any set of n points in the plane, not all on a line, has an element incident to at least $\frac{n}{2} - c$ connecting lines.*

It was believed that, apart from some small examples listed in [**394**], this statement is true with $c = 0$, until Felsner exhibited an infinite series of configurations, showing that $c \geq 3/2$.

In Section 3, we reduce Theorem 2.4 to a statement (Theorem 3.2) showing that under some further restrictions the surplus is indeed bounded. More precisely, if there is no connecting line whose leftmost and rightmost points are both red, then we have $|G| \geq 2\lfloor |R|/2 \rfloor$, so in particular $|R| - |G| \leq 1$.

Another way of rephrasing Ungar's theorem is that from all closed segments whose endpoints belong to a noncollinear set of n points in the plane, one can always select at least $2\lfloor n/2 \rfloor$ such that no two of them are parallel. Unless we explicitly state it otherwise, every *segment* used in this chapter is assumed to be *closed*. Our proof of Theorem 2.4 is based on a far-reaching generalization of Ungar's result. To formulate this statement, we need to relax the condition of two segments being *parallel*.

DEFINITION 2.8. *Two segments belonging to distinct lines are called* avoiding *if one of the following two conditions is satisfied:*
  (i) *they are parallel, or*
  (ii) *the intersection of their supporting lines does not belong to any of the segments.*

An alternative definition is that two segments are avoiding if and only if they are disjoint and their convex hull is a quadrilateral.

The following strengthening of Ungar's theorem, which is of independent interest, implies Theorem 2.4 (and Theorem 3.2 stated in the next section).

THEOREM 2.9 (Generalized Ungar theorem). *From all closed segments determined by a set of n noncollinear points in the plane, one can always select at least $2\lfloor n/2 \rfloor$ pairwise nonavoiding ones, lying on distinct lines.*

Theorem 2.9 is established in Section 4.

3. Directions in Space vs. Points in the Plane

In this section, we reduce Theorem 2.4 to Theorem 2.9.

Let P be a set of n points in $\mathbb{R}^3$ such that not all of them lie in a common plane and no three of them are collinear. Let p_0 be an *extreme* point of P, i.e., a vertex of the convex hull of P. Consider a supporting plane to P at p_0, and translate it to the side that contains P. Let π denote the resulting plane. Project from p_0 all points of $P \setminus \{p_0\}$ onto π. We obtain a set R of $n-1$ *distinct* points in π, not all on a line, and we will refer to the elements of R as *red* points. Each red point corresponds to a direction determined by p_0 and some other point of P.

For each pair of elements $p, p' \in P \setminus \{p_0\}$, take a line parallel to pp' that passes through p_0. Color with *green* the intersection point of this line with π, unless it has already been colored red. The set of all green points is denoted by G. By definition, we have $R \cap G = \emptyset$.

We need the following simple property of the sets R and G, which implies that along every line passing through at least two red points either the leftmost or the rightmost point belonging to $R \cup G$ is green.

LEMMA 3.1. *Every line connecting two red points $r, r' \in R$ passes through at least one green point $g \in G$ that does not belong to the (closed) segment rr'.*

PROOF. Let ℓ be a line in π passing through at least two red points $r, r' \in R$. Assume without loss of generality that r and r' are the leftmost and rightmost red points along ℓ. Let p and p' denote those elements of P whose projections to π are r and r', respectively. Observe that in the plane induced by p_0 and ℓ, the direction of pp' does not belong to the convex cone enclosed by the rays $p_0 p$ and $p_0 p'$, so the line through p_0 parallel to pp' will cross ℓ in a green point g meeting the requirements. □

To establish Theorem 2.4, it is sufficient to verify the following result.

THEOREM 3.2. *Let R be a set of n red points in the plane, not all collinear, and let G be a set of m green points such that $R \cap G = \emptyset$ and every line ℓ connecting at least two red points in R passes through a green point $g \in G$ that does not belong to any segment rr', for $r, r' \in R \cap \ell$.*
Then we have $m \geq 2\lfloor n/2 \rfloor$.

Indeed, to prove Theorem 2.4, it is enough to notice that in our setting we have $|R| = n - 1$ and that the number of different directions determined by P is equal to
$$|R| + |G| \geq n - 1 + 2 \left\lfloor \frac{n-1}{2} \right\rfloor = n + 2 \left\lceil \frac{n}{2} \right\rceil - 3.$$
Thus, applying Theorem 3.2, Theorem 2.4 immediately follows.

It is interesting to note that Theorem 3.2 also implies Ungar's above-mentioned theorem. To see this, regard the elements of our given planar point set as *red*, and the directions determined by them as green points on the line at infinity, and apply Theorem 3.2. (If we wish, we can perform a projective transformation and avoid the use of points at infinity.)

It remains to prove Theorem 3.2. However, as mentioned before, this result can be easily deduced from Theorem 2.9, which is a further extension of Ungar's theorem.

PROOF. (Proof of Theorem 3.2 using Theorem 2.9.) Applying Theorem 2.9 to the set R, we obtain $2\lfloor n/2 \rfloor$ segments with red endpoints that lie in distinct lines and no pair of them are avoiding. By the condition in Theorem 3.2, the continuation of each of these segments passes through a green point. Assign such a green point to each segment. Observe that these points are all distinct. Indeed, if we can assign the same green point to two different segments, then they must be avoiding, by definition. This completes the proof of Theorem 3.2 and hence of Theorem 2.4. □

4. Proof of the Generalized Ungar Theorem

The aim of this section is to prove Theorem 2.9, the generalized Ungar theorem.

Fix an (x, y)-coordinate system in the plane. We apply a standard duality transform that maps a point $p = (p_1, p_2)$ to the line p^* with equation $y + p_1 x + p_2 = 0$. Vice versa, a nonvertical line l with equation $y + l_1 x + l_2 = 0$ is mapped to the point $l^* = (l_1, l_2)$. Consequently, any two parallel lines are mapped into points having the same x-coordinate. It is often convenient to imagine that the dual picture lies in another, so-called *dual*, plane, different from the original one, which is referred to as the *primal* plane.

The above mapping is incidence and order preserving, in the sense that p lies above, on, or below ℓ if and only ℓ^* lies above, on, or below p^*, respectively. The points of a segment $e = ab$ in the primal plane are mapped to the set of all lines in the closed *double wedge* e^*, which is bounded by a^* and b^* and does not contain the vertical direction. All of these lines pass through the point $q = a^* \cap b^*$, which is called the *apex* of the double wedge e^*. All double wedges used here are assumed to be closed, and they never contain the vertical direction.

DEFINITION 4.1. *We call two double wedges* avoiding *if their apices are distinct and the apex of neither of them is contained in the other.*

It is easy to see that, according to this definition, two noncollinear segments in the primal plane are avoiding if and only if they are mapped to avoiding double wedges.

Switching to the dual plane, Theorem 2.9 can now be reformulated as follows.

THEOREM 4.2. *Let L be a set of n pairwise nonparallel lines in the plane, not all of which pass through the same point. Then the set of all double wedges bounded by pairs of lines in L has at least $2\lfloor n/2 \rfloor$ pairwise nonavoiding elements with different apices.*

Note that the definition of double wedges depends on the choice of the coordinate system, so *a priori* Theorem 4.2 gives a separate statement in each coordinate frame. However, each of these statements is equivalent to Theorem 2.9, and that result does not depend on coordinates. Therefore, we are free to use whatever coordinate system we like. In the final part of the analysis, we have to exploit this property. But until then, no restriction on the coordinate system is imposed.

Suppose that a set of $2\lfloor n/2 \rfloor$ double wedges meets the conditions in Theorem 4.2. Clearly, we can replace each element of this set, bounded by a pair of lines $\ell_1, \ell_2 \in L$, by the *maximal* double wedge with the same apex, i.e., the double wedge bounded by those lines through $\ell_1 \cap \ell_2$ which have the *smallest* and *largest* slopes. If every pair of double wedges in the original set were nonavoiding, then this property remains valid after the replacement.

It is sufficient to prove Theorem 4.2 for the case when n is even, because for odd n the statement trivially follows.

The proof is constructive. Let $\mathcal{A}(L)$ denote the *arrangement* of L, consisting of all vertices, edges, and faces of the planar map induced by L (see Chapter 2). We will construct a set of n vertices of $\mathcal{A}(L)$ with distinct x-coordinates, and show that the maximal double wedges whose apices belong to this set are pairwise nonavoiding.

We start by defining a sequence J of vertices $v_1, v_2, \ldots$, which will be referred to as *junctions*. Let L^- (resp., L^+) denote the subset of L consisting of the $n/2$ lines with the smallest (resp., largest) slopes. If we wish to simplify the picture, we can apply an affine transformation that keeps the vertical direction fixed and carries the elements of L^- and L^+ to lines of negative and positive slopes, respectively (whence the choice of notation). However, we will never use this property explicitly.

The construction proceeds as follows.

STEP 1: Set $i := 1$ and $L_1^- := L^-$, $L_1^+ := L^+$.

STEP 2: If $L_i^- = L_i^+ = \emptyset$, the construction of J terminates. Otherwise, as we will see, neither set is empty. Let v_i be the *leftmost* intersection point between a line in L_i^- and a line in L_i^+. Let d_i^- (and d_i^+) denote the number of elements of L_i^- (and L_i^+, respectively) incident to v_i, and put $d_i = \min\{d_i^-, d_i^+\}$. Define L_{i+1}^- (and L_{i+1}^+) as the set of lines obtained from L_i^- (resp., L_i^+) by deleting from it the d_i elements that are incident to v_i and have the smallest (resp., largest) slopes among those incident lines. (That is, if $d_i^- = d_i^+$, then all lines incident to v_i are deleted; otherwise, if, say, $d_i^- > d_i^+$, we are left with $d_i^- - d_i^+$ lines through v_i that belong to L_i^- and separate the deleted elements of L_i^- from the deleted elements of L_i^+.) Set $i := i + 1$, and repeat Step 2.

Let $J = \langle v_1, v_2, \ldots, v_k \rangle$ denote the resulting sequence.

It is easy to verify the following properties of this construction.

4. PROOF OF THE GENERALIZED UNGAR THEOREM

CLAIM 4.3. (i) $|L_i^-| = |L_i^+|$, for each $i = 1, \ldots, k$.

(ii) For every $1 \leq i < j \leq k$, the junction v_i lies in the left unbounded face f_j of $\mathcal{A}(L_j^- \cup L_j^+)$ which separates L_j^- and L_j^+ at $x = -\infty$ (whose rightmost vertex is v_j). v_i lies in the interior of f_j if $d_i^- = d_i^+$; otherwise it may lie on the boundary of f_j.

(iii) $\sum_{i=1}^{k} d_i = n/2$. □

Next, between any two consecutive junctions v_i and v_{i+1}, for $1 \leq i < k$, we specify $d_i + d_{i+1} - 1$ further vertices of $\mathcal{A}(L)$, called *stations*.

Fix an index $1 \leq i < k$, and consider the vertical slab between v_i and v_{i+1}. By Claim 4.3 (ii), v_i lies inside or on the boundary of the face f_{i+1} of $\mathcal{A}(L_{i+1}^- \cup L_{i+1}^+)$, whose rightmost vertex is v_{i+1}. Hence, the segment $e = v_i v_{i+1}$ is contained in the closure of f_{i+1}. Now at least one of the following two conditions is satisfied: (a) all the d_i lines removed from L_i^+ and all the d_{i+1} lines removed from L_{i+1}^- pass above e, or (b) all the d_i lines removed from L_i^- and all the d_{i+1} lines removed from L_{i+1}^+ pass below e. (We caution the reader that this statement is not totally obvious when e belongs to the boundary of f_{i+1}.)

Assume, by symmetry, that (a) holds. Denote the lines removed from L_i^+ by $\ell_1^+, \ldots, \ell_{d_i}^+$, listed according to increasing slopes, and those removed from L_{i+1}^- by $\ell_1^-, \ldots, \ell_{d_{i+1}}^-$, listed according to decreasing slopes. Define the set of *stations* S_i in the vertical slab between v_i and v_{i+1} as the collection of all intersection points of $\ell_{d_i}^+$ with the lines $\ell_1^-, \ldots, \ell_{d_{i+1}}^-$, and all intersection points of $\ell_{d_{i+1}}^-$ with the lines $\ell_1^+, \ldots, \ell_{d_i}^+$. Clearly, we have $|S_i| = d_i + d_{i+1} - 1$ such points.

See Figure 3 for an illustration.

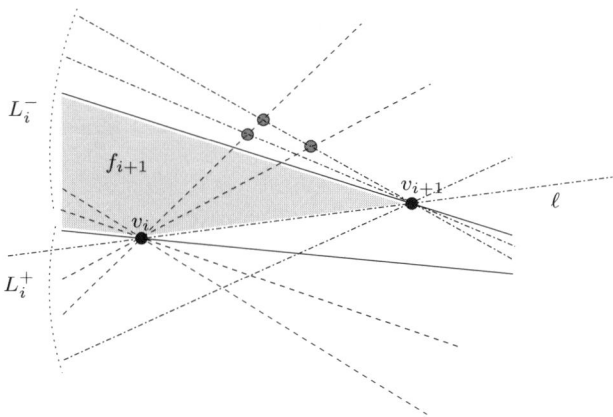

FIGURE 3. Constructing stations between consecutive junctions.

Finally, we have to consider the portions of the plane to the left of v_1 and to the right of v_k and collect there a set S_k of $d_k + d_1 - 1$ additional *stations*. Actually, exploiting the fact that we can (almost) freely select the coordinate system used for the duality transform, it is possible to select $d_k + d_1 - 1$ suitable stations, so that all of them, or all but one, lie to the left of v_1. The proper choice of the coordinate system as well as the details of the construction of S_k are described in [**588**].

Let $Q = J \cup \left(\cup_{i=1}^{k} S_i\right)$. In view of Claim 4.3 (iii), the total number $|Q|$ of junctions and stations equals

$$|Q| = |J| + \sum_{i=1}^{k} |S_i| = k + \sum_{i=1}^{k-1}(d_i + d_{i+1} - 1) + (d_k + d_1 - 1) = 2\sum_{i=1}^{k} d_i = n.$$

All elements of Q are distinct. To complete the proof of Theorem 4.2 (and hence of Theorem 2.9), we need to verify the following

CLAIM 4.4. *Associate with each element $q \in Q$ the maximal* double wedge $W(q)$ *(not containing the vertical line through q), which is bounded by a pair of lines passing through q. Then the resulting set of n double wedges has no two avoiding elements.*

5. Colored Versions of the Sylvester–Gallai Theorem

Da Silva and Fukuda [**257**] suggested the following attractive bipartite version (and strengthening) of the Sylvester–Gallai theorem.

CONJECTURE 5.1. *Let R be a set of red points and let B be a set of blue points in the plane, not all on a straight line. Assume that*
  (i) *R and B are separated by a straight line, and*
  (ii) *$|R|$ and $|B|$ differ by at most one.*
Then there exists a bichromatic *ordinary line, i.e., a line passing through precisely one red and one blue point.*

It is not hard to see that this statement does not remain true if we drop any of the above assumptions. For instance, it is easy to see that the conjecture is false if we drop condition (i).

Suppose that n is even, say, $n = 2k$. Let P be the vertex set of a regular n-gon, and let Q be the set of intersection points of the line at infinity with all lines determined by two elements of P. Clearly, we have $|P| = |Q| = n$. Color $P \cup Q$ with two colors, red and blue, so that the number of red points equals the number of blue points and every *ordinary line* is *monochromatic*.

Notice that every ordinary line determined by $P \cup Q$ passes through one element of Q. Furthermore, for every pair of opposite vertices $p_1, p_2 \in P$, there is a unique point $q \in Q$, such that $p_1 q$ and $p_2 q$ are ordinary lines. Pick $\lfloor \frac{k}{2} \rfloor$ pairs of opposite vertices of P and the corresponding points on the line at infinity, and color them red. Color the remaining $\lceil \frac{k}{2} \rceil$ pairs of opposite vertices of P and the corresponding points on the line at infinity blue. Finally, color the k uncolored points of Q so as to balance the number of red and blue points. Obviously, all ordinary lines determined by $P \cup Q$ are monochromatic. If we wish to avoid using points at infinity, we can modify this construction by applying a suitable projective transformation. Similar constructions can be given in the case when $n > 3$ is odd.

By generating all possible combinatorial types of small point configurations, Finschi and Fukuda [**359**] managed to find a nine-element counterexample to Conjecture 5.1; see Figure 4. However, it is still possible that the conjecture is true for every sufficiently large n, or for all even values of n.

Pach and Pinchasi [**587**] showed that without making any special assumption it is still true that there always exist bichromatic lines containing relatively few points of $R \cup B$.

5. COLORED VERSIONS OF THE SYLVESTER–GALLAI THEOREM

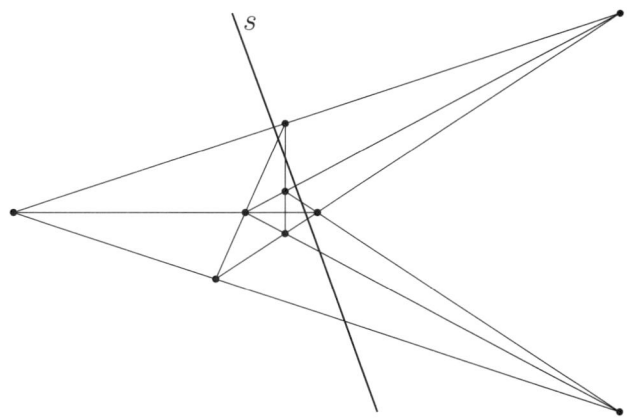

FIGURE 4. The Finschi–Fukuda nine-element counterexample to Conjecture 5.1.

THEOREM 5.2 ([**587**]). *Given n red and n blue points in the plane, not all on a line,*
  (i) *there exist more than $n/2$ bichromatic lines that pass through at most two red points and at most two blue points;*
  (ii) *the number of bichromatic lines passing through at most six points is at least one tenth of the total number of connecting lines.*

The proof is based on the following simple consequence of Euler's Polyhedral Formula (see e.g. [**60**]), which immediately implies the Sylvester–Gallai theorem.

LEMMA 5.3 (Melchior's inequality). *Let P be a finite noncollinear point set in the plane, and let l_i $(i = 2, 3, \ldots)$ denote the number of lines passing through precisely i elements of P. Then we have*

$$\sum_{i=2}^{n-1}(i-3)l_i \leq -3.$$

Now we prove Theorem 5.2.

PROOF. Let R and B be two disjoint n-element point sets in the plane, and assume that not all elements of $R \cup B$ are on the same line. We will refer to the elements of R and B, as *red* points and *blue* points, respectively.

For any ordered pair of nonnegative integers (i, j), $i + j \geq 2$, let l_{ij} denote the number of lines passing through precisely i red and j blue points. In particular, the number of bichromatic lines is $\sum_{i,j\geq 1} l_{ij}$. Set $l_{ij} := 0$, whenever $i + j \leq 1$.

The number of monochromatic point pairs is equal to

$$\sum_{i,j\geq 0}\left[\binom{i}{2} + \binom{j}{2}\right] l_{ij} = 2\binom{n}{2} = n^2 - n.$$

The number of bichromatic point pairs is $\sum_{i,j\geq 0} ij l_{ij} = n^2$. Thus, we have

(5.1)
$$\sum_{\substack{i,j \geq 0 \\ i+j \geq 2}} \left[\binom{i}{2} + \binom{j}{2} - ij\right] l_{ij} = -n.$$

Lemma 5.3 implies that

(5.2)
$$\sum_{\substack{i,j \geq 0 \\ i+j \geq 2}} (i+j-3)l_{ij} \leq -3.$$

Adding up twice (5.1) and $1+\varepsilon$ times (5.2), for some positive ε, we obtain

(5.3)
$$\sum_{\substack{i,j \geq 0 \\ i+j \geq 2}} \left[(i-j)^2 + \varepsilon(i+j-3) - 3\right] l_{ij} \leq -2n - 3(1+\varepsilon).$$

For any $i,j \geq 0$ $(i+j \geq 2)$, let γ_{ij} denote the coefficient of l_{ij} in the above inequality, so that $\sum_{i,j} \gamma_{ij} l_{ij} < 0$.

First, set $\varepsilon = 1$. It is easy to verify that $\gamma_{11} = -4$, $\gamma_{12} = \gamma_{21} = \gamma_{22} = -2$, and that all other coefficients γ_{ij} are nonnegative. Therefore, (5.3) yields that

$$-4l_{11} - 2l_{12} - 2l_{21} - 2l_{22} \leq -2n - 6.$$

Consequently,
$$2l_{11} + l_{12} + l_{21} + l_{22} \geq n + 3,$$
which proves part (i).

To establish part (ii), set $\varepsilon = 3/5$. Then
$$\gamma_{11} = -\frac{18}{5}, \; \gamma_{12} = \gamma_{21} = -2, \; \gamma_{22} = -\frac{12}{5}, \; \gamma_{23} = \gamma_{32} = -\frac{4}{5}, \; \gamma_{33} = -\frac{6}{5},$$
and all other coefficients are at least $2/5$. Hence,

$$\sum_{\substack{i,j \geq 0 \\ i+j \geq 2}} \gamma_{ij} l_{ij} + 4l_{11} + \frac{12}{5}l_{12} + \frac{12}{5}l_{21} + \frac{14}{5}l_{22} + \frac{6}{5}l_{23} + \frac{6}{5}l_{32} + \frac{8}{5}l_{33} \geq \frac{2}{5} \cdot \sum_{\substack{i,j \geq 0 \\ i+j \geq 2}} l_{ij}.$$

Comparing the last inequality with (5.3), we obtain

$$-2n - 3\left(1+\frac{3}{5}\right) + 4l_{11} + \frac{12}{5}l_{12} + \frac{12}{5}l_{21} + \frac{14}{5}l_{22} + \frac{6}{5}l_{23} + \frac{6}{5}l_{32} + \frac{8}{5}l_{33} \geq \frac{2}{5} \cdot \sum_{\substack{i,j \geq 0 \\ i+j \geq 2}} l_{ij}.$$

That is,

$$\sum_{\substack{i,j \geq 0 \\ 2 \leq i+j \leq 6}} l_{ij} \geq l_{11} + \frac{3}{5}l_{12} + \frac{3}{5}l_{21} + \frac{7}{10}l_{22} + \frac{3}{10}l_{23} + \frac{3}{10}l_{32} + \frac{2}{5}l_{33}$$

$$\geq \frac{1}{10} \cdot \sum_{i,j \geq 0} l_{ij} + \frac{n}{2} + \frac{6}{5},$$

which completes the proof of part (ii) of the theorem. $\square$

CHAPTER 2

Arrangements of Surfaces: Evolution of the Basic Theory

1. Introduction

In this chapter we introduce one of the most basic constructs in discrete and computational geometry—arrangements of curves or of surfaces. As we will see, in this and in subsequent chapters, arrangements play a key role in many of the problems that we will consider.

Formally, the *arrangement* of a finite collection Γ of geometric objects in $\mathbb{R}^d$, denoted as $\mathcal{A}(\Gamma)$, is the decomposition of the space into relatively open connected cells of dimensions $0, \ldots, d$ induced by Γ, where each cell is a maximal connected set of points lying in the intersection of a fixed subset of Γ; see Figure 1 for an illustration.

Besides being interesting in their own right, because of the rich geometric, combinatorial, algebraic, and topological structures that arrangements possess, they also lie at the heart of several geometric problems arising in a wide range of applications including robotics, computer graphics, molecular modeling, and computer vision. Before proceeding further, we present a few such examples.

(a) Assume that we have a robot system B with d degrees of freedom, i.e., we can represent each placement of B as a point in $\mathbb{R}^d$, and we call the space of all placements the *configuration space* of B. Suppose the three-dimensional workspace of B is cluttered with, say, polyhedral obstacles, whose shapes and locations are known. B is allowed to move freely in a motion that traces a continuous path in the configuration space, but B has to avoid collision with the obstacles. For each combination of a geometric feature (vertex, edge, face) of an obstacle and a similar feature (face, edge, vertex) of B, define their *contact surface* as the set of all points in $\mathbb{R}^d$ that represent placements of B at which contact is made between

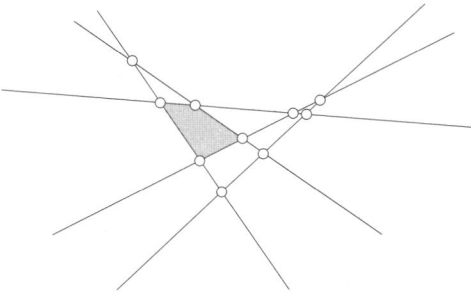

FIGURE 1. An arrangement of lines.

these specific features. Let Γ be the set of all contact surfaces. Let Z be a point corresponding to a given initial *free* placement of B, i.e., a placement at which it does not intersect any obstacle. Then the set of all free placements of B that can be reached from Z via a collision-free continuous motion corresponds to the cell containing Z in the arrangement of the contact surfaces. Thus, the problem of determining whether there exists a collision-free path from an initial configuration I to a final configuration F is equivalent to determining whether I and F lie in the same cell of $\mathcal{A}(\Gamma)$. This close relationship between arrangements and motion planning has led to considerable work on arrangements; see, for example, [**45, 104, 412, 422, 472, 473, 507, 512, 681**]. If we want to compute the set of all placements reachable from the initial placement I, the *combinatorial complexity* of the cell in $\mathcal{A}(\Gamma)$ containing I, i.e., the total number of lower-dimensional faces appearing on its boundary, serves as a trivial lower bound for the running time. It turns out that in many instances one can design motion-planning algorithms whose performance almost matches this bound.

(b) A molecule can be modeled as an arrangement of spheres, where the radius of each sphere depends on the atom that it models and the position of each sphere depends on the molecular structure. In the *Van der Waals model*, a molecule is a family of possibly overlapping spheres, where the radius of each sphere is determined by the van der Waals radius of the corresponding atom in the molecule. Lee and Richards [**508**] proposed another model, called *solvent accessible* model, which is used to study the interaction between the protein and solvent molecules. In this model too, a molecule is modeled as a sphere, but the balls representing the solvent molecules are shrunk to points and the balls representing atoms in the protein are inflated by the same amount [**648**]. Even though these models ignore various properties (e.g., chemical or electrical) of molecules, they have been useful in a variety of applications. Many problems in molecular modeling can be formulated as problems related to arrangements of spheres. For example, computing the "outer surface" of the molecule corresponds to computing the unbounded cell of the corresponding arrangement of spheres, or, more generally, the (complement of the) union of the balls (atoms) bounded by the given spheres. See [**291, 292, 417, 423, 548**] for more details on applications of arrangements in molecular biology.

(c) Arrangements are also attractive because of their relationship with several other structures. For example, using the *duality* transform, a point $p = (p_1, \ldots, p_d)$ in $\mathbb{R}^d$ can be mapped to the hyperplane $\sum_{i=1}^{d} x_i p_i = 1$, and vice versa. This facilitates the formulation of several problems related to point configurations in terms of arrangements of hyperplanes. See [**224, 288**] for a small sample of such problems. The Grassmann–Plücker relation transforms[1] k-flats in $\mathbb{R}^d$ to hyperplanes or points in $\mathbb{R}^u$, for $u = \binom{d+1}{k+1} - 1$ [**152, 429**]; e.g., lines in $\mathbb{R}^3$ can be mapped to hyperplanes or points in $\mathbb{R}^5$. Therefore many problems involving lines in $\mathbb{R}^3$ have been solved using hyperplane arrangements in $\mathbb{R}^5$ [**220, 270, 622, 698**]. Moreover, since lines in space have four degrees of freedom, it is better, in some cases, to represent lines in space as points in $\mathbb{R}^4$. Typically, one then obtains arrangements of nonlinear surfaces. Chapter 7 deals exclusively with problems involving lines in space, and many more details can be found there. The well-known combinatorial structures called *oriented*

---

[1] With some care, excluding some lower-dimensional variety of flats, one can achieve representation in real, rather than projective, space.

matroids of rank $k+1$ are closely related to arrangements of pseudo-hyperplanes in $\mathbb{R}^k$ [**152, 650**], and *zonotopes* in $\mathbb{R}^d$ correspond to hyperplane arrangements in $\mathbb{R}^{d-1}$ [**288, 759**]. Several applications of arrangements in singularity theory, algebraic group theory, and other fields of mathematics can be found in [**572, 574, 575**].

The study of arrangements of lines and hyperplanes has a long, rich history. The first paper on this topic is perhaps by J. Steiner in 1826 [**702**], in which he obtained bounds on the number of cells in arrangements of lines and circles in the plane and of planes and spheres in $\mathbb{R}^3$. His results have since been extended in several ways [**66, 67, 68, 185, 651, 746, 755, 756**]. A summary of early work on arrangements of hyperplanes can be found in the monograph and the survey paper by Grünbaum [**393, 394**]. Most of the work on hyperplane arrangements until the 1980s dealt with the combinatorial structure of the entire arrangement or of a single cell in the arrangement (i.e., a convex polyhedron), and with the algebraic and topological properties of the arrangement [**348, 573, 574, 575**]. Various substructures and algorithmic issues of hyperplane arrangements, motivated by problems in computational and combinatorial geometry, have received attention mostly during the last twenty years.

Although hyperplane arrangements possess a rich structure, many applications (e.g., the motion-planning problem and the molecular models described above) call for a systematic study of higher-dimensional arrangements of patches of algebraic surfaces. For more than a century, researchers in algebraic geometry have studied arrangements of algebraic surfaces, but their focus has largely been on algebraic and combinatorial issues rather than on algorithmic ones. Considerable progress has been made on all fronts during the last twenty years.

It is beyond the scope of a survey chapter, or even of a book, to cover all aspects of arrangements. In this chapter we will survey combinatorial and algorithmic problems on arrangements of (hyper)surfaces (or of surface patches) in real affine space $\mathbb{R}^d$. (Hyperplane arrangements in complex space have also been studied; see, e.g., [**153, 575**].) We will assume that d is a small constant, that the surfaces are algebraic and their degree is bounded by a constant, and that any input surface patch is a semialgebraic set defined by a Boolean combination of a constant number of polynomial equalities and inequalities of constant maximum degree. We refer to such a collection of surfaces or of surface patches as having *constant description complexity*. There has also been some recent work on combinatorial and algorithmic issues involving arrangements of more general surfaces, known as *semi-pfaffian* sets, which include graphs of trigonometric, exponential, or logarithmic functions on bounded domains [**373, 479**]. We also note that a study of algebraic and topological problems on arrangements of algebraic surfaces can be found in [**156**]. In this chapter we will mostly review the known results on the combinatorial complexity of various substructures of arrangements, the known algorithms for computing these substructures, and the geometric applications that benefit from these results. Many other combinatorial problems related to arrangements are discussed in [**152, 358, 385, 584, 585, 690, 759**]. An excellent source on combinatorial and algorithmic results on arrangements of hyperplanes is the 1987 book by Edelsbrunner [**288**]. The 1995 book of Sharir and Agarwal [**681**] covers some of the topics discussed here in more detail. Other survey papers on arrangements include [**401, 414, 421,**

677], as well as the earlier versions of this chapter and Chapter 3 (Agarwal and Sharir [**51, 52**]).

In this chapter we review the basic theory of arrangements in higher dimensions, where the front of the research nowadays lies. However, readers unfamiliar with the earlier reults on planar arrangements and Davenport–Schinzel sequences should read first Chapter 3 (mainly Section 6 of that chapter), which reviews those earlier results.

This chapter is organized as follows. In Section 2 we define arrangements formally, state the assumptions we will be making in this chapter, and discuss the known bounds on the complexity of entire arrangements. Sections 3–10 discuss combinatorial complexities of various substructures of arrangements. Section 11 discusses several methods for representing arrangements. Section 12 describes algorithms for computing the entire arrangement, and Section 13 reviews algorithms for computing various substructures of arrangements. We discuss a few applications of arrangements in Section 14.

2. Preliminaries

Let $\Gamma = \{\gamma_1, \ldots, \gamma_n\}$ be a collection of n (hyper)surfaces or surface patches in $\mathbb{R}^d$. The set Γ induces a decomposition of $\mathbb{R}^d$ into connected cells (or faces), called the *arrangement* of Γ and denoted $\mathcal{A}(\Gamma)$, so that each cell is a maximal connected subset of the intersection of a fixed (possibly empty) subset of surface patches that avoids all other surface patches. Thus a d-dimensional cell, for example, is a maximal connected region that does not meet any member of Γ. The *combinatorial complexity* of $\mathcal{A}(\Gamma)$ is the total number of cells, of all dimensions, in $\mathcal{A}(\Gamma)$. The combinatorial complexity of a k-dimensional cell C in $\mathcal{A}(\Gamma)$ is the number of cells of $\mathcal{A}(\Gamma)$ of dimension less than k that are contained in the boundary of C.

We assume that Γ satisfies the following properties.

(A1) Each $\gamma_i \in \Gamma$ is a semialgebraic set of constant description complexity.[2] The local dimension of every point in γ_i is $d-1$.

(A2) Each $\gamma_i \in \Gamma$ is of the form $(Q_i = 0) \wedge F_i(P_{i_1} \geq 0, P_{i_2} \geq 0, \ldots, P_{i_u} \geq 0)$. Here u is a constant; F_i is a Boolean formula; $Q_i, P_{i_1}, \ldots, P_{i_u} \in \mathbb{R}[x_1, \ldots, x_d]$; and the degrees of Q_i, P_{i_j} are at most b, for some constant b. Let $\mathcal{Q} = \{Q_1, \ldots, Q_n\}$.

Note that (A2) implies (A1), but we will mention both assumption for the sake of clarity. We will refer to a semialgebraic set satisfying (A1) and (A2) a *(hyper)surface patch* in $\mathbb{R}^d$ (of constant description complexity). If γ_i is simply the zero set of Q_i, we will call γ_i a *(hyper)surface*. Using a stratification algorithm [**156, 432**], we can decompose each γ_i into a constant number connected surface patches so that the interior of each patch is *smooth* and each of them satisfies (A1) and (A2) with a different, possibly larger, value of b. We can also assume that each resulting patch is *monotone* in $x_1, \ldots, x_{d-1}$ (i.e., any line parallel to the x_d-axis intersects the patch in at most one point). In some cases, the resulting collection may also include *vertical* surface patches, namely, patches whose projection on the

---

[2]A subset of $\mathbb{R}^d$ is called a *real semialgebraic set* if it is obtained as a finite Boolean combination of sets of the form $\{f = 0\}$ or $\{f > 0\}$ for d-variate polynomials f. As already defined, a semialgebraic set has *constant description complexity* if it can be described in terms of a constant number of polynomials, with a constant bound on the degrees of the corresponding polynomials.

hyperplane $x_d = 0$ has dimension $\leq d-2$. However, in most of the presentation we will assume that no vertical patches exist.

A collection of hyperplanes is called *simple* if any d of the hyperplanes intersect in exactly one point, and no $d+1$ of them have a nonempty intersection. In a simple arrangement of hyperplanes, a k-dimensional cell is contained in exactly $d-k$ hyperplanes. We will also need a similar concept for arrangements of surface patches. A collection Γ satisfying assumptions (A1) and (A2) is said to be in *general position* if the coefficients of the polynomials defining the surface patches in Γ and their boundaries are algebraically independent[3] over the rationals; otherwise, Γ is called *degenerate*. This condition ensures that no degeneracy occurs among the surface patches, such as too many surface patches with a common point, tangencies or overlaps between different intersections of subsets of the surface patches, etc. We note that this definition of general position is quite strong (e.g., surfaces defined by polynomials with integer coefficients are not in general position in this strong sense). In all the applications much weaker versions of general position are required, which rule out a specific list of forbidden degenerate situations. If Γ is in general position, then any d surface patches of Γ intersect in at most s points for some constant s depending on d and b. By Bezout's theorem [**429**], $s \leq b^d$.

If Γ is degenerate, we can perturb the coefficients of the polynomials in $\mathcal{Q}$ by various infinitesimals, so that the coefficients of the perturbed polynomials are in extension fields of the reals that are fields of *Puiseux Series* in these infinitesimals, and so that the resulting surface patches are in general position. Moreover, it can be shown that, as far as worst-case bounds are concerned, the perturbation may reduce the combinatorial complexity of any cell of the arrangement by at most a constant factor [**634, 678, 681**]. Actually, in many cases the size of a substructure of Γ has maximum complexity when Γ is in general position. This observation allows us to restrict our attention to arrangements in general position while investigating the combinatorial complexity of substructures of arrangements.

However, in order to achieve the general position defined above, the perturbation scheme has to introduce a different infinitesimal for each coefficient, which makes any algorithm based on this perturbation scheme impractical. Fortunately, most of the algorithms involving arrangements either work for any degenerate arrangement or require a considerably weaker definition of general position, e.g., the intersection of any k surface patches is either empty or a $(d-k)$-dimensional set, all surface patches intersect "properly," etc. The perturbation scheme required by an algorithm depends on the degenerate situations that it wants to rule out. Several constructive perturbation schemes have been proposed that use only a few infinitesimals [**133, 300, 320, 321, 361**]. Although these schemes cannot handle all the cases, they work for a wide range of applications. The paper by Seidel [**671**] contains a detailed discussion on "linear" perturbation and its applications in geometric algorithms. A few algorithms have also been proposed to handle degeneracies directly without resorting to perturbations; see e.g. [**117, 186**]. We will, nevertheless, use the strong definition of general position, defined above, in order to simplify the exposition, and refer the reader to the original papers for specific *general-position assumptions* required by different algorithms.

---

[3] A set $\{x_1, \ldots, x_k\}$ of real numbers is *algebraically independent* (over the rationals) if no k-variate polynomial with integer coefficients vanishes at $(x_1, x_2, \ldots, x_k)$.

In the light of the preceding discussion, and since we are mainly interested in asymptotic bounds, we will make the following additional assumptions on the surface patches in Γ, without any real loss of generality, whenever required.

(A3) Each surface patch in Γ is connected and monotone in $x_1, \ldots, x_{d-1}$, and its relative interior is smooth.
(A4) The surface patches in Γ are in general position.
(A5) Any d surface patches in Γ intersect in at most s points, for some constant s. (This is a consequence of the preceding assumptions, but is stated to introduce s explicitly.)

Generally, we will be stating assumptions (A1) and (A2), but most of the proofs and algorithms sketched in the chapter will also make assumptions (A3)–(A5).

Assumptions (A1)–(A3) imply that we can regard each surface patch γ as the graph of a partially defined $(d-1)$-variate function $x_d = \gamma(x_1, \ldots, x_{d-1})$ of constant description complexity. We will refer to the projection of γ onto the hyperplane $x_d = 0$ as the *domain*, denoted γ^*, of γ (over which the function γ is defined). The boundary of γ^*, called the *domain boundary*, is a collection of $O(1)$ $(d-2)$-dimensional surface patches in $\mathbb{R}^{d-1}$ satisfying assumptions corresponding to (A1)–(A2). Abusing the notation slightly, we will not distinguish between the surface patch γ and the underlying function $\gamma(x_1, \ldots, x_{d-1})$.

The most fundamental question in the combinatorial study of an arrangement $\mathcal{A}(\Gamma)$ of surfaces is to prove a bound on the combinatorial complexity, $f(\Gamma)$, of $\mathcal{A}(\Gamma)$.

In 1826, Steiner [**702**] studied the complexity of arrangements of lines and circles in $\mathbb{R}^2$ and of planes and spheres in $\mathbb{R}^3$. In particular, he showed that an arrangement of n planes in $\mathbb{R}^3$ in general position has $\binom{n}{3}$ vertices, $\binom{n}{2} + 3\binom{n}{3}$ edges, $n^2 + 3\binom{n}{3}$ 2-faces, and $1 + n + \binom{n}{2} + \binom{n}{3}$ 3-cells. Later Roberts [**651**] extended Steiner's formula to count the number faces in arbitrary arrangements of planes (allowing all kinds of degeneracies) in $\mathbb{R}^3$, using the inclusion-exclusion principle. Brousseau [**182**] used a plane-sweep argument to count the number of faces in arrangements of planes in $\mathbb{R}^3$. (A similar argument was used by Hadwiger [**408**] to derive Euler's formula for convex polytopes.) His method was later extended by Alexanderson and Wetzel [**68**].

Buck [**185**] was the first to bound the combinatorial complexity of hyperplane arrangements in higher dimensions. In more recent work, Zaslavsky [**755, 756**] studied hyperplane arrangements; he used the Möbius inversion formula and lattice theory to count the number of cells of all dimensions in (possibly degenerate) hyperplane arrangements. Let Γ be a set of n hyperplanes in $\mathbb{R}^d$. Let $\varphi_k(\Gamma)$ denote the number of k-cells in $\mathcal{A}(\Gamma)$. Zaslavsky [**755**] and Las Vergnas [**506**] proved that for nonsimple arrangements, $\varphi_k(\Gamma)$ depends on the underlying matroid structure. There are several results on bounding $\varphi_k(\Gamma)$ in nonsimple hyperplane arrangements. For example, Fukuda *et al.* [**371**] proved that the mean number of $(k-1)$-cells bounding a k-cell in an arrangement of n hyperplanes is less than $2k$, which implies that $\varphi_k(\Gamma) \leq \binom{d}{k}\varphi_d(\Gamma)$. See [**371, 544, 659**] for some other results of this type.

In summary, the following theorem gives a bound on the combinatorial complexity of hyperplane arrangements. (See [**51**] for a proof.)

THEOREM 2.1 (Buck [**185**]). *Let Γ be a set of n hyperplanes in $\mathbb{R}^d$. For any $0 \le k \le d$,*

$$\varphi_k(\Gamma) \le \binom{n}{d-k} \sum_{i=0}^{k} \binom{n-d+k}{i}.$$

Equality holds when $\mathcal{A}(\Gamma)$ is simple.

For arrangements $\mathcal{A}(\Gamma)$ of a set Γ of surfaces satisfying assumptions (A1) and (A2), obtaining a sharp bound on $f(\Gamma)$, the combinatorial complexity of $\mathcal{A}(\Gamma)$, is not easy. If the surface patches are in general position, in the sense defined above, it is obvious that $f(\Gamma) = O(n^d)$. However, it is not easy to argue that the arrangements have maximum complexity when the surface patches are in general position (this is due to the complicated algebraic structures that can arise in degenerate settings). Heintz et al. [**435**] proved that $f(\Gamma) = (nb)^{O(d)}$. A lower bound of $\Omega((nb/d)^d)$ is not difficult to prove. Warren [**740**] had proved that the number of d-dimensional cells in an arrangement of n hypersurfaces, each of degree $\le b$, in $\mathbb{R}^d$ is $O((nb/d)^d)$. This bound also follows from the results by Milnor [**551**], Petrovskiĭ and Oleĭnik [**628**], and Thom [**720**]. Using a perturbation argument, Pollack and Roy [**634**] generalized Warren's result and proved that the number of cells of all dimensions in an arrangement of n hypersurfaces is $(O(nb)/d)^d$. An easy consequence of their result is the following theorem.

THEOREM 2.2. *Let Γ be a set of n surface patches in $\mathbb{R}^d$ satisfying assumptions (A1) and (A2). Then*

$$f(\Gamma) = \left(\frac{O(nb)}{d}\right)^d.$$

An extension of this theorem by Basu et al. [**132**] shows that if Σ is a k-dimensional algebraic variety of degree at most b in $\mathbb{R}^d$, then the number of cells in the subdivision of Σ induced by Γ is at most $O((n/k)^k b^d)$.

Improved bounds on the complexity of the arrangement can be proved in some special cases. For example, if Γ is a set of n $(d-1)$-simplices in $\mathbb{R}^d$ that form the boundaries of k convex polytopes, then $f(\Gamma) = O(n^{\lfloor d/2 \rfloor} k^{\lceil d/2 \rceil})$ [**90**]. See [**144**] for improved bounds in a few other cases. A concept closely related to the combinatorial complexity of arrangements is the number of *realizable sign sequences* of a family of polynomials. Let $\mathcal{Q} = \{Q_1, \ldots, Q_n\}$ be a set of d-variate polynomials as defined above, and let Γ be the family of the zero-sets of the polynomials in $\mathcal{Q}$. We can define $\sigma_i(\mathbf{x})$, for a point $\mathbf{x} \in \mathbb{R}^d$, as follows.

$$\sigma_i(\mathbf{x}) = \begin{cases} -1 & Q_i(\mathbf{x}) < 0, \\ 0 & Q_i(\mathbf{x}) = 0, \\ +1 & Q_i(\mathbf{x}) > 0. \end{cases}$$

Since $\sigma_i(\mathbf{x})$ remains the same for all points $\mathbf{x}$ in a single cell of $\mathcal{A}(\Gamma)$, we can define the *sign sequence* for each cell $\sigma(C) = \langle \sigma_1(\mathbf{x}), \sigma_2(\mathbf{x}), \ldots, \sigma_n(\mathbf{x}) \rangle$ for any point $\mathbf{x} \in C$. A sign sequence σ is *realized* by $\mathcal{A}(\Gamma)$ if there is a cell $C \in \mathcal{A}(\Gamma)$ with $\sigma = \sigma(C)$. A well-studied question in algebraic geometry is to bound the number of sign sequences that can be realized by a set of polynomials [**70**]. Obviously, $f(\Gamma)$ is an upper bound on this quantity, but the actual bound could be smaller, since the same sign sequence can be attained by more than one cell.

3. Lower Envelopes

Definitions and preliminary results. In Chapter 3 we will review lower envelopes of arcs in the plane, and will show the relationship between such envelopes and *Davenport–Schinzel sequences*, a powerful combinatorial construct with rather surprising properties. Here we assume familiarity of the reader with the basic results concerning Davenport–Schinzel sequences and univariate lower envelopes, but we will not need too much of that theory. We use the standard notation $\lambda_q(n)$ for the maximum length of Davenport–Schinzel sequences of order q on n symbols, and recall (see Chapter 3) that $\lambda_q(n)$, for $q \geq 3$, is nearly linear (ever so slightly super-linear) in n for any fixed q. Specifically, $\lambda_q(n) = n\beta_q(n)$, where $\beta_q(n)$ is an extremely slowly growing function, expressed in terms of the inverse Ackermann function $\alpha(n)$; the precise known bounds are stated in Chapter 3. We also recall that the complexity of the lower envelope of n univariate continuous fully defined (resp., partially defined) functions, each pair of which intersect in at most q points, is at most $\lambda_q(n)$ (resp., $\lambda_{q+2}(n)$). We refer the reader to Chapter 3 for many additional details, which can also be found in [**52, 536, 681**].

In this section we study lower envelopes of surface patches in higher dimensions. Let $\Gamma = \{\gamma_1, \ldots, \gamma_n\}$ be a collection of surface patches in $\mathbb{R}^d$ satisfying assumptions (A1)–(A3). If we regard each surface patch as the graph of a partially defined function, the *lower envelope* of Γ, denoted $L(\Gamma)$ (or L for brevity), is defined as the graph of the partially defined function

$$L_\Gamma(\mathbf{x}) = \min_{1 \leq i \leq n} \gamma_i(\mathbf{x}), \quad \mathbf{x} \in \mathbb{R}^{d-1};$$

$\gamma_i(\mathbf{x})$ is set to $+\infty$ if $\mathbf{x} \notin \gamma_i^*$; see Figure 2. The *upper envelope* $U(\Gamma)$ of Γ is defined as the graph of the partially defined function

$$U_\Gamma(\mathbf{x}) = \max_{1 \leq i \leq n} \gamma_i(\mathbf{x}), \quad \mathbf{x} \in \mathbb{R}^{d-1};$$

$\gamma_i(\mathbf{x})$ is set to $-\infty$ if $\mathbf{x} \notin \gamma_i^*$. We can extend the definitions of lower and upper envelopes even if Γ satisfies only (A1) and (A2). We can decompose each γ_i into $O(1)$ connected patches, each of which is monotone in the $x_1, \ldots, x_{d-1}$ directions and satisfies (A1) and (A2). Let Γ' denote the resulting set of surface patches. We define $L(\Gamma) = L(\Gamma')$ and $U(\Gamma) = U(\Gamma')$.

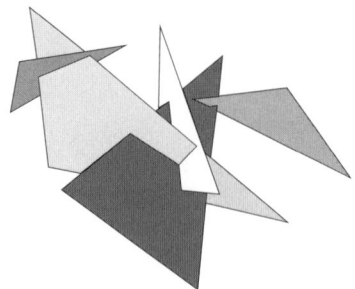

FIGURE 2. Lower envelope of triangles in $\mathbb{R}^3$, as viewed from below.

$L(\Gamma)$ induces a partition of $\mathbb{R}^{d-1}$ into maximal connected $((d-1)$-dimensional) regions so that $L(\Gamma)$ is attained by a fixed (possibly empty) subset of Γ over the interior of each such region (in general position, it is attained either by one surface or by no surface at all). The boundary of such a region consists of points at which $L(\Gamma)$ is attained by at least two of the surface patches or by the relative boundary of at least one surface. Let $\mathcal{M}(\Gamma)$ denote this subdivision of $\mathbb{R}^{d-1}$, which we call the *minimization diagram* for the collection Γ. A *face* of $\mathcal{M}(\Gamma)$ is a maximal connected region over which $L(\Gamma)$ is attained by the same set of functions and/or relative boundaries of function graphs in Γ. Note that if a face $f \in \mathcal{M}(\Gamma)$ lies on the boundary of the domain of a surface in Γ, then f may not correspond to any face of $L(\Gamma)$. However, if f lies in the relative interior of the domains of all the relevant surface patches, f is the orthogonal projection of a face $\hat{f}$ of $L(\Gamma)$. The *combinatorial complexity* of $L(\Gamma)$, denoted $\kappa(\Gamma)$, is the number of faces of all dimensions in $\mathcal{M}(\Gamma)$. For an infinite family $\mathbf{G}$ of surface patches satisfying assumptions (A1)–(A2), we define $\kappa(n, d, \mathbf{G}) = \max \kappa(\Gamma)$, where the maximum is taken over all subsets Γ of $\mathbf{G}$ of size n. If $\mathbf{G}$ is the set of all surface patches satisfying (A1)–(A2) (with some fixed choices of the parameters), or if $\mathbf{G}$ is obvious from the context, we will simply write $\kappa(n, d)$. The *maximization diagram* is defined as the subdivision of $\mathbb{R}^{d-1}$ induced, in the same manner, by the upper envelope $U(\Gamma)$ of Γ.

As will be discussed in Chapter 3, the complexity of the lower envelope of n arcs in the plane, each pair of which intersects in at most s points, is at most $\lambda_{s+2}(n)$, the maximum length of an $(n, s+2)$-Davenport–Schinzel sequence (see also [**681**]). Extending to higher dimensions, it was conjectured that the complexity of the lower envelope of a family of n surface patches satisfying (A1)–(A2) is $O(n^{d-2}\lambda_q(n))$, for some constant $q \geq 0$. If Γ is a set of n hyperplanes in $\mathbb{R}^d$, then the Upper Bound Theorem implies that the worst-case complexity of $L(\Gamma)$ is $\Theta(n^{\lfloor d/2 \rfloor})$ (see [**543**, **759**]). Let Δ be the set of all $(d-1)$-simplices in $\mathbb{R}^d$. Extending the lower-bound construction by Wiernik and Sharir [**747**] to higher dimensions, one can prove that $\kappa(n, d, \Delta) = \Omega(n^{d-1}\alpha(n))$. This indicates that we cannot hope to aim for an $o(n^{d-1})$ bound on $\kappa(n, d)$ for general surface patches. At the end of this section we will discuss some cases, in addition to the case of hyperplanes, in which better bounds on $\kappa(n, d)$ can be proved.

Using a divide-and-conquer approach, Pach and Sharir [**595**] proved that, for a set Γ of n simplices in $\mathbb{R}^d$, the number of $(d-1)$-dimensional faces in $\mathcal{M}(\Gamma)$ is $O(n^{d-1}\alpha(n))$. Roughly speaking, they divide Γ into subsets Γ_1, Γ_2, each of size at most $\lceil n/2 \rceil$, and bound the number of $(d-1)$-dimensional faces of $\mathcal{M}(\Gamma_1), \mathcal{M}(\Gamma_2)$ recursively. They prove that the number of $(d-1)$-dimensional faces in $\mathcal{M}(\Gamma)$ is $|\mathcal{M}(\Gamma_1)| + |\mathcal{M}(\Gamma_2)| + O(n^{d-1}\alpha(n))$, thereby obtaining the claimed bound. Edelsbrunner [**289**] extended their result to give the same asymptotic bound for the number of faces of all dimensions. Simpler proofs for this bound were proposed by Sharir and Agarwal [**681**] and Tagansky [**714**]. Roughly speaking, both proofs proceed by induction on d, and they bound the change in the complexity of the minimization diagram as a simplex is inserted into Γ.

The main complexity bound. All the aforementioned proofs rely crucially on the fact that if Γ is a set of surface patches in general position, then any d-tuple of surface patches intersect in at most one point. These proofs do not extend to the case when a d-tuple intersect in $s \geq 2$ points. Halperin and Sharir [**419**] proved a near-quadratic bound on $\kappa(n, 3)$ for the case when $s \leq 2$. Sharir [**678**] extended their approach to higher values of s and d. Their results are stated in the following theorem.

THEOREM 3.1 (Halperin and Sharir [**419**]; Sharir [**678**]). *Let Γ be a set of n surface patches in $\mathbb{R}^d$ satisfying assumptions (A1)–(A2). Then $\kappa(n,d) = O(n^{d-1+\varepsilon})$, for any $\varepsilon > 0$. The constant of proportionality depends on ε, d, b (and s).*

PROOF. We will sketch the proof for a set of bivariate surface patches in $\mathbb{R}^3$ satisfying assumptions (A1)–(A5) with $s = 2$, i.e., a triple of surface patches intersect in at most two points. For a pair of surface patches $\gamma_i, \gamma_j \in \Gamma$, let β_{ij} denote the intersection arc $\gamma_i \cap \gamma_j$. If β_{ij} is not x_1-monotone, we decompose it at its x_1-extremal points; each intersection arc is thereby decomposed into $O(1)$ pieces. If any of these points appears on the lower envelope, we regard it as a vertex on the envelope and its projection as a vertex on the minimization diagram.

Since Γ is in general position, it suffices to bound the number of vertices in $\mathcal{M}(\Gamma)$. Indeed, after the aforementioned decomposition of the intersection arcs, a higher-dimensional face f of $\mathcal{M}(\Gamma)$ must be incident to a vertex v of $\mathcal{M}(\Gamma)$, and we can charge f to v. By the general-position assumption, each vertex is charged only a constant number of times. For a subset $R \subseteq \Gamma$, let $\phi^*(R)$ denote the number of vertices in $\mathcal{M}(R)$; set $\phi^*(r) = \max \phi^*(R)$, where the maximum is taken over all subsets R of Γ of size r.

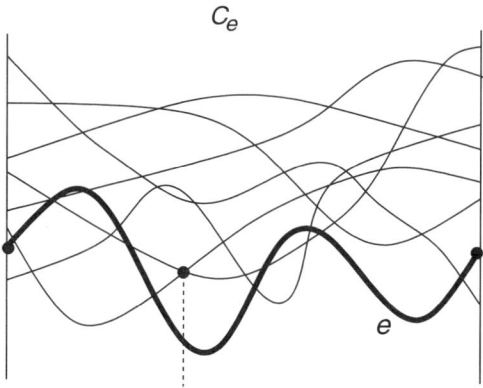

FIGURE 3. Vertical cylinder C_e and the vertical cross-section Γ_e of Γ.

We call a vertex of $\mathcal{M}(\Gamma)$ a *boundary* vertex if it lies on the boundary $\partial \gamma_i^*$ of the domain of a surface γ_i; otherwise, we call it an *inner* vertex. The number of boundary vertices is $O(n\lambda_q(n))$, where q is a constant depending on b, the maximum degree of the surface patches and their boundaries. Indeed, let C_e be the vertical cylinder erected on an edge e of the boundary $\partial \gamma_i^*$, i.e., $C_e = e \times \mathbb{R}$. Define $\Gamma_e = \{\gamma \cap C_e \mid \gamma \in \Gamma \setminus \{\gamma_i\}\}$, which is a collection of $O(n)$ arcs; see Figure 3. Each arc in Γ_e satisfies assumptions (A1)–(A3) (with $d = 2$, and with larger, but still constant, parameters b and s). It is easily seen that a boundary vertex of $\mathcal{M}(\Gamma)$ appearing on e is a vertex of $\mathcal{M}(\Gamma_e)$. If the arcs in Γ_e intersect in at most $q - 2$ points, $O(\lambda_q(n))$ boundary vertices lie on e. Summing over all $O(n)$ edges of domain boundaries of Γ, we obtain the desired bound on the number of boundary vertices.

We call an inner vertex *regular* if it is not an x_1-extremal vertex of any of the three intersection curves. The number of irregular vertices is obviously $O(n^2)$. For a subset $R \subseteq \Gamma$, let $\phi(R)$ denote the number of regular (inner) vertices in $\mathcal{M}(R)$, and let $\phi(r) = \max_{|R|=r} \phi(R)$. The above discussion implies that

$$\phi^*(\Gamma) \leq \phi(\Gamma) + O(n\lambda_q(n)).$$

Next, we derive a recurrence for $\phi(n)$, which will solve to $O(n^{2+\varepsilon})$, for any $\varepsilon > 0$. Fix a regular vertex v of $\mathcal{M}(\Gamma)$. Let $\hat{v}$ be the corresponding vertex of $L(\Gamma)$. Since v is an inner vertex, $\hat{v}$ is well defined, and is one of the two intersection points of three surface patches, say, $\gamma_1, \gamma_2, \gamma_3$. Assume, without loss of generality, that if $|\gamma_1 \cap \gamma_2 \cap \gamma_3| = 2$, then the x_1-coordinate of the other intersection point of γ_1, γ_2, and γ_3 is larger than that of $\hat{v}$. Since $\hat{v}$ is a regular vertex, one of the three pairwise-intersection curves β_{ij}, say β_{12}, lies above $L(\Gamma)$ in the halfspace $x_1 < x_1(v)$ in a sufficiently small neighborhood of $\hat{v}$. We mark on β_{12} the intersection points of β_{12} with other surface patches of Γ and the points that lie above the boundaries of other surface patches in Γ.

We fix a parameter $t = t(\varepsilon)$ and follow β_{12} in the $(-x_1)$-direction, starting from $\hat{v}$, until one of the following three events occurs:[4]

(C1) we reach the left endpoint of β_{12} (including the case of reaching $x_1 = -\infty$);
(C2) β_{12} appears on $L(\Gamma)$; or
(C3) we crossed t marked points on β_{12}.

We call v a vertex of type (Ci), for $i = 1, 2, 3$, if we first reach an event of type (Ci). If (C1) occurs, we charge v to the left endpoint of β_{12}. Since each endpoint is charged at most twice, the total number of regular vertices of type (C1) is $O(n^2)$. (It is easily verified that this bound also holds for the cases where we reach $x_1 = -\infty$.) If (C2) occurs, then we must have passed above the boundary of γ_3 while following β_{12} because β_{12} lies strictly above γ_3 in the halfspace $x_1 < x_1(v)$. Let w be the marked point on β_{12} lying above $\partial \gamma_3^*$ that we have encountered. We charge v to w. Suppose w lies above an edge e of $\partial \gamma_3^*$. We can define C_e and Γ_e as above; then w is a vertex of $\mathcal{A}(\Gamma_e)$. Since (C2) occurred before (C3), at most t marked points lie on β_{12} between v and w, which implies that fewer than t arcs of Γ_e lie below w. As shown in [676], and follows from the Clarkson–Shor probabilistic technique [246], the number of vertices of $\mathcal{A}(\Gamma_e)$ that lie above at most t arcs is $O(t\lambda_q(n))$. Summing over all edges of domain boundaries, the number of marked points on intersection arcs to which a vertex of type (C2) is charged is $O(nt\lambda_q(n))$. Since each marked point is charged $O(1)$ times, the number of type (C2) vertices is $O(nt\lambda_q(n))$.

Finally, if (C3) occurs, then we charge $1/t$ to each marked point on β_{12} that we visited. Each marked point will be charged only $O(1/t)$ units, and each such marked point lies above at most t surface patches of Γ. Theorem 6.1 in Section 6, based on the Clarkson–Shor probabilistic technique [246], implies that the number of such marked points, summed over all intersection curves, is $O(t^3\phi^*(n/t))$. The total number of vertices of type (C3) is thus

$$O(1/t) \cdot O(t^3\phi^*(n/t)) = O(t^2\phi(n/t) + n\lambda_q(n)).$$

---

[4]If the x_1-coordinate of the other intersection point of γ_1, γ_2, and γ_3 were smaller than that of $\hat{v}$, we would have traced β_{12} in the $(+x_1)$-direction.

Hence, we obtain the following recurrence for $\phi(n)$:
$$\phi(n) \leq At^2\phi\left(\frac{n}{t}\right) + Btn\lambda_q(n),$$
where A and B are constants (depending on b). The solution of the above recurrence is
$$\phi(n) = O(tn^{1+\log_t A}\lambda_q(n)).$$
For any fixed $\varepsilon > 0$, if $t = t(\varepsilon)$ is chosen sufficiently large, then $\phi(n) = O(n^{2+\varepsilon})$, where the constant of proportionality depends on ε (and on the other constant parameters). This proves the theorem for $d = 3, s = 2$.

For $s > 2$, Sharir [**678**] introduces the notion of *index* of a regular vertex. The index of a vertex $v \in \bigcap_{i=1}^{3} \gamma_i$ is the number of points of $\bigcap_{i=1}^{3} \gamma_i$ whose x_1-coordinates are smaller than that of v. For $0 \leq j < s$, let $\phi^{(j)}(\Gamma)$ be the number of regular vertices in $L(\Gamma)$ of index j. Then $\phi(\Gamma) = \sum_{j=0}^{s-1} \phi^{(j)}(\Gamma)$.

Modifying the above argument slightly, Sharir derived a system of recurrences for the quantities $\phi^{(j)}(\Gamma)$, for $j < s$. There are three main differences. First, the tracing of β_{12} is always done in the decreasing x_1-direction. Second, the value of the parameter t now depends on j and is denoted by t_j. Third, there is one more stopping criterion:

(C4) β_{12} intersects γ_3; let z be the (first) intersection point.

Using the fact that the index of z is $\leq j - 1$ and that at most t_j surface patches lie below z, Sharir derives the following recurrence for $\phi^{(j)}(n) = \max_{|\Gamma|=n} \phi^{(j)}(\Gamma)$.

$$\phi^{(j)}(n) \leq A_j t_j^2 \phi^*\left(\frac{n}{t_j}\right) + B_j\left(t_j^3 \phi^{(j-1)}\left(\frac{n}{t_j}\right) + nt_j\lambda_q(n)\right).$$

By expanding this system of recurrences and by choosing the values of t_j carefully (in terms of a pre-specified $\varepsilon > 0$), Sharir proved that the solution of this system satisfies
$$\phi^*(n) = O(n^{2+\varepsilon}),$$
for any $\varepsilon > 0$, where the constant of proportionality depends on ε (and on the other constant parameters). The theorem is proved in higher dimensions by induction on d, using a similar charging scheme. See the original paper [**678**] for details. □

PROBLEM 3.2. *Let Γ be a set of n surface patches in $\mathbb{R}^d$ satisfying assumptions (A1) and (A2). Is $\kappa(n, d) = O(n^{d-2}\lambda_q(n))$, where q is a constant depending on d and b?*

We remark that this question has recently been answered in the affirmative for $d = 3, s = 2$, where the surfaces of Γ are graphs of totally defined functions [**344**].

Bounds in special cases. As noted above, sharper bounds are known on the complexity of lower envelopes in some special cases; see [**663, 681**]. For example, if Γ is a set of pseudo-planes in $\mathbb{R}^3$, i.e., each triple of surfaces intersects in at most one point and the intersection of a pair of surfaces in Γ is a single (closed or unbounded) Jordan curve, then $\kappa(\Gamma) = O(n)$. On the other hand, if Γ is a set of pseudo-spheres, i.e., each triple intersects in at most two points and the intersection curve of any pair is a single Jordan curve, then $\kappa(\Gamma) = O(n^2)$. If Γ is a family of hypersurfaces in $\mathbb{R}^d$, a sharper bound on $\kappa(\Gamma)$ can sometimes be proved using the so-called *linearization* technique. Here is a sketch of this technique.

Let $\Gamma = \{\gamma_1, \ldots, \gamma_n\}$ be a collection of hypersurfaces of degree at most b, i.e., each γ_i is the zero set of a d-variate polynomial Q_i of degree at most b. Let $\mathcal{Q} = \{Q_1, \ldots, Q_n\}$. We say that Γ admits a *linearization* of dimension k if, for some $p > 0$, there exists a $(d+p)$-variate polynomial of the form

$$g(\mathbf{x}, \mathbf{a}) = \psi_0(\mathbf{a}) + \psi_1(\mathbf{a})\varphi_1(\mathbf{x}) + \psi_2(\mathbf{a})\varphi_2(\mathbf{x}) + \cdots + \psi_{k-1}(\mathbf{a})\varphi_{k-1}(\mathbf{x}) + \varphi_k(\mathbf{x}),$$

for $\mathbf{x} \in \mathbb{R}^d$, $\mathbf{a} \in \mathbb{R}^p$, so that, for each $1 \leq i \leq n$, we have $Q_i(\mathbf{x}) = g(\mathbf{x}, \mathbf{a}_i)$ for some $\mathbf{a}_i \in \mathbb{R}^p$. Here each $\psi_j(\mathbf{a})$, for $0 \leq j \leq k$, is a p-variate polynomial, and each $\varphi_j(\mathbf{x})$, for $1 \leq j \leq k+1$, is a d-variate polynomial. It is easily seen that such a polynomial representation always exists for $p \leq d^{b+1}$—let the φ's be the monomials that appear in at least one of the polynomials of $\mathcal{Q}$, and let $\psi_j(\mathbf{a}) = a_j$ (where we think of $\mathbf{a}$ as the vector of coefficients of the monomials).

We define a transform $\varphi : \mathbb{R}^d \mapsto \mathbb{R}^k$ that maps each point in $\mathbb{R}^d$ to the point

$$\varphi(\mathbf{x}) = (\varphi_1(\mathbf{x}), \varphi_2(\mathbf{x}), \ldots, \varphi_k(\mathbf{x}));$$

the image $\varphi(\mathbb{R}^d)$ is a d-dimensional algebraic surface Σ in $\mathbb{R}^k$. For each function $Q_i(\mathbf{x}) = g(\mathbf{x}, \mathbf{a}_i)$, we define a k-variate linear function

$$h_i(\mathbf{y}) = \psi_0(\mathbf{a}_i) + \psi_1(\mathbf{a}_i)y_1 + \cdots \psi_{k-1}(\mathbf{a}_i)y_{k-1} + y_k.$$

Let $H = \{h_i = 0 \mid 1 \leq i \leq n\}$ be a set of n hyperplanes in $\mathbb{R}^k$. Let ξ be a vertex of $L(\Gamma)$. If ξ is incident to $\gamma_1, \ldots, \gamma_d$, then $Q_1(\xi) = \cdots = Q_d(\xi) = 0$ and $Q_{d+1}(\xi)\sigma_{d+1}0, \ldots, Q_n(\xi)\sigma_n 0$, where $\sigma_i \in \{>, <\}$. By construction, $Q_i(\xi) = h_i(\varphi(\xi))$. Let $Q \in \mathbb{R}[x_1, \ldots, x_d]$ be a d-variate polynomial. If we regard Q as a univariate polynomial in x_d and the coefficient of the leading term in Q is a positive constant, then we call Q a *positive polynomial*. If all Q_i's are positive, then, by the definition of lower envelopes, $Q_i(\xi) < 0$ for every $i > d$. Hence, $h_1(\varphi(\xi)) = \cdots h_d(\varphi(\xi)) = 0$ and $h_{d+1}(\varphi(\xi)) < 0, \ldots, h_n(\varphi(\xi)) < 0$. That is, $\varphi(\xi)$ is a vertex of $L(H) \cap \Sigma$. Since each h_i is a hyperplane in $\mathbb{R}^k$ and the degree of Σ depends only on d and b, the Upper Bound Theorem for convex polyhedra (see McMullen and Shephard [**543**] and Ziegler [**759**]) implies that the number of vertices in $\Sigma \cap L(H)$ is $O(n^{\lfloor k/2 \rfloor})$. Hence, we obtain the following result.

THEOREM 3.3. *Let Γ be a collection of n algebraic hypersurfaces in $\mathbb{R}^d$, of constant maximum degree b. If Γ admits a linearization of dimension k and each surface Γ is the zero set of a positive polynomial, then $\kappa(\Gamma) = O(n^{\lfloor k/2 \rfloor})$, where the constant of proportionality depends on k, d, and b.*

We illustrate the linearization technique by giving an example. A sphere in $\mathbb{R}^d$ with center $(a_1, \ldots, a_d)$ and radius a_{d+1} can be regarded as the zero set of the polynomial $g(\mathbf{x}, \mathbf{a})$, where

$$\begin{aligned} g(\mathbf{x}, a_1, \ldots, a_{d+1}) &= [a_1^2 + \cdots + a_d^2 - a_{d+1}^2] - [2a_1 \cdot x_1] - [2a_2 \cdot x_2] - \cdots - \\ & \quad [2a_d \cdot x_d] + [x_1^2 + \cdots x_d^2]. \end{aligned}$$

Thus, setting

$$\psi_0(\mathbf{a}) = \sum_{i=1}^d a_i^2 - a_{d+1}^2, \quad \psi_1(\mathbf{a}) = -2a_1, \quad \cdots \quad \psi_d(\mathbf{a}) = -2a_d, \quad \psi_{d+1}(\mathbf{a}) = 1,$$

$$\varphi_1(\mathbf{x}) = x_1, \quad \cdots \quad \varphi_d(\mathbf{x}) = x_d, \quad \varphi_{d+1}(\mathbf{x}) = \sum_{i=1}^d x_i^2,$$

we obtain a linearization of dimension $d+1$. We thus obtain the following corollary.

COROLLARY 3.4. *Let Γ be a set of n spheres in $\mathbb{R}^d$. Then $\kappa(\Gamma) = O(n^{\lceil d/2 \rceil})$.*

REMARK 3.5. *An intriguing challenge is to find scenarios where one can prove that the actual complexity of $\Sigma \cap L(H)$ is smaller than $O(n^{\lfloor k/2 \rfloor})$. This is clearly the case when $\lfloor k/2 \rfloor \geq d$ (there is no point in using linearization for such values of k). See Theorem 5.3 below for the context related to this question.*

The overlay of minimization diagrams. Motivated by several applications, researchers have studied the complexity of the overlay of two minimization diagrams. That is, let Γ and Γ' be two families of surface patches satisfying assumptions (A1)–(A2); set $n = |\Gamma| + |\Gamma'|$. The *overlay* of $\mathcal{M}(\Gamma)$ and $\mathcal{M}(\Gamma')$ is the decomposition of $\mathbb{R}^{d-1}$ into maximal connected regions so that each region lies within a fixed pair of faces of $\mathcal{M}(\Gamma)$ and $\mathcal{M}(\Gamma')$. It is conjectured that the complexity of the overlay of the two diagrams is also close to $O(n^{d-1})$; that is, the bound on the complexity of the overlay is asymptotically similar to that for a single diagram. Although this conjecture is obviously true for the minimization diagrams of arcs in the plane, it is not intuitive even in $\mathbb{R}^3$ because the overlay of two planar maps with m edges each may have $\Omega(m^2)$ vertices. Edelsbrunner et al. [**296**] proved an $O(n^{d-1}\alpha(n))$ upper bound if Γ and Γ' are sets of a total of n $(d-1)$-simplices in $\mathbb{R}^d$.

Agarwal *et al.* [**43**] proved that the overlay of two minimization diagrams, defined for a total of n surface patches, in $\mathbb{R}^3$ has $O(n^{2+\varepsilon})$ complexity, for any $\varepsilon > 0$. Note that in $\mathbb{R}^3$, each vertex of the overlay is a vertex of $\mathcal{M}(\Gamma)$, a vertex of $\mathcal{M}(\Gamma')$, or an intersection point of an edge of $\mathcal{M}(\Gamma)$ with an edge of $\mathcal{M}(\Gamma')$. The proof in [**43**] establishes an upper bound on the number of intersection points by generalizing the proof technique of Theorem 3.1.

More recently, Koltun and Sharir [**495**] have extended the result to $\mathbb{R}^4$, showing that the complexity of the overlay of two or three minimization diagrams of a total of n surface patches in $\mathbb{R}^4$ satisfying assumptions (A1)–(A5) is $O(n^{3+\varepsilon})$, for any $\varepsilon > 0$. They also gave a simpler proof for the near-quadratic bound in $\mathbb{R}^3$ of [**43**]. A simple recent observation by Koltun and Sharir [**497**] reduces the overlay of k minimization diagrams in $\mathbb{R}^d$, for $k \leq d-1$, into a substructure of a *single* minimization diagram in $\mathbb{R}^{d+k-1}$, implying that the complexity of the overlay is $O(n^{d+k-2+\varepsilon})$, for any $\varepsilon > 0$. While these bounds are not sharp (compare, e.g., with the bounds cited above for $d = 3$, $k = 2$), they improve upon previous naive estimates. Moreover, for the case of minimization diagrams of *hyperplanes*, Koltun and Sharir obtain improved bounds that turn out to be tight in the worst case. Their results are related to, and inspired by, a similar observation due to Chan [**203**].

PROBLEM 3.6. *What is the complexity of the overlay of two (or any number up to $d-1$) minimization diagrams of collections of surface patches in $\mathbb{R}^d$, for $d \geq 5$?*

The following problem is closely related to the overlay of minimization diagrams. Let Γ, Γ' be two sets of surface patches in $\mathbb{R}^d$ satisfying (A1)–(A5). Regarding each surface as the graph of a partially defined function, define

$$\mathcal{S}(\Gamma, \Gamma') = \left\{ \mathbf{x} \;\middle|\; L_\Gamma(x_1, \ldots, x_{d-1}) \geq x_d \geq U_{\Gamma'}(x_1, \ldots, x_{d-1}) \right\},$$

i.e., $\mathcal{S}(\Gamma, \Gamma')$, called the *sandwich region*, is the set of points lying above all surface patches of Γ' and below all surface patches of Γ; see Figure 4. It can be shown that the combinatorial complexity of $\mathcal{S}(\Gamma, \Gamma')$ is at most proportional to the complexity

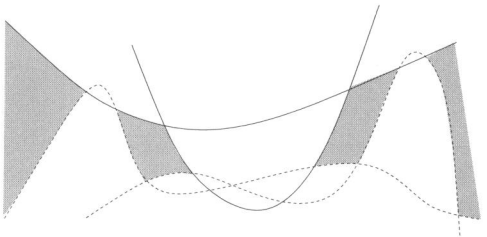

FIGURE 4. The sandwich region $\mathcal{S}(\Gamma, \Gamma')$; solid arcs are in Γ, and dashed arcs are in Γ'.

of the overlay of the minimization diagram of Γ and the maximization diagram of Γ'. The results of Agarwal *et al.* [**43**] and of Koltun and Sharir [**495**] imply that the complexity of $\mathcal{S}(\Gamma, \Gamma')$ is $O(n^{2+\varepsilon})$ in 3-space, and $O(n^{3+\varepsilon})$ in 4-space, for any $\varepsilon > 0$. In general, the complexity of the overlay of the minimization diagram of Γ and the maximization diagram of Γ' may be larger than that of $\mathcal{S}(\Gamma, \Gamma')$. As an application, which also illustrates this discrepancy, consider the following example. Let $S = \{S_1, \ldots, S_n\}$ be a set of n spheres in $\mathbb{R}^3$. A line in $\mathbb{R}^3$ can be parameterized by four real parameters. We can therefore define the set of lines tangent to a sphere S_i and lying above (resp., below) S_i as a surface patch γ_i (resp., γ_i') in $\mathbb{R}^4$. Define $\Gamma = \{\gamma_i \mid 1 \leq i \leq n\}$ and $\Gamma' = \{\gamma_i' \mid 1 \leq i \leq n\}$. Agarwal *et al.* [**21**] showed that, if the lines are parameterized carefully, then $\mathcal{S}(\Gamma, \Gamma')$ is the set of lines intersecting all the spheres of S and that the combinatorial complexity of $\mathcal{S}(\Gamma, \Gamma')$ is $O(n^{3+\varepsilon})$, for any $\varepsilon > 0$. However, a construction of Pellegrini [**621**] implies that the combinatorial complexity of the overlay of the two corresponding minimization diagrams can be $\Omega(n^4)$.

For the special case of arrangements of $(d-1)$-simplices in $\mathbb{R}^d$, the proofs of [**289, 595**] actually imply that the complexity of the sandwich region is also $O(n^{d-1}\alpha(n))$. A recent result of Koltun and Wenk [**498**] establishes a slightly weaker bound on the complexity of the overlay of minimization diagrams in arrangements of $(d-1)$-simplices in $\mathbb{R}^d$.

4. Single Cells

Lower envelopes are closely related to other substructures in arrangements, notably *cells* and *zones*. See Figure 5 for an illustration of a cell in a planar arrangement. The lower envelope is a portion of the boundary of the bottommost cell of the arrangement, though the worst-case complexity of $L(\Gamma)$ can be larger than that of the bottommost cell of $\mathcal{A}(\Gamma)$. In two dimensions, it was shown by Guibas *et al.* [**402**] (see also Chapter 3) that the complexity of a single face in an arrangement of n arcs, each pair of which intersect in at most s points, is $O(\lambda_{s+2}(n))$, and so has the same asymptotic bound as the complexity of the lower envelope of such a collection of arcs. The same holds more or less in higher dimensions, although it did take some time to get there: The complexity of a single cell in an arrangement of n surface patches in $\mathbb{R}^d$ satisfying the assumptions (A1) and (A2) is $O(n^{d-1+\varepsilon})$, for any $\varepsilon > 0$—see below. The Upper Bound Theorem implies that the complexity of a single cell in an arrangement of hyperplanes in $\mathbb{R}^d$ is $O(n^{\lfloor d/2 \rfloor})$, and the linearization technique described in Section 3 implies that the complexity of a single cell in an arrangement of n spheres is $O(n^{\lceil d/2 \rceil})$. However, the lower-bound construction for lower envelopes implies a lower bound of $\Omega(n^{d-1}\alpha(n))$ for the complexity of a single cell for arrangements of simplices.

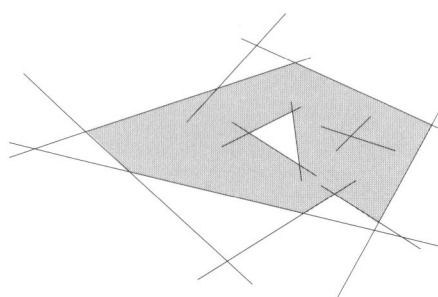

FIGURE 5. A single cell in an arrangement of segments.

Pach and Sharir [**595**] were the first to prove a subcubic upper bound on the complexity of a single cell in arrangements of triangles in $\mathbb{R}^3$. This bound was improved by Aronov and Sharir [**102**] to $O(n^{7/3})$, and subsequently to $O(n^2 \log n)$ [**104**]. The latter approach extends to higher dimensions; that is, the complexity of a single cell in an arrangement of n $(d-1)$-simplices in $\mathbb{R}^d$ is $O(n^{d-1} \log n)$. A simpler proof was given by Tagansky [**714**]. These approaches, however, do not extend to nonlinear surfaces even in $\mathbb{R}^3$.

Halperin [**412, 413**] proved near-quadratic bounds on the complexity of a single cell in arrangement of certain classes of n bivariate surface patches, which arise in motion-planning applications. Halperin and Sharir [**422**] proved a near-quadratic bound on the complexity of a single cell in an arrangement of the contact surfaces that arise in a rigid motion of a simple polygon amid polygonal obstacles in the plane, i.e., the surfaces that represent the placements of the polygon at which it touches one of the obstacles. The proof borrows ideas from the proof of Theorem 3.1.

A near-optimal bound on the complexity of a single cell in the arrangement of an arbitrary collection of surface patches in $\mathbb{R}^3$ satisfying assumptions (A1) and (A2) was finally proved by Halperin and Sharir [**420**]. It took some time to extend to higher dimensions the main topological property that is needed in the proof. This has been done by Basu [**131**], who has thus obtained the general result:

THEOREM 4.1 (Halperin and Sharir [**420**], Basu [**131**]). *Let Γ be a set of surface patches in $\mathbb{R}^d$ satisfying assumptions (A1) and (A2). Then the complexity of a single cell in $\mathcal{A}(\Gamma)$ is $O(n^{d-1+\varepsilon})$, for any $\varepsilon > 0$, where the constant of proportionality depends on ε, d, and on the maximum degree of the surface patches and of their boundaries.*

The proof (in $\mathbb{R}^3$, say) proceeds along the same lines as the proof of Theorem 3.1. However, the following two additional results are established to "bootstrap" the recurrences that the proof derives. Let C be the cell of $\mathcal{A}(\Gamma)$ whose complexity we want to bound.

  (a) There are only $O(n^2)$ vertices v of the cell C that are locally x-extreme (that is, there is a neighborhood N of v and a connected component C' of the intersection of N with the interior of C, such that v lies to the left

(in the x-direction) of all points of C', or v lies to the right of all these points).

  (b) There are only $O(n^{2+\varepsilon})$ vertices, for any $\varepsilon > 0$, on *popular faces* of C, that is, 2-faces f for which C lies locally near f on both sides of f.

Property (a) is proved by an appropriate decomposition of C into $O(n^2)$ subcells, in the style of a *Morse decomposition* of C (see [**550**]), so that each subcell has at most two points that are locally x-extreme in C. Property (b) is proved by applying the machinery of the proof of Theorem 3.1, where the quantity to be analyzed is the number of vertices of popular faces of C, rather than all inner vertices. Once these two results are available, the proof of Theorem 3.1 can be carried through, with appropriate modifications, to yield a recurrence for the number of vertices of C, whose solution is $O(n^{2+\varepsilon})$, for any $\varepsilon > 0$. We refer the reader to the original paper [**420**] for more details. Basu's paper [**131**] follows the same recurrence scheme, and provides an appropriate extension of (a) and (b) to higher dimensions.

The linearization technique in the previous section can be extended to bound the complexity of a cell as well, namely, one can prove the following result.

THEOREM 4.2. *Let Γ be a collection of n hypersurfaces in $\mathbb{R}^d$, of constant maximum degree b. If Γ admits a linearization of dimension k, then the combinatorial complexity of a cell of $\mathcal{A}(\Gamma)$ is $O(n^{\lfloor k/2 \rfloor})$, where the constant of proportionality depends on k, d, and b.*

5. Zones

Let Γ be a set of n surfaces in $\mathbb{R}^d$. The *zone* of a variety σ (not belonging to Γ), denoted as $zone(\sigma; \Gamma)$, is defined to be the set of d-dimensional cells in $\mathcal{A}(\Gamma)$ that intersect σ. The complexity of $zone(\sigma; \Gamma)$ is defined to be the sum of the complexities of the cells of $\mathcal{A}(\Gamma)$ that belong to $zone(\sigma; \Gamma)$, where the complexity of a cell in $\mathcal{A}(\Gamma)$ is the number of faces of all dimensions that are contained in the closure of the cell.

The complexity of a zone was first studied by Edelsbrunner *et al.* [**301**]; see also [**224**]. The "classical" zone theorem [**288, 303**] asserts that the maximum complexity of the zone of a hyperplane in an arrangement of n hyperplanes in $\mathbb{R}^d$ is $\Theta(n^{d-1})$, where the constant of proportionality depends on d. The original proof given by Edelsbrunner *et al.* [**301**] had some technical problems. A correct, and simpler, proof was given by Edelsbrunner *et al.* [**303**]. Their technique is actually quite general and can also be applied to obtain several other interesting combinatorial bounds involving arrangements. For example, the proof by Aronov and Sharir for the complexity of a single cell in arrangements of simplices [**104**] used a similar approach. Other results based on this technique can be found in [**12, 98, 100**]. We therefore describe the technique, as applied in the proof of the zone theorem:

THEOREM 5.1 (Zone theorem [**303**]). *The maximum complexity of the zone of a hyperplane in an arrangement of n hyperplanes in $\mathbb{R}^d$ is $\Theta(n^{d-1})$.*

This result is easy to prove for $d = 2$; see Chapter 3 for details. For a set Γ of n hyperplanes in $\mathbb{R}^d$ and another hyperplane b, let $\tau_k(b; \Gamma)$ denote the total number of k-faces contained on the boundary of cells in $zone(b; \Gamma)$; each such k-face is counted once for each cell that it bounds. Let

$$\tau_k(n,d) = \max \tau_k(b;\Gamma),$$

where the maximum is taken over all hyperplanes b and all sets Γ of n hyperplanes in $\mathbb{R}^d$. The maximum complexity of $zone(b;\Gamma)$ is at most $\sum_{k=0}^{d} \tau_k(n,d)$. Thus the following lemma immediately implies the upper bound in Theorem 5.1.

LEMMA 5.2. *For each d and $0 \leq k \leq d$,*

$$\tau_k(n,d) = O(n^{d-1}),$$

where the constants of proportionality depend on d and k.

PROOF. We use induction on d. As just noted, the claim holds for $d = 2$. Assume that the claim holds for all $d' < d$, let Γ be a set of n hyperplanes in $\mathbb{R}^d$, and let b be some other hyperplane. Without loss of generality, we can assume that the hyperplanes in $\Gamma \cup \{b\}$ are in general position. We define a k-*border* to be a pair (f, C), where f is a k-face incident to a d-dimensional cell C of $\mathcal{A}(\Gamma)$. Thus $\tau_k(b;\Gamma)$ is the total number of k-borders (f, C) for which $C \in zone(b;\Gamma)$.

We pick a hyperplane $\gamma \in \Gamma$ and count the number of all k-borders (f, C) in $zone(b;\Gamma)$ such that f is not contained in γ. If we remove γ from Γ, then any such k-border is contained in a k-border (f', C') of $zone(b;\Gamma \setminus \{\gamma\})$ (i.e., $f \subseteq f'$ and $C \subseteq C'$). Our strategy is thus to consider the collection of k-borders in $zone(b;\Gamma \setminus \{\gamma\})$, and to estimate the increase in the number of k-borders as we add γ back to Γ. Observe that we do not count k-borders that lie in γ.

Let $\Gamma|_\gamma = \{\gamma' \cap \gamma \mid \gamma' \in \Gamma \setminus \{\gamma\}\}$; the set $\Gamma|_\gamma$ forms a $(d-1)$-dimensional arrangement of $n-1$ hyperplanes within γ. Let (f, C) be a k-border of $zone(b;\Gamma \setminus \{\gamma\})$, and consider what happens to it when we reinsert γ. The following cases may occur:

- $\gamma \cap C = \emptyset$: In this case the k-border (f, C) gives rise to exactly one k-border in $zone(b;\Gamma)$, namely itself.
- $\gamma \cap C \neq \emptyset$, $\gamma \cap f = \emptyset$: Let γ^+ be the open halfspace bounded by γ that contains f, and let $C^+ = C \cap \gamma^+$. If C^+ intersects b, then (f, C) gives rise to one k-border in $zone(b;\Gamma)$, namely (f, C^+) (this is the case for the edge $f = e$ in Figure 6); otherwise it gives rise to no k-border in $zone(b;\Gamma)$.
- $\gamma \cap f \neq \emptyset$: Let γ^+ and γ^- be the two open halfspaces bounded by γ and let $C^+ = C \cap \gamma^+$ and $C^- = C \cap \gamma^-$. If the closure of only one of C^+ and C^- intersects b, say, C^+, then (f, C) gives rise to only one k-border in $zone(b;\Gamma)$, namely $(f \cap \gamma^+, C^+)$ (this is the case for the edge $f = e'$ in Figure 6). If both C^+ and C^- intersect b, then (f, C) gives rise to two k-borders in $zone(b;\Gamma)$, namely $(f \cap \gamma^+, C^+)$ and $(f \cap \gamma^-, C^-)$ (this is the case for the edge $f = e''$ in Figure 6). In this case, however, we can charge uniquely this increase in the number of k-borders to $(f \cap \gamma, C \cap \gamma)$, which, as easily seen, is a $(k-1)$-border in $zone(b \cap \gamma; \Gamma|_\gamma)$.

If we repeat this process over all k-borders of $zone(b;\Gamma \setminus \{\gamma\})$, we obtain that the total number of k-borders (f, C) in $zone(b;\Gamma)$, for f not contained in γ, is at most

$$\begin{aligned}
\tau_k(b;\Gamma \setminus \{\gamma\}) + \tau_{k-1}(b \cap \gamma; \Gamma|_\gamma) &\leq \tau_k(n-1,d) + \tau_{k-1}(n-1,d-1) \\
&= \tau_k(n-1,d) + O(n^{d-2}),
\end{aligned}$$

5. ZONES

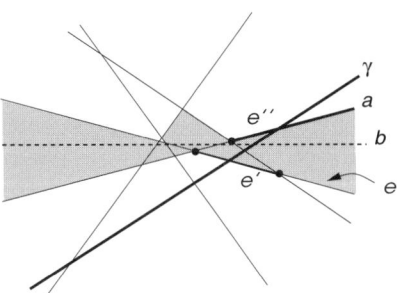

FIGURE 6. Inserting γ into $zone(b; \Gamma \setminus \{\gamma\})$.

where the last inequality follows from the induction hypothesis. Repeating this analysis for all hyperplanes $\gamma \in \Gamma$, summing up the resulting bounds, and observing that each k-border of $zone(b; \Gamma)$ is counted exactly $n - d + k$ times, we obtain

$$\tau_k(n, d) \leq \frac{n}{n - d + k} \left(\tau_k(n - 1, d) + O(n^{d-2}) \right).$$

Edelsbrunner et al. [303] showed that this recurrence solves to $O(n^{d-1})$ for $k \geq 2$. Using Euler's formula for cell complexes, one can show that $\tau_k(n, d) = O(n^{d-1})$ for $k = 0, 1$ as well. This completes the proof of the theorem. For the lower bound, it is easily checked that the complexity of the zone of a hyperplane b in an arrangement of n hyperplanes in $\mathbb{R}^d$ in general position is $\Omega(n^{d-1})$. In fact, the complexity of the cross-section of the arrangement within b is already $\Omega(n^{d-1})$. □

The above technique can be extended to bound the quantity $\sum_{C \in \mathcal{A}(\Gamma)} |C|^2$, where Γ is a set of hyperplanes, C ranges over all d-dimensional cells of the arrangement, and $|C|$ denotes the number of lower-dimensional faces incident to C. For $d \leq 3$, an easy application of the zone theorem (Theorem 5.1) implies that $\sum_C |C|^2 = O(n^d)$; this bound is obviously tight if the lines or planes of Γ are in general position. For $d > 3$, the same application of the zone theorem yields only $\sum_C |C| f_C = O(n^d)$, where f_C is the number of hyperplanes of Γ meeting the boundary of C. Using the same induction scheme as in the proof of Theorem 5.1, Aronov et al. [98] showed that

$$\sum_{C \in \mathcal{A}(\Gamma)} |C|^2 = O(n^d \log^{\lfloor d/2 \rfloor - 1} n).$$

A different proof and a slight strengthening of the bound has subsequently been given by Aronov and Sharir [108]. It is believed that the right bound is $O(n^d)$. Note that such a result does not hold for arrangements of simplices or of surfaces because the complexity of single cell can be $\Omega(n^{d-1})$.

The zone theorem for hyperplane arrangements can be extended as follows.

THEOREM 5.3 (Aronov, Pellegrini, and Sharir [100]). Let Γ be a set of n hyperplanes in $\mathbb{R}^d$. Let σ be a p-dimensional algebraic variety of some fixed degree, or the relative boundary of any convex set with affine dimension $p + 1$, for $0 \leq p \leq d$. The complexity of the $zone(\sigma; \Gamma)$ is $O(n^{\lfloor (d+p)/2 \rfloor} \log^\beta n)$, where $\beta = d + p \pmod{2}$, and the bound is almost tight (up to the logarithmic factor) in the worst case.

In particular, for $p = d - 1$, the complexity of the zone is $O(n^{d-1} \log n)$, which is almost the same as the complexity of the zone of a hyperplane in such an arrangement.

The proof proceeds along the same lines of the inductive proof of Theorem 5.1. However, we face the following technical difference. When we remove and re-insert a hyperplane $\gamma \in \Gamma$, a face f of a cell C of $zone(\sigma; \Gamma \setminus \{\gamma\})$ may be split into two subfaces, both lying in $zone(\sigma; \Gamma)$, but γ need not meet σ within C (as it has to do when σ is a hyperplane). Therefore, the charging scheme used in the proof of Theorem 5.1 becomes inadequate, because $f \cap \gamma$ need not belong to the zone of $\sigma \cap \gamma$ in the $(d-1)$-dimensional cross-section of $\mathcal{A}(\Gamma)$ along γ. What is true, however, is that $f \cap \gamma$ is a face incident to a *popular facet* of $zone(\sigma; \Gamma)$ along γ, that is, a facet $g \subseteq \gamma$ whose two incident cells belong to the (d-dimensional) zone. Thus the induction proceeds not by decreasing the dimension of the arrangement (as was done in the proof of Theorem 5.1), but by reapplying the same machinery to bound the number of vertices of popular facets of the original $zone(\sigma; \Gamma)$. This in turn requires similar bounds on the number of vertices of lower-dimensional popular faces. We refer the reader to Aronov *et al.* [**100**] for more details.

In general, the zone of a surface in an arrangement of n surfaces in $\mathbb{R}^d$ can be transformed to a single cell in another arrangement of $O(n)$ surface patches in $\mathbb{R}^d$. For example, Let Γ be a set of n $(d-1)$-simplices in $\mathbb{R}^d$, and let σ be a hyperplane. We split each $\gamma \in \Gamma$ into two polyhedra at the intersection of γ and σ (if the intersection is nonempty), push these two polyhedra slightly away from each other, and, if necessary, retriangulate each polyhedron into a constant number of simplices. In this manner, we obtain a collection Γ' of $O(n)$ simplices, and all cells of the zone of σ in $\mathcal{A}(\Gamma)$ now fuse into a single cell of $\mathcal{A}(\Gamma')$. Moreover, by the general position assumption, the complexity of the zone of σ in Γ is easily seen to be dominated by the complexity of the new single cell of $\mathcal{A}(\Gamma')$. (The same technique has been used earlier in [**295**], to obtain a near-linear bound on the complexity of the zone of an arc in a two-dimensional arrangement of arcs.) Hence, the following theorem is an easy consequence of the result by Aronov and Sharir [**104**].

THEOREM 5.4. *The complexity of the zone of a hyperplane in an arrangement of n $(d-1)$-simplices in $\mathbb{R}^d$ is $O(n^{d-1} \log n)$.*

Using a similar argument one can prove the following.

THEOREM 5.5 (Halperin and Sharir [**420**], Basu [**131**]). *Let Γ be a collection of n surface patches in $\mathbb{R}^d$, satisfying assumptions (A1) and (A2). The combinatorial complexity of the zone in $\mathcal{A}(\Gamma)$ of an algebraic surface σ of some fixed degree is $O(n^{d-1+\varepsilon})$, for any $\varepsilon > 0$, where the constant of proportionality depends on ε, d, on the maximum degree of the given surfaces and their boundaries, and on the degree of σ.*

6. Levels

The *level* of a point $p \in \mathbb{R}^d$ in an arrangement $\mathcal{A}(\Gamma)$ of a set Γ of surface patches satisfying (A1)–(A3) is the number of surfaces of Γ lying vertically below p. For $0 \leq k < n$, we define the *k-level* (resp., *$(\leq k)$-level*), denoted by $\mathcal{A}_k(\Gamma)$ (resp., $\mathcal{A}_{\leq k}(\Gamma)$), to be the closure of all points on the surfaces of Γ whose level is k (resp., at most k). A face of $\mathcal{A}_k(\Gamma)$ or $\mathcal{A}_{\leq k}(\Gamma)$ is a maximal connected portion of a face of $\mathcal{A}(\Gamma)$ consisting of points having a fixed subset of surfaces lying below them. For totally defined functions, any such face coincides with a face of $\mathcal{A}(\Gamma)$. Note that $\mathcal{A}_0(\Gamma)$ is the same as $L(\Gamma)$. If the surfaces of Γ are graphs of totally defined functions, then the level of all points on a face of $\mathcal{A}(\Gamma)$ is the same and $\mathcal{A}_k(\Gamma)$ is the graph of a totally defined continuous function; otherwise $\mathcal{A}_k(\Gamma)$ may have discontinuities. See Figure 7 for an example of levels in arrangements of lines and of segments.

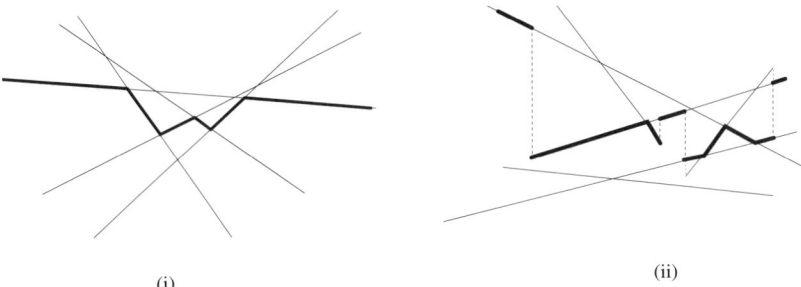

FIGURE 7. The 2-level in (i) an arrangement of lines, and (ii) in an arrangement of segments.

Levels in hyperplane arrangements in $\mathbb{R}^d$ are closely related to *k-sets* of point sets in $\mathbb{R}^d$. Let S be a set of n points in $\mathbb{R}^d$, and let S^* be the set of hyperplanes dual to S (recall example (c) in the introduction). A subset $A \subset S$ is called a *k-set* (resp., $\leq k$-*set*) if $|A| = k$ (resp., $|A| \leq k$) and A can be strictly separated from $S \setminus A$ by a hyperplane h. The level of the point h^*, dual to h, in $\mathcal{A}(S^*)$ is k or $n-k$. The *k-set problem* is to bound the maximum number of k-sets of S (in terms of k and n). It is easy to see that the maximum number of k-sets in a set of n points in $\mathbb{R}^d$ is bounded by the maximum number of facets in the k-level and the $(n-k)$-level in an arrangement of n hyperplanes in $\mathbb{R}^d$.

Let $\psi_k(\Gamma)$ (resp., $\psi_{\leq k}(\Gamma)$) be the total number of faces in $\mathcal{A}_k(\Gamma)$ (resp., $\mathcal{A}_{\leq k}(\Gamma)$). Let $\mathbf{G}$ be a (possibly infinite) family of surfaces in $\mathbb{R}^d$ satisfying assumptions (A1)–(A3). We define $\psi_k(n, d, \mathbf{G}) = \max \psi_k(\Gamma)$ and $\psi_{\leq k}(n, d, \mathbf{G}) = \max \psi_{\leq k}(\Gamma)$, where the maximum in both cases is taken over all subsets $\Gamma \subseteq \mathbf{G}$ of size n. If $\mathbf{G}$ is not important or follows from the context, we will omit the argument $\mathbf{G}$.

The following theorem follows from a result by Clarkson and Shor [**246**], already used earlier for the analysis of the complexity of lower envelopes.

THEOREM 6.1 (Clarkson and Shor [**246**]). *Let $\mathbf{G}$ be an infinite family of surfaces satisfying assumptions (A1)–(A3). Then for any $0 \leq k < n - d$,*

$$\psi_{\leq k}(n, d, \mathbf{G}) = O\left((k+1)^d \kappa\left(\frac{n}{k+1}, d, \mathbf{G}\right)\right),$$

where $\kappa(n, d, \mathbf{G}) = \psi_0(n, d, \mathbf{G})$ is the maximum complexity of the lower envelope of n surfaces in $\mathbf{G}$.

PROOF. Let $\Gamma \subset \mathbf{G}$ be a set of n surface patches satisfying assumptions (A1)–(A5). For a subset $X \subseteq \Gamma$ and an integer $0 \leq k \leq |X| - d$, let $V_k(X)$ denote the set of vertices at level k in $\mathcal{A}(X)$. As is easily seen, $\psi_{\leq k}(\Gamma)$ is proportional to $\sum_{j=0}^{k} |V_j(\Gamma)|$, which we thus proceed to bound. We bound below only the number of vertices in the first k levels that are incident to the relative interiors of d surface patches; the other types of vertices are easier to analyze, and the same bound applies to them as well. We choose a random subset $R \subseteq \Gamma$ of size $r = \lfloor n/(k+1) \rfloor$ and bound the expected number of vertices in $V_0(R)$. Assuming general position, a vertex $v \in V_j(\Gamma)$ is in $V_0(R)$ if and only if the d surfaces incident to v are in R and none of the j surfaces of Γ lying below v is chosen in R, so the probability that $v \in V_0(R)$ is $\binom{n-j-d}{r-d}/\binom{n}{r}$. Hence, easy manipulation of binomial coefficients implies that

$$\begin{aligned} \mathbf{E}\{|V_0(R)|\} &= \sum_{j=0}^{n-d} |V_j(\Gamma)| \frac{\binom{n-j-d}{r-d}}{\binom{n}{r}} \\ &\geq \sum_{j=0}^{k} |V_j(\Gamma)| \frac{\binom{n-j-d}{r-d}}{\binom{n}{r}} \\ &= \Omega\left(\frac{1}{(k+1)^d}\right) \sum_{j=0}^{k} |V_j(\Gamma)|. \end{aligned}$$

Thus

(6.4) $$\sum_{j=0}^{k} |V_j(\Gamma)| \leq c(k+1)^d \mathbf{E}\{|V_0(R)|\},$$

for some constant c. Since every vertex in $V_0(R)$ lies on the lower envelope of R, the assertion now follows from the definition of κ. $\square$

COROLLARY 6.2.
(i) $\psi_{\leq k}(n, d) = O((k+1)^{1-\varepsilon} n^{d-1+\varepsilon})$.
(ii) Let $\mathbf{H}$ be the set of all hyperplanes in $\mathbb{R}^d$. Then,

$$\psi_{\leq k}(n, d, \mathbf{H}) = \Theta\bigl(n^{\lfloor d/2 \rfloor}(k+1)^{\lceil d/2 \rceil}\bigr).$$

PROOF. Part (i) follows from Theorems 3.1 and 6.1. Part (ii) follows from the fact that $\kappa(n, d, \mathbf{H}) = \Theta(n^{\lfloor d/2 \rfloor})$. $\square$

There is even a tighter upper bound of $n(k+1)$ on the number of $\leq k$-sets of n points in the plane, $k \leq n/2$ [**73**, **620**]; see also [**383**].

In contrast to these bounds on the complexity of the ($\leq k$)-level, which are tight or almost tight in the worst case, much less is known about the complexity of a single k-level, even for the simplest case of arrangements of lines in the plane. For example, Corollary 6.2(ii), for $d = 2$, implies that the complexity of an average level in an arrangement of lines in the plane is linear, but no upper bound that is even close is known. For a set Γ of n lines in the plane, Lovász[5] [**518**] proved that $\psi_{\lfloor n/2 \rfloor}(\Gamma) = O(n^{3/2})$. Erdős et al. [**330**] extended his argument to prove that $\psi_k(\Gamma) = O(n\sqrt{k+1})$. Since the original proof, many different proofs have been proposed for obtaining the same bound on $\psi_k(\Gamma)$ [**18**, **405**]. Goodman and Pollack [**384**] proved a similar bound on the maximum complexity of the k-level

---

[5]According to L. Lovász [**518**], the $(n/2)$-set problem was originally posed by A. Simmons, and E. Strauss had constructed a set of points in the plane in which the number of $(n/2)$-sets was $\Omega(n \log n)$.

in an arrangement of pseudo-lines. Erdős et al.'s bound was slightly improved by Pach et al. [**603**] to $o(n\sqrt{k+1})$, using a rather complicated argument.

A major breakthrough in this direction was made in the late 1990s by Dey, who obtained the following improvement.

THEOREM 6.3 (Dey [**264**]). *Let Γ be a set of n lines in the plane. Then for any $0 \leq k < n$, $\psi_k(\Gamma) = O(n(k+1)^{1/3})$.*

Dey's proof is quite simple and elegant. It uses the following result on geometric graphs, which was independently proved by Ajtai et al. [**61**] and by Leighton [**509**].[6] We will later discuss this result in detail in Chapter 5 (and also in Chapter 4).

LEMMA 6.4 (Crossing Lemma). *Let G be a geometric graph with n vertices and $m \geq 4n$ edges. Then there are $\Omega(m^3/n^2)$ pairs of edges in G whose relative interiors cross.*

PROOF OF THEOREM 6.3. For simplicity we assume that n is even and prove the bound for $k = n/2$. We argue in the dual plane, where we have a set S of n points in general position and we wish to establish the asserted bound for the number of *halving segments* of S, where a halving segment is a straight segment connecting a pair of points $u, v \in S$ so that the line passing through u and v has exactly $(n/2)-1$ points of S below it. Let H denote the set of halving segments.

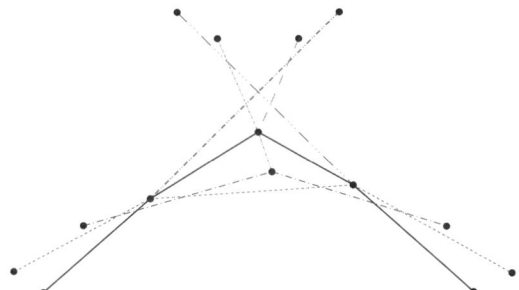

FIGURE 8. A set of 14 points with 14 halving segments, split into 7 convex x-monotone chains.

The segments in H can be decomposed into $n/2$ convex x-monotone chains as follows. Let uv be an edge of H, with v lying to the right of u. We rotate the line that passes through u and v counterclockwise about v and stop as soon as the line overlaps another halving segment vw incident to v, or reaches the vertical direction. It is easy to check that, if w exists, it lies to the right of v and that uvw is a left turn.[7] We now rotate about w, and continue in this manner until our line becomes vertical. We apply the same procedure "backwards" by turning the line uv clockwise around u, and keep iterating until the line becomes vertical. The halving segments that we have encountered during the whole process constitute one convex polygonal chain. By applying this procedure repeatedly, we obtain the desired decomposition of the entire H into convex x-monotone polygonal chains.

---

[6]A *geometric graph* $G = (V, E)$ is a graph drawn in the plane so that its vertices are points and its edges are straight segments connecting pairs of these points. A geometric graph need not be planar.

[7]This property is called the *antipodality* property of the graph of halving segments.

Using the properties of halving segments proved by Lovász [**518**], we can conclude that the segments are partitioned into $n/2$ convex chains. (These convex chains are in a certain sense dual to the concave chains in the dual line arrangement that were defined by Agarwal et al. [**18**]; see also [**405**].)

The number of crossing points between two convex chains is bounded by the number of upper common tangents between the same two chains. The construction implies that any line passing through two points of S is an upper common tangent of at most one pair of chains. Thus there are $O(n^2)$ crossings between the segments in H. By Lemma 6.4, any graph with n vertices and crossing number $O(n^2)$ has at most $O(n^{4/3})$ edges, so S has at most $O(n^{4/3})$ halving segments. A similar, slightly more detailed, argument proves the bound for arbitrary values of k. □

Tamaki and Tokuyama generalized Dey's proof to prove a similar bound on the complexity of the k-level in arrangements of pseudo-lines [**716**]. A simplified proof of this latter result has been given in [**684**]. Combining the ideas from an old result of Welzl [**743**] with Dey's proof technique, one can obtain the following generalization. See also [**83**] for some other generalizations of Dey's result.

COROLLARY 6.5. *Let Γ be a set of n lines in the plane. Then for any $0 \le k < n$, $0 < j < n - k$, we have*

$$\sum_{t=k}^{k+j} \psi_t(\Gamma) = O(n(k+1)^{1/3} j^{2/3}).$$

Bárány and Steiger proved a linear upper bound on the expected number of k-sets in a random planar point set [**128**], where the points are chosen uniformly from a convex region. Edelsbrunner et al. [**307**] proved that if S is a set of points in the plane so that the ratio of the maximum and the minimum distance in S is at most $c\sqrt{n}$ (a so-called *dense* set), then the number of k-sets in S is $O(c\sqrt{n}\psi_k(c\sqrt{n}))$. Applying Dey's result, the number of k-sets in a dense point set is $O(n^{7/6})$. Alt et al. [**78**] have proved that if the points in S lie on a constant number of pairwise disjoint convex curves, then the number of k-sets in S is $O(n)$.

Lower bounds. Erdős et al. [**330**] constructed, for any n and $0 \le k < n$, a set Γ of n lines so that $\psi_k(\Gamma) = \Omega(n \log(k+1))$; see Edelsbrunner and Welzl for another construction that gives the same lower bound [**308**]. A fairly old unpublished result of Klawe et al. [**482**] constructs a set Γ of n x-monotone *pseudo-lines* so that $\psi_{n/2}(\Gamma)$ has $n2^{\Omega(\sqrt{\log n})}$ vertices. In a fairly recent breakthrough development, Tóth [**722**] has established the same asymptotic bound for the case of lines (i.e., for the number of halving sets for a set of points). A recent work of Nivasch [**569**] simplifies the construction and improves the constant in the exponent.

Extensions; higher dimensions. The following question is related to the complexity of levels in arrangements of lines: Let Γ be a set of n lines in the plane. Let Π be an x-monotone polygonal path whose vertices (the points at which Π bends) are a subset of the vertices of $\mathcal{A}(\Gamma)$ and whose edges are contained in the lines of Γ. What is the maximum number of vertices in Π? Matoušek [**527**] proved that there exists a set Γ of n lines in the plane whose arrangement contains an x-monotone path with $\Omega(n^{5/3})$ vertices. This has been considerably strengthened recently by Balogh et al. [**122**], who present such a construction with $\Omega(n^2/2^{c\sqrt{\log n}})$ vertices, for some constant $c > 0$.

Agarwal et al. [**18**] proved a nontrivial upper bound on the complexity of the k-level in an arrangement of segments. Combining their argument with that of Dey's, it follows that the maximum complexity of the k-level in a planar arrangement of n segments is $O(n(k+1)^{1/3}\alpha(n/(k+1)))$. Recent significant progress has been made on the analysis of the complexity of a single level in an arrangement of n arcs in the plane. Tamaki and Tokuyama [**715**] proved that the complexity of any level in an arrangement of parabolas, each with a vertical axis, is $O(n^{23/12})$. (Their bound actually applies to *pseudo-parabolas*, i.e., graphs of continuous, totally defined, univariate functions, each pair of which intersect at most twice.) This has later been improved by Agarwal et al. [**41**], and further by Chan [**204, 205**]. The best bound [**205**] is $\tilde{O}(n^{3/2})$ (where $\tilde{O}(\cdot)$ hides polylogarithmic factors). Chan's analysis yields many other nontrivial bounds. In particular, the complexity of a single level in an arrangement of arcs, each pair of which intersects at most s times, is $O(n^{2-\frac{1}{2s}})$, for $s \geq 3$ odd, and $O(n^{2-\frac{1}{2(s-1)}})$, for $s \geq 4$ even. In spite of this impressive progress, none of these bounds is known (or suspected) to be tight.

PROBLEM 6.6.
  (i) *What is the maximum complexity of a level in an arrangement of n lines in the plane?*
  (ii) *What is the maximum complexity of a level in an arrangement of n x-monotone Jordan arcs, each pair of which intersect in at most s points, for some constant $s > 1$?*

Bárány et al. [**127**] proved an $O(n^{3-\gamma})$ bound on the complexity of the k-level in arrangements of n planes in $\mathbb{R}^3$, for any k, for some (small) absolute constant $\gamma > 0$. The bound was improved by Aronov et al. [**91**], Eppstein [**322**], and Nivasch and Sharir [**571**] to $O(n^{8/3}\text{polylog}\,n)$, and then by Dey and Edelsbrunner [**265**] to $O(n^{8/3})$. The best bound known, due to Sharir et al. [**685**], is $O(n(k+1)^{3/2})$. Agarwal et al. [**18**] established a weaker bound on the complexity of the k-level for arrangements of triangles in $\mathbb{R}^3$. A nontrivial bound on the complexity of the k-level in an arrangement of n hyperplanes in $d > 3$ dimensions, of the form $O(n^{d-\varepsilon_d})$, for some constant ε_d that decreases exponentially with d, was obtained in [**71, 760**]. This has later been slightly improved to $O(n^{\lfloor d/2 \rfloor}k^{\lceil d/2 \rceil - \varepsilon_d})$ in [**18**]. Recently, Matoušek et al. [**540**] have improved the bound in $\mathbb{R}^4$ to $O(n^{4-2/45})$, with an elementary proof (the proof of the general bounds uses tools from algebraic topology).

As mentioned above, Dey's bound for k-levels in arrangements of lines also applies to arrangements of pseudo-lines, but the above-mentioned bounds for levels in arrangements of planes in three dimensions are not known to hold for *pseudo-planes*, namely, a collection of xy-monotone surfaces, each pair of which intersects in an unbounded arc, and each triple of which intersects in a single point. The first (ever so slightly) subcubic bound on the complexity of a single level in such an arrangement was recently established by Chan [**207**].

Table 1 summarizes the known upper bounds on k-levels.

7. Many Cells and Related Problems

In the previous two sections we bounded the complexity of families of d-dimensional cells in a d-dimensional arrangement $\mathcal{A}(\Gamma)$ which satisfied certain conditions (e.g., cells intersected by a surface, or the cells at level at most k). We

TABLE 1. Upper bounds on k-levels.

Objects	Bound	Source
Lines in $\mathbb{R}^2$	$O(n(k+1)^{1/3})$	[264]
Segments in $\mathbb{R}^2$	$O(n(k+1)^{1/3}\alpha(n/(k+1)))$	[18, 264]
Planes in $\mathbb{R}^3$	$O(n(k+1)^{3/2})$	[685]
Triangles in $\mathbb{R}^3$	$O(n^2(k+1)^{5/6}\alpha(n/(k+1)))$	[18]
Hyperplanes in $\mathbb{R}^4$	$O(n^{4-2/45})$	[540]
Hyperplanes in $\mathbb{R}^d$	$O(n^{\lfloor d/2 \rfloor} k^{\lceil d/2 \rceil - \varepsilon_d})$	[760]
Pseudo-parabolas in $\mathbb{R}^2$	$\tilde{O}(n^{3/2})$	[205]
s-intersecting curve segments in $\mathbb{R}^2$, $s \geq 3$ odd	$O(n^{2-\frac{1}{2s}})$	[205]
s-intersecting curve segments in $\mathbb{R}^2$, $s \geq 4$ even	$O(n^{2-\frac{1}{2(s-1)}})$	[205]

can ask a more general question: *What is the complexity of any m distinct (full-dimensional) cells in $\mathcal{A}(\Gamma)$?* A single cell in an arrangement of lines in the plane can have up to n edges, but can the total complexity of m cells in an arrangement of lines be $\Omega(mn)$? This is certainly false for the entire collection of cells, where $m = \Omega(n^2)$.

We can also formulate the above problem as follows: Let P be a set of m points and Γ a set of n surfaces in $\mathbb{R}^d$ satisfying assumptions (A1) and (A2). Assume that no point of P lies on any surface of Γ, and define $\mathcal{C}(P,\Gamma)$ to be the set of cells in $\mathcal{A}(\Gamma)$ that contain at least one point of P. Define $\mu(P,\Gamma) = \sum_{C \in \mathcal{C}(P,\Gamma)} |C|$, and $\mu(m,n,\mathbf{G}) = \max \mu(P,\Gamma)$, where the maximum is taken over all sets P of m points and over all sets Γ of n surfaces in a given class $\mathbf{G}$.

We next review the known bounds on $\mu(m,n,\mathbf{G})$. They are summarized in Table 2 below. Let $\mathbf{L}$ be the set of all lines in the plane. An old result of Canham [191] asserts that $\mu(m,n,\mathbf{L}) = O(m^2 + n)$, from which it easily follows that $\mu(m,n,\mathbf{L}) = O(m\sqrt{n} + n)$. Although this bound is optimal for $m \leq \sqrt{n}$, it is weak for larger values of m (again, try $m = \Theta(n^2)$). Clarkson et al. [245] proved that $\mu(m,n,\mathbf{L}) = \Theta(m^{2/3}n^{2/3} + n)$. Their technique, based on random sampling, is general and constructive, and has led to several important combinatorial and algorithmic results on arrangements [245, 400, 401]; see also Chapter 4. For example, following a similar, but considerably more involved, approach, Aronov et al. [92] proved that $\mu(m,n,\mathbf{E}) = O(m^{2/3}n^{2/3} + m\log n + n\alpha(n))$, where $\mathbf{E}$ is the set of all line segments in the plane; see also Agarwal et al. [23]. An improved bound can be attained if the number of vertices in the arrangement of segments is small. Hershberger and Snoeyink [438] proved an $O(m^{2/3}n^{2/3} + n)$ upper bound on the complexity of m distinct cells in the arrangements of n segments in the plane, when the segments satisfy certain additional conditions.

The old bound of Clarkson et al. [245] on the complexity of m distinct cells in an arrangement of n circles (see Table 2 below) has recently been improved by Agarwal et al. [23] to $O(m^{2/3}n^{2/3} + m^{6/11}n^{9/11}\text{polylog } n + n\log n)$ (this is slightly better than the bound as stated in [23, Theorem 4.4]); see Chapter 4 for related results and more details. Still, when m is relatively small, no tight bounds are known.

PROBLEM 7.1. *What is the maximum complexity of m distinct cells in an arrangement of n circles in the plane?*

TABLE 2. Complexity of many cells.

Objects	Complexity	Source
Lines in $\mathbb{R}^2$	$\Theta(m^{2/3}n^{2/3} + n)$	[245]
Segments in $\mathbb{R}^2$	$O(m^{2/3}n^{2/3} + n\alpha(n) + n\log m)$	[92]
Unit circles in $\mathbb{R}^2$	$O(m^{2/3}n^{2/3}\alpha^{1/3}(n) + n)$	[245]
Circles in $\mathbb{R}^2$	$O(m^{2/3}n^{2/3} + m^{6/11}n^{9/11}\text{polylog }n + n\log n)$	[23]
Arcs in $\mathbb{R}^2$	$O(\sqrt{m}\lambda_q(n))$	[295]
Planes in $\mathbb{R}^3$	$\Theta(m^{2/3}n + n^2)$	[17]
Hyperplanes in $\mathbb{R}^d$, $d \geq 4$	$O(m^{1/2}n^{d/2}\log^\beta n)$ $\beta = (\lfloor d/2 \rfloor - 2)/2$	[108]

Complexity of many cells in hyperplane arrangements in higher dimensions was first studied by Edelsbrunner and Haussler [299]. Let **H** be the set of all hyperplanes in $\mathbb{R}^d$. They proved that the maximum number of $(d-1)$-dimensional faces in m distinct cells in an arrangement of n hyperplanes in $\mathbb{R}^d$ is $O(m^{1/2}n^{d/2} + n^{d-1})$. Refining an argument by Edelsbrunner *et al.* [298], Agarwal and Aronov [17] improved this bound to $O(m^{2/3}n^{d/3} + n^{d-1})$. As follows from [299], this bound is tight in the worst case. Aronov and Sharir [108], slightly improving upon an earlier result in [98], proved that $\mu(m, n, \mathbf{H}) = O(m^{1/2}n^{d/2}\log^\beta n)$, where $\beta = (\lfloor d/2 \rfloor - 2)/2$. Aronov *et al.* [98] also proved several lower bounds on $\mu(m, n, \mathbf{H})$: For odd values of d and for $m \leq n$, $\mu(m, n, \mathbf{H}) = \Theta(mn^{\lfloor d/2 \rfloor})$; for m of the form $\Theta(n^{d-2k})$ where $0 \leq k \leq \lfloor d/2 \rfloor$ is an integer, $\mu(m, n, \mathbf{H}) = \Omega(m^{1/2}n^{\lfloor d/2 \rfloor})$; and for arbitrary values of m, $\mu(m, n, \mathbf{H}) = \Omega(m^{1/2}n^{d/2-1/4})$. Agarwal [12], Guibas *et al.* [397], and Halperin and Sharir [418] obtained bounds on "special" subsets of cells in hyperplane arrangements.

A problem closely related to, but somewhat simpler than, the many-cells problem is the *incidence problem*. Here is a simple instance of this problem: Let Γ be a set of n lines and P a set of m points in the plane. Define $I(P, \Gamma) = \sum_{\ell \in \Gamma} |P \cap \ell|$; set $I(m, n) = \max I(P, \Gamma)$, where the maximum is taken over all sets P of m distinct points and over all sets Γ of n distinct lines in the plane. Of course, this problem is interesting only when the lines in Γ are in highly degenerate position. If $n = m^2 + m + 1$, then a finite projective plane of order m has n points and n lines and each line contains $m + 1 = \Omega(n^{1/2})$ points, so the number of incidences between n points and n lines is $\Omega(n^{3/2})$. Szemerédi and Trotter [711] proved that such a construction is impossible in $\mathbb{R}^2$. In a subsequent paper, Szemerédi and Trotter [712] proved that $I(m, n) = O(m^{2/3}n^{2/3} + m + n)$. We will devote a full chapter (Chapter 4) to the topic of incidences and of many related problems and results, where the many-cells problem will also be addressed.

8. Generalized Voronoi Diagrams

An interesting application of the new bounds on the complexity of lower envelopes is to generalized Voronoi diagrams in higher dimensions. Let S be a set of n pairwise-disjoint convex objects in $\mathbb{R}^d$, each of constant description complexity, and let ρ be some metric in $\mathbb{R}^d$. For a point $\mathbf{x} \in \mathbb{R}^d$, let $\Phi(\mathbf{x})$ denote the set of objects nearest to $\mathbf{x}$, i.e.,

$$\Phi(\mathbf{x}) = \{s \in S \mid \rho(\mathbf{x}, s) \leq \rho(\mathbf{x}, s') \; \forall s' \in S\}.$$

The *Voronoi diagram* $\text{Vor}_\rho(S)$ of S under the metric ρ (sometimes also simply denoted as $\text{Vor}(S)$) is a partition of $\mathbb{R}^d$ into maximal connected regions C of various dimensions, so that, for each C, the set $\Phi(\mathbf{x})$ is the same for all $\mathbf{x} \in C$. Let γ_i be the graph of the function $x_{d+1} = \rho(\mathbf{x}, s_i)$. Set $\Gamma = \{\gamma_i \mid 1 \leq i \leq n\}$. Edelsbrunner and Seidel [**302**] observed that $\text{Vor}_\rho(S)$ is the minimization diagram of Γ, a fact which follows by definition.

In the classical case, in which ρ is the Euclidean metric and the objects in S are singletons (points), the graphs of these distance functions can be replaced by a collection of n hyperplanes in $\mathbb{R}^{d+1}$, using the linearization technique, without affecting the minimization (Voronoi) diagram. Hence the maximum possible complexity of $\text{Vor}(S)$ is $O(n^{\lceil d/2 \rceil})$, which actually can be achieved (see, e.g., [**489, 667**]). In more general settings, though, this reduction is not possible. Nevertheless, the bounds on the complexity of lower envelopes imply that, under reasonable assumption on ρ and on the objects in S, the complexity of the diagram is $O(n^{d+\varepsilon})$, for any $\varepsilon > 0$. While this bound is nontrivial, it is conjectured to be too weak. For example, this bound is near-quadratic for planar Voronoi diagrams, but the complexity of almost every planar Voronoi diagram is only $O(n)$, although there are certain distance functions for which the corresponding planar Voronoi diagram can have quadratic complexity (this is the case for multiplicatively weighted Voronoi diagrams; see [**115**]).

In three dimensions, the above-mentioned bound for point sites and Euclidean metric is $\Theta(n^2)$. It has been a long-standing open problem to determine whether a similar quadratic or near-quadratic bound holds in $\mathbb{R}^3$ for more general objects and metrics (here the bounds on lower envelopes only give an upper bound of $O(n^{3+\varepsilon})$). The problem stated above calls for improving this bound by roughly another factor of n. Since we are aiming for a bound that is "two orders of magnitude" better than the complexity of $\mathcal{A}(\Gamma)$, it appears to be a considerably more difficult problem than that of general lower envelopes. The only hope of making progress here is to exploit the special structure of the distance functions $\rho(x, s)$.

Fortunately, some progress on this problem was made recently. It was shown by Chew *et al.* [**236**] that the complexity of the Voronoi diagram is $O(n^2 \alpha(n) \log n)$ for the case where the objects of S are *lines* in $\mathbb{R}^3$ and the metric ρ is a *convex distance function* induced by a convex polytope with a constant number of facets (see [**236**] for more details). Note that such a distance function is not necessarily a metric, because it will fail to be symmetric if the defining polytope is not centrally symmetric. The L_1 and L_∞ metrics are special cases of such distance functions. The best known lower bound for the complexity of the diagram in this special case is $\Omega(n^2 \alpha(n))$. Recently, Koltun and Sharir [**496**] have extended this result to an arbitrary collection of pairwise disjoint *line segments* or *triangles*; the bound for the case of line segments is $O(n^2\alpha(n)\log n)$, and, for the case of triangles, is $O(n^{2+\varepsilon})$, for any $\varepsilon > 0$.

Dwyer [**286**] has shown that the expected complexity of the Voronoi diagram of a set of n random lines in $\mathbb{R}^3$ is $O(n^{3/2})$. Boissonnat *et al.* [**160**] have shown that the maximum complexity of the L_1-Voronoi diagram of a set of n points in $\mathbb{R}^3$ is $\Theta(n^2)$. Finally, it is shown in [**713**] that the complexity of the three-dimensional Voronoi diagram of point sites under a general polyhedral convex distance function (induced by a polytope with $O(1)$ facets) is $O(n^2 \log n)$.

In spite of all this progress, the case of lines in $\mathbb{R}^3$ under the Euclidean metric is still open:

PROBLEM 8.1. *Is the complexity of the Voronoi diagram of a set S of n lines under the Euclidean metric in $\mathbb{R}^3$ close to n^2?*

A partial affirmative answer has recently been given by Koltun and Sharir [**494**], for the special case where the lines have a constant number of distinct orientations. Everett *et al.* [**340**] have recently made further progress on this problem, by analyzing the topological structure of the *trisector* of three mutually skew lines in $\mathbb{R}^3$, i.e., the 1-dimensional locus of all points equidistant from the three lines.

An interesting special case of these problems involves *dynamic Voronoi diagrams* for moving points in the plane. Let S be a set of n points in the plane, each moving along some line at some fixed velocity. The goal is to bound the number of combinatorial changes of the Euclidean $\mathrm{Vor}(S)$ over time. This dynamic Voronoi diagram can easily be transformed into a three-dimensional Voronoi diagram, by adding the time t as a third coordinate. The points become lines in $\mathbb{R}^3$, and the metric is a (somewhat unorthodox) distance function induced by a horizontal disk (that is, the distance from a point $p(x_0, y_0, t_0)$ to a line ℓ is the Euclidean distance from p to the point of intersection of ℓ with the horizontal plane $t = t_0$). Here too the open problem is to derive a near-quadratic bound on the complexity of the diagram. Cubic or near-cubic bounds are known for this problem, even under more general settings [**367, 399, 678**], but subcubic bounds are known only in some very special cases [**235**].

PROBLEM 8.2. *Is the complexity of the dynamic Euclidean Voronoi diagram of a set S of n moving points in the plane close to n^2? Is this true at least in the special case where the points have linear trajectories with constant common speed?*

Next, consider the problem of bounding the complexity of generalized Voronoi diagrams in higher dimensions. As mentioned above, when the objects in S are n points in $\mathbb{R}^d$ and the metric is Euclidean, the complexity of $\mathrm{Vor}(S)$ is $O(n^{\lceil d/2 \rceil})$. As d increases, this becomes significantly smaller than the naive $O(n^{d+1})$ bound or the improved bound, $O(n^{d+\varepsilon})$, obtained by viewing the Voronoi diagram as a lower envelope in $\mathbb{R}^{d+1}$. The same bound of $O(n^{\lceil d/2 \rceil})$ has been obtained in [**160**] for the complexity of the L_∞-diagram of n points in $\mathbb{R}^d$ (it was also shown that this bound is tight in the worst case). It is thus tempting to conjecture that the maximum complexity of generalized Voronoi diagrams in higher dimensions is close to this bound. Unfortunately, this was shown by Aronov to be false [**89**], by presenting a lower bound of $\Omega(n^{d-1})$. The sites used in this construction are lower-dimensional flats, and the distance is either Euclidean or a polyhedral convex distance function. (It is interesting that the lower bound in Aronov's construction depends on the affine dimension $0 \leq k \leq d-2$ of the sites: It is $\Omega(\max\{n^{k+1}, n^{\lceil (d-k)/2 \rceil}\})$.) Thus, for $d = 3$, this lower bound does not contradict the conjecture made above, that the complexity of generalized Voronoi diagrams should be at most near-quadratic in this case. Also, in higher dimensions, the conjecture mentioned above is still not refuted when the sites are singleton points. Finally, for the general case, the construction by Aronov still leaves a gap of roughly a factor of n between the known upper and lower bounds, in any dimension.

Another special case of generalized Voronoi diagrams are *power diagrams*. Here we have a set $\mathcal{B}$ of n balls in $\mathbb{R}^d$, and the distance from a point $x \in \mathbb{R}^d$ to a ball $B \in \mathcal{B}$ is $\text{power}(x, B) = $ length of a tangent from x to B; that is, if B is centered at p and has radius r, then $\text{power}(x, B) = d^2(x, p) - r^2$. It is an easy exercise to see that power diagrams in $\mathbb{R}^d$ are minimization diagrams of lower envelopes of hyperplanes in $\mathbb{R}^{d+1}$; see, e.g., Boissonnat and Yvinec [**162**]. Hence their complexity has the same asymptotic bound $O(n^{\lceil d/2 \rceil})$ as standard Voronoi diagrams.

Another special case are *additive weight Voronoi diagrams*: Let P be a set of n points in $\mathbb{R}^d$, such that each $p \in P$ is associated with a weight $w_p \in \mathbb{R}$. The additively weighted distance from a point $x \in \mathbb{R}^d$ to $p \in P$ is $d_w(x,p) := d(x,p) + w_p$, and the additive weight Voronoi diagram is the partition of d-space into Voronoi cells, where the cell $V_w(p)$ of a point $p \in P$ is

$$V_w(p) = \{x \in \mathbb{R}^d \mid d_w(x,p) \leq d_w(x,q) \text{ for } q \in P\}.$$

Putting $a := \max_{p \in P} w_p$, we can interpret this diagram as the standard Euclidean Voronoi diagram of the n (not necessarily disjoint) balls $B(p, a - w_p)$, for $p \in P$. An elegant construction (see [**162**, Theorem 18.3.1]) shows that an additive weight Voronoi diagram in $\mathbb{R}^d$ can be embedded as a substructure of a power diagram in $\mathbb{R}^{d+1}$. Hence the complexity of an additive weight Voronoi diagram of n points (or, rather, the standard diagram of balls) in $\mathbb{R}^d$ is $O(n^{\lfloor d/2 \rfloor + 1})$.

9. Union of Geometric Objects

An expanded version of this section can be found in the recent survey by Agarwal *et al.* [**42**].

Let $\mathcal{K}$ be a family of n bodies in $\mathbb{R}^d$, whose boundaries satisfy assumptions (A1)–(A2), and let $\mathcal{U}$ denote the union of $\mathcal{K}$. In general, we can think of the arrangement $\mathcal{A}(\mathcal{K})$ of the boundaries of the objects of $\mathcal{K}$, and observe that the boundary of $\mathcal{U}$ is a substructure of $\mathcal{A}(\mathcal{K})$—each face of $\partial \mathcal{U}$ is a face of $\mathcal{A}(\mathcal{K})$. The combinatorial complexity of $\mathcal{U}$ is the number of these boundary faces, of all dimensions. The goal is to look for special properties of $\mathcal{K}$ that would ensure that the complexity of $\mathcal{U}$ is small. In the worst case, that complexity might be proportional to that of the entire arrangement $\mathcal{A}(\mathcal{K})$ and thus might be $\Theta(n^d)$. A simple example is a collection of long and thin triangles in $\mathbb{R}^2$, where it is easy to arrange them so that the complexity of their union is $\Theta(n^2)$; see Figure 9 for a similar scenario.

However, in many favorable cases, the complexity of $\mathcal{U}$ is considerably smaller. Here is an easy example. Let $\mathcal{K}$ be a collection of n halfspaces in $\mathbb{R}^d$, each bounded by a hyperplane. Then the union of $\mathcal{K}$ is the complement of a convex polyhedron with at most n facets. By the Upper Bound Theorem [**543**, **759**], this complexity is only $O(n^{\lfloor d/2 \rfloor})$. In this section we will see many additional examples of this phenomenon.

Pseudo-disks in the plane. A family $\mathcal{K}$ of simply connected regions in the plane is said to be a family of *pseudo-disks* if the boundaries of each pair of objects in $\mathcal{K}$

9. UNION OF GEOMETRIC OBJECTS

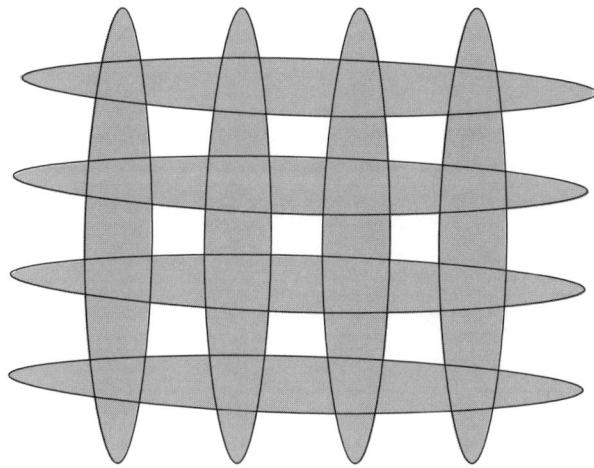

FIGURE 9. Union of Jordan regions.

intersect at most twice. Disks are obvious examples of pseudo-disks, but there are many other natural examples. An interesting case is that of *Minkowski sums* of a fixed convex object B with n pairwise disjoint convex sets $A_1, A_2, \ldots, A_n$. Recall that the Minkowski sum $A \oplus B$ of two sets A, B is the set of pointwise sums of their elements:

$$A \oplus B = \{x + y \mid x \in A, \ y \in B\}.$$

Put $K_i = A_i \oplus B$. If ∂K_i and ∂K_j intersected at four or more points, then they would have to have at least four distinct common outer tangents. As is well known, if a line at orientation θ supports $A \oplus B$ at a point z, then there are points $x \in \partial A$, $y \in \partial B$, such that $z = x + y$, and the line with orientation θ that supports A (resp., B) passes through x (resp., y). It follows that each common outer tangent to ∂K_i and ∂K_j can be translated to become a common outer tangent to A_i and A_j. However, since these sets are disjoint, they have only two common outer tangents, showing that K_i and K_j are indeed (convex) pseudo-disks. See Figure 10 for an illustration of this argument; the figure actually shows Minkowski sums with $-B$, to fit into the motion planning application discussed in the next paragraph.

This situation arises in the motion planning problem for a convex rigid robot B that is free to translate in the plane amid a collection of convex pairwise disjoint obstacles $A_1, \ldots, A_n$, which it must avoid. Assume that B is initially given at a placement that contains the origin o, and represent any translated position of B by the point $z \in \mathbb{R}^2$ to which o is translated (that is, this placement is $B + z$). Then $B + z$ intersects an obstacle A_i if there are points $x \in A_i$, $y \in B$, such that $x = y + z$, or $z = x - y$. Hence, denoting by $-B$ the reflection $\{-y \mid y \in B\}$ of B, it follows that $B + z$ collides with A_i if and only if $z \in A_i \oplus (-B)$. In other words, the *free configuration space* $\mathcal{F}$ of B, namely, the locus of all translations of B at which it does not collide with any obstacle, is the complement of the union $\mathcal{U} = \bigcup_i (A_i \oplus (-B))$.

Hence, to solve this translational motion planning problem, we need to compute $\mathcal{U}$. Before doing so, we need to determine its combinatorial complexity: How many vertices and edges of the arrangement $\mathcal{A}(\mathcal{K})$, where $\mathcal{K} = \{A_i \oplus (-B)\}_{i=1}^n$, appear on

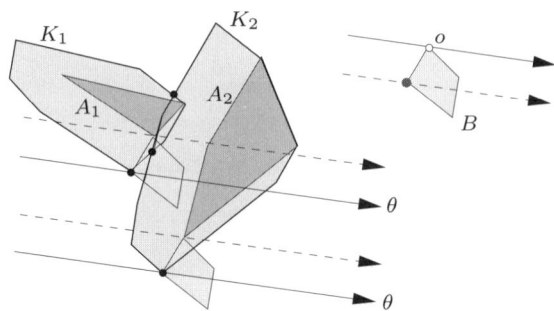

FIGURE 10. Illustrating the proof that $K_1 = A_1 \oplus (-B)$ and $K_2 = A_2 \oplus (-B)$ can have at most two common outer tangents.

$\partial \mathcal{U}$? The surprising answer is that this complexity is linear in n; it was shown by Kedem et al. [**471**], one of the oldest results concerning unions of geometric objects:

THEOREM 9.1 (Kedem et al. [**471**]). *The number of vertices of the boundary of the union of n pseudo-disks in the plane is at most $6n - 12$, for $n \geq 3$. This bound is sharp.*

PROOF. Let $\mathcal{K}$ be a family of n pseudo-disks, and let $\mathcal{U}$ denote its union. The proof is an application of a beautiful result of Hanani and Tutte [**238, 727**], stating that a graph drawn in the plane, so that each pair of its edges that do not have a common endpoint cross an even number of times, is planar. We construct such a graph G: We may assume that the boundary of each $K \in \mathcal{K}$ appears on $\partial \mathcal{U}$; indeed, by removing all other sets from $\mathcal{K}$, we do not lose any feature of $\partial \mathcal{U}$. For each $K \in \mathcal{K}$, choose an arbitrary fixed point $p_K \in \partial K \cap \partial \mathcal{U}$, and make these points the vertices of G. If $K, K' \in \mathcal{K}$ are such that their boundaries intersect at a point q that is a vertex of $\mathcal{U}$, draw an edge that connects p_K to $p_{K'}$ by following ∂K from p_K to q (in either of the two possible ways) and then following $\partial K'$ from q to $p_{K'}$. See Figure 11. (As stated, edges of G may overlap each other, so we perturb them slightly to make sure this does not happen.)

We claim that each pair of edges of G intersect an even number of times. In fact, each pair of "half-edges", one connecting some point p_K to an intersection point v, and the other connecting another point p_L to another intersection point u, are either disjoint or intersect exactly twice. This is easy to show, because each of these half-edges starts and ends on the boundary of the union, and the boundaries of the two corresponding pseudo-disks intersect at most twice; see Figure 11. Hence, by the Hanani–Tutte theorem, G is planar, and thus has at most $3n - 6$ edges, implying that $\partial \mathcal{U}$ can have at most twice that many vertices. □

3-intersecting curves. A natural question that arises is what happens if any two boundaries intersect in at most three points. Notice that in general this question makes no sense, since any two closed curves must intersect in an even number of points (assuming nondegenerate configurations). To make the problem interesting, let Γ be a collection of n Jordan arcs, such that both endpoints of each arc $\gamma_i \in$

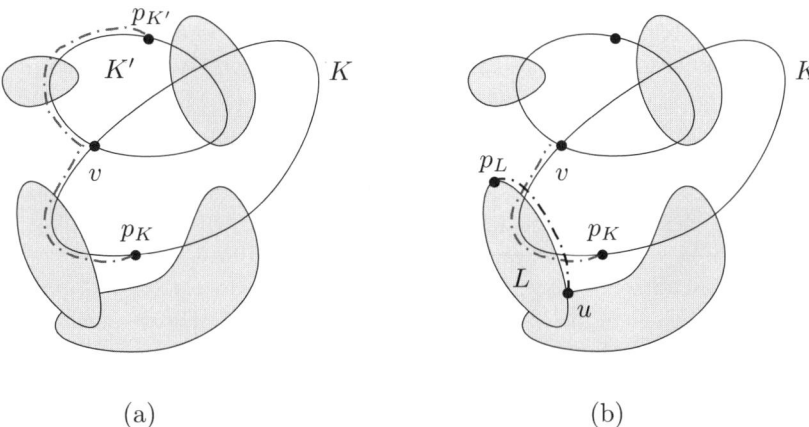

FIGURE 11. The union of pseudo-disks via a planarity argument. (a) Drawing an edge of G. (b) Two half-edges are disjoint or intersect twice.

Γ lie on the x-axis, and such that K_i is the region between γ_i and the x-axis. Edelsbrunner et al. [**294**] have shown that if each pair of arcs in Γ intersect in at most three points then the maximum combinatorial complexity of the union K is $\Theta(n\alpha(n))$. The upper bound requires a rather sophisticated analysis of the topological structure of K, and the lower bound follows from the construction by Wiernik and Sharir for lower envelopes of segments [**747**]; see also Shor [**689, 681**] and Chapter 3.

Fat objects in the plane. Define a convex planar body K to be δ-*fat*, if there exist two disks D, D', such that $D \subseteq K \subseteq D'$, and the ratio between the radii of D', D is at most δ. Let $\mathcal{K}$ be a family of n convex δ-fat objects in the plane, whose boundaries satisfy assumptions (A1)–(A2). Intuitively, we should expect their union $\mathcal{U}$ to have small complexity, because the simple examples where the union has quadratic complexity involve long and skinny objects.

This has been shown in a series of papers. The simplest example of fat objects are fat triangles, where fatness can also be defined in terms of their angles, so that a triangle is δ-fat (with a slight "abuse" of the parameter δ) if each of its angles is at least δ. Matoušek et al. [**537**] have shown that the union of n δ-fat triangles has only $O(n)$ holes (connected components of its complement) and its complexity is $O(n \log \log n)$ (and can be $\Omega(n\alpha(n))$ in the worst case). The constants of proportionality depend on δ, where the best bound, due to Pach and Tardos [**605**], is $O(\frac{1}{\delta} \log \frac{1}{\delta})$. Earlier works by Alt et al. [**79**] and by Efrat et al. [**315**] consider the case of *fat wedges* (whose angle is at least some $\delta > 0$), and show that the complexity of the union of such wedges is linear.

For arbitrary convex fat objects, Efrat and Sharir [**317**] have shown that the complexity of the union is $O(n^{1+\varepsilon})$, for any $\varepsilon > 0$, where the constant of proportionality depends on δ, ε, and the degree b in (A2).

Several refinements of this result have also been established. Call a (not necessarily convex) object K κ-*curved* if for each point $p \in \partial K$ there exists a disk D

that passes through p, is contained in K, and its radius is κ times the diameter of K. Informally, κ-curved objects cannot have convex corners, nor can they have very thin bottlenecks. Efrat and Katz [**314**] have shown that the complexity of the union of n κ-curved objects satisfying (A1)–(A2) is $O(\lambda_s(n)\log n)$, where s is a constant that depends on the description complexity of the input objects. This has further been extended by Efrat [**313**], to so-called (δ,β)-*covered objects*, which are objects K with the property that for each point $p \in \partial K$ there exists a δ-fat triangle T that has p as a vertex, is contained in K, and each of its edges is at least β times the diameter of K. Efrat has shown that if $\mathcal{K}$ is a collection of n (δ,β)-covered objects, each pair of whose boundaries intersect in at most $s = O(1)$ points, then the complexity of their union is $O(\lambda_{s+2}(n)\log^2 n \log\log n)$, where s is the maximum number of intersections between the boundaries of any pair of the given objects. A recent paper by de Berg [**140**] gives a simple proof of similar bounds.

Regular vertices and related problems. The analysis of the complexity bounds just cited requires as an initial but important substep an analysis of the number of *regular vertices* of the union: these are vertices of the union that are incident to two boundaries that intersect exactly twice. In fact, the analysis can only handle directly the irregular vertices of the union. Motivated by this problem, Pach and Sharir [**599**] have shown that, for an arbitrary collection of n convex regions, each pair of whose boundaries cross in a constant number of points, one has $R \leq 2I + 6n - 12$, where R (resp., I) is the number of regular (resp., irregular) vertices on the boundary of the union. This result has been used in [**317**] and the other works to obtain near-linear bounds. It is curious that the proof works only when all the regions are convex; one would expect the inequality to hold in more general settings too.

Nevertheless, regular vertices are interesting in their own right. Some additional results concerning them have been obtained by Aronov *et al.* [**93**], and some of their bounds have recently been refined by Ezra *et al.* [**343**]. First, if there are only regular vertices (i.e., every pair of boundaries intersect at most twice), then the inequality obtained by [**599**] implies that the complexity of the union in this case is at most $6n - 12$, so the result by Pach and Sharir extends the older result of Kedem *et al.* [**471**]. In general, though, I can be quadratic, so the above inequality only yields a quadratic upper bound on the number of regular vertices of the union. However, it was shown in [**93, 343**] that in many cases R is subquadratic. This is the case when the given regions are such that every pair of their boundaries cross at most a constant number of times. If in addition all the regions are convex, the upper bound is close to $O(n^{4/3})$ [**343**].

As a final, somewhat unrelated result, we mention the work of Aronov and Sharir [**105**], who proved that the complexity of the union of n convex polygons in $\mathbb{R}^2$ with a total of s vertices is $O(n^2 + s\alpha(n))$.

Union of objects in higher dimensions. The analysis of the complexity of the union in three (and higher) dimensions has been lagging behind the extensive research on the union of objects in $\mathbb{R}^2$, but in the past several years significant progress was made on this front too. Of course, the union of n thin plate-like objects that form a three-dimensional grid can have complexity $\Omega(n^3)$. However, as in the plane, it is important, and motivated by a line of practical applications, to consider realistic input models, and in particular to study the complexity of the union of *fat* objects. A prevailing conjecture is that the complexity of the union of fat objects (say, under the definition of δ-fatness, requiring the ratio between the radii of the smallest enclosing ball and the largest inscribed ball of each object to be at most some fixed constant δ) is near-quadratic. Such a bound has however proved quite elusive to obtain for general fat objects, and this has been recognized as one of the major open problems in computational geometry [**262**, Problem 4].

Let us first consider the case of convex polyhedra. Aronov et al. [**109**] (see also [**103**]) proved that the complexity of the union of n (not necessarily fat) convex polyhedra in $\mathbb{R}^3$ with a total of s faces is $O(n^3 + sn \log n)$. This bound, which is nearly tight in the worst case, is interesting, because it is cubic only in the number of polyhedra, and not in the overall number of their facets.

Subcubic (in most cases, nearly-quadratic) bounds on the complexity of the union do exist in certain special cases. The complexity of the union of axis-parallel cubes is $O(n^2)$, and it drops to $O(n)$ if the cubes have the same size [**160**]. It was shown by Pach et al. [**593**] that the complexity of the union of n cubes, which are not axis-parallel but have equal (or "almost equal") size, is $O(n^{2+\varepsilon})$ for any $\varepsilon > 0$. The main technical result established in [**593**], which is interesting in its own right, is that the complexity of the union of n δ-fat *dihedral wedges* (wedges bounded by a pair of halfplanes which form an angle of at least δ between them) is $O(n^{2+\varepsilon})$ for any $\varepsilon > 0$. This result has been used in a recent work of Ezra and Sharir [**346**] to show that the complexity of the union of n δ-fat tetrahedra in $\mathbb{R}^3$ (of arbitrary sizes) is also $O(n^{2+\varepsilon})$ for any $\varepsilon > 0$.

This result essentially settles the case of fat convex polyhedra, but still leaves open many other instances of the general conjecture mentioned above (see below for a list of some specific open problems).

Minkowski sums in 3-space. Unions of objects also arise as subproblems in the study of generalized Voronoi diagrams, as follows. Let S and ρ be as in the previous section (say, for the 3-dimensional case). Let K denote the region consisting of all points $x \in \mathbb{R}^3$ whose smallest distance from a site in S is at most r, for some fixed parameter $r > 0$. Then $K = \bigcup_{s \in S} B(s, r)$, where $B(s, r) = \{x \in \mathbb{R}^3 \mid \rho(x, s) \leq r\}$. We thus face the problem of bounding the combinatorial complexity of the union of n objects in $\mathbb{R}^3$ (of some special type). For example, if S is a set of lines and ρ is the Euclidean distance, the objects are n congruent infinite cylinders in $\mathbb{R}^3$. In general, if the metric ρ is a distance function induced by some convex body P, the resulting objects are the *Minkowski sums* $s \oplus (-rP)$, for $s \in S$, where, as above, $A \oplus B = \{x + y \mid x \in A, y \in B\}$. Of course, this problem can also be stated in any higher dimension.

Since it has been conjectured that the complexity of the whole Voronoi diagram in $\mathbb{R}^3$ should be near-quadratic, the same conjecture should apply to the (simpler) structure K (whose boundary can be regarded as a *level curve* of the diagram at *height* r; it does indeed correspond to the cross-section at height r of the lower envelope in $\mathbb{R}^4$ that represents the diagram). This conjecture was confirmed by Aronov and Sharir in [**106**], in the special case where both P and the objects of S are convex polyhedra. They specialized their analysis of the union of convex polytopes to obtain an improved bound in the special case in which the polyhedra in question are Minkowski sums of the form $R_i \oplus P$, where the R_i's are n pairwise-disjoint convex polyhedra, P is a convex polyhedron, and the total number of faces of these Minkowski sums is s. The improved bounds are $O(ns \log n)$ and $\Omega(ns\alpha(n))$. If P is a cube, then the complexity of the Minkowski sum is $O(n^2\alpha(n))$ [**424**].

Agarwal and Sharir [**53**] have shown that the union of Minkowski sums of disjoint polyhedra of overall complexity n with a ball has complexity $O(n^{2+\varepsilon})$ for any $\varepsilon > 0$. As a special case (in which the given polyhedra are lines), the complexity of the union of n congruent infinite cylinders in $\mathbb{R}^3$ is nearly quadratic.

The case of the union of cylinders of different radii has recently been settled by Ezra [**342**], who showed that in this case too, the complexity is only $O(n^{2+\varepsilon})$, for any $\varepsilon > 0$.

Other bounds in three and higher dimensions. The first rather general result on the complexity of the union of fat objects in 3-space stems from the analysis technique of Agarwal and Sharir [**53**] and appears in their paper: The complexity of the union $\mathcal{U}$ is $O(n^{2+\varepsilon})$ for any $\varepsilon > 0$ if $\mathcal{K}$ consists of n convex objects of near-equal size, with C^2-continuous boundaries, bounded mean curvature, and constant description complexity (that is, satisfying assumptions (A1)–(A2)).

In $d \geq 4$ dimensions, the results become even more scarce. As already mentioned, the complexity of the union of n halfspaces (each bounded by a hyperplane) in $\mathbb{R}^d$ is $O(n^{\lfloor d/2 \rfloor})$, as follows from the Upper Bound Theorem [**543, 759**]. The complexity of the union of n balls in d-space is $O(n^{\lceil d/2 \rceil})$, as follows by lifting them to hyperplanes in $\mathbb{R}^{d+1}$. Boissonnat et al. [**160**] provide an upper bound of $O(n^{\lceil d/2 \rceil})$ for the union of n axis-parallel cubes in $\mathbb{R}^d$, which improves to $O(n^{\lfloor d/2 \rfloor})$ when the cubes have equal (or nearly equal) size. The union complexity of n convex bodies in $\mathbb{R}^d$ satisfying (A1)–(A2) *with a common interior point o* is $O(n^{d-1+\varepsilon})$ for any $\varepsilon > 0$, which follows from the observation that the boundary of their union can be interpreted as the upper envelope of $(d-1)$-variate functions (in spherical coordinates around o). See also a refined bound for polyhedra in $\mathbb{R}^3$ in [**450**]. Koltun and Sharir [**495**] extend the above-mentioned result of [**53**] to four dimensions, and proved that the complexity of the union of n convex objects of near-equal size, with C^2-continuous boundaries, bounded mean curvature, and constant description complexity in $\mathbb{R}^4$ is $O(n^{3+\varepsilon})$ for any $\varepsilon > 0$.

A more general result has later been established in three and four dimensions. In complete analogy with the definition of κ-curved objects in two dimensions, a compact body c is κ-*round* (for a fixed $\kappa > 0$) if for every point $p \subset \partial c$ there exists a closed ball $B(p, c, \kappa)$ of radius $\kappa \cdot \text{diam}(c)$, which contains p and is contained in c. (If c is convex, $B(p, c, \kappa)$ is unique. The definition, though, allows c to be non-convex and to have *reflex* edges and vertices, although c cannot have any *convex* edge or vertex.) See Figure 12. Aronov et al. [**94**] have shown that the complexity of the union of a collection $\mathcal{K}$ of n κ-round (not necessarily convex) bodies in $\mathbb{R}^3$ or in $\mathbb{R}^4$, satisfying assumptions (A1)–(A2), is $O(n^{2+\varepsilon})$ in $\mathbb{R}^3$ and $O(n^{3+\varepsilon})$ in $\mathbb{R}^4$, for any $\varepsilon > 0$, where the constant of proportionality depends on ε, κ and the maximum degree b in (A2). In principle, the analysis of [**94**] can be applied in any dimension $d \geq 3$, except for its last step, which reduces the problem to that of bounding the number of vertices of the *sandwich region* between the upper envelope of a collection of $(d-1)$-variate functions and the lower envelope of another such collection. As already mentioned, sharp bounds on the number of such vertices are known only for $d = 3$ and $d = 4$, which is the only reason for the present inability to extend the above bounds to $d > 4$.

10. Decomposition of Arrangements

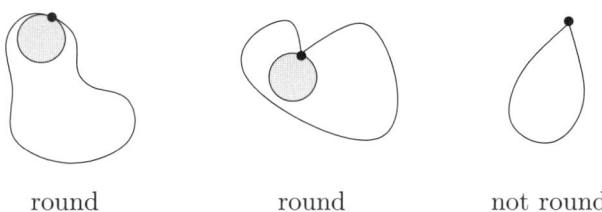

round   round   not round

FIGURE 12. Examples of (planar analogues of) κ-round and non-κ-round bodies.

We summarize some of the open problems involving unions in three and higher dimensions:

PROBLEM 9.2.
  (i) *What is the complexity of the union of n κ-round objects (satisfying (A1)–(A2)) in $\mathbb{R}^d$, for $d \geq 5$?*
  (ii) *What is the complexity of the union of n Minkowski sums of the form $A_i \oplus B$, where the A_i's and B are compact convex objects (satisfying (A1)–(A2)) in $\mathbb{R}^3$, and the A_i's are pairwise disjoint?*
  (iii) *What is the complexity of the union of n δ-fat objects (satisfying (A1)–(A2)) in $\mathbb{R}^d$, for $d \geq 3$?*

10. Decomposition of Arrangements

Many applications call for decomposing each cell of an arrangement into subcells of constant complexity; see Sections 12 and 13 for a sample of such applications. In this section we describe a few general schemes that have been proposed for decomposition of arrangements.

10.1. Triangulating hyperplane arrangements. Each k-dimensional cell in an arrangement of hyperplanes is a convex polyhedron, so we can triangulate it into k-simplices. If the cell is unbounded, some of the simplices in the triangulation will be unbounded. A commonly used scheme to triangulate a convex polytope P is the so-called *bottom-vertex triangulation*, denoted P^∇. It recursively triangulates every face of P as follows. An edge is a one-dimensional simplex, so there is nothing to do. Suppose we have triangulated all j-dimensional cells of P, for $j < k$. We now triangulate a k-dimensional cell C as follows. Let v be the vertex of C with the minimum x_d-coordinate. For each $(k-1)$-dimensional simplex Δ lying on the boundary of C but not containing v (Δ was constructed while triangulating a $(k-1)$-dimensional cell incident to C), we extend Δ to a k-dimensional simplex by taking the convex hull of Δ and v; see Figure 13(i). (Unbounded cells require some care in this definition; see [**242**]). The number of simplices in P^∇ is proportional to the number of vertices in P.

If we want to triangulate the entire arrangement or more than one of its cells, we compute the bottom-vertex triangulation f^∇ for each face f in the increasing order of their dimension. Let $\mathcal{A}^\nabla(\Gamma)$ denote the bottom-vertex triangulation of $\mathcal{A}(\Gamma)$. A useful property of $\mathcal{A}^\nabla(\Gamma)$ is that each simplex $\Delta \in \mathcal{A}^\nabla(\Gamma)$ is defined by a set $D(\Delta)$ of at most $d(d+3)/2$ hyperplanes of Γ, in the sense that $\Delta \in \mathcal{A}^\nabla(D(\Delta))$. Moreover, if $\mathcal{K}(\Delta) \subseteq \Gamma$ is the subset of hyperplanes intersecting Δ, then $\Delta \in \mathcal{A}^\nabla(R)$, for a subset

$R \subseteq \Gamma$, if and only if $D(\Delta) \subseteq R$ and $\mathcal{K}(\Delta) \cap R = \emptyset$. A disadvantage of bottom-vertex triangulation is that some vertices may have large degree. Methods for obtaining low-degree triangulations have been proposed in two and three dimensions [**271**].

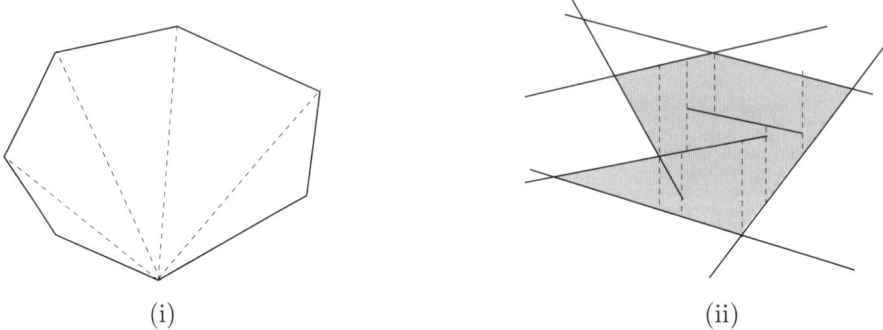

FIGURE 13. (i) Bottom-vertex triangulation of a convex polygon; (ii) vertical decomposition of a cell in an arrangement of segments.

10.2. Vertical decomposition. Unfortunately, the bottom-vertex triangulation scheme does not work for arrangements of surfaces. Collins [**250**] described a general decomposition scheme, called *cylindrical algebraic decomposition*, that decomposes $\mathcal{A}(\Gamma)$ into $(bn)^{2^{O(d)}}$ cells, each semialgebraic of constant description complexity (however, the maximum algebraic degree involved in defining a cell grows exponentially with d), and homeomorphic to a ball of the appropriate dimension. Moreover, Collins's algorithm produces a cell complex, i.e., closures of any two cells are either disjoint or their intersection is the closure of another lower-dimensional cell of the decomposition. This bound is quite far from the known trivial lower bound of $\Omega(n^d)$, which is a lower bound on the size of the (undecomposed) arrangement. A significantly better scheme for decomposing arrangements of general surfaces is their *vertical decomposition*. Although vertical decompositions of polygons in the plane have been in use for a long time, they were extended to higher dimensions only in the late 1980s. We describe this method briefly.

Let C be a d-dimensional cell in $\mathcal{A}(\Gamma)$. The vertical decomposition of C, denoted $C^{\|}$, is computed as follows. We erect a vertical "wall" up and down (in the x_d-direction) within C from each $(d-2)$-dimensional face of C and from points of vertical tangencies (i.e., the points at which the tangent hyperplanes are parallel to the x_d-direction), and extend these walls until they hit another surface (or, failing this, all the way to $\pm\infty$). This results in a decomposition of C into "prism-like" subcells, so that each subcell has a unique top facet and a unique bottom facet, and every vertical line cuts the subcell in a connected (possibly empty) interval. We next project each resulting subcell τ on the hyperplane $x_d = 0$. Let C_τ be the projected cell. We apply recursively the same technique to C_τ and compute its vertical decomposition $C_\tau^{\|}$. (We continue the recursion in this manner until we reach $d = 1$.) We then "lift" $C_\tau^{\|}$ back into $\mathbb{R}^d$, by extending each subcell $c \in C_\tau^{\|}$ into the vertical cylinder $c \times \mathbb{R}$, and by clipping the cylinder within τ; see Figure 13(ii). A standard argument shows that each cell of $C^{\|}$ is a semialgebraic set of constant description complexity. In fact, these cells have the same structure as the Collins cells, but the number of subcells in $C^{\|}$ is much smaller than that in the Collins decomposition of C. Applying the above step to each cell of $\mathcal{A}(\Gamma)$, we obtain the vertical decomposition $\mathcal{A}^{\|}(\Gamma)$ of $\mathcal{A}(\Gamma)$. Note, though, that $\mathcal{A}^{\|}(\Gamma)$ is not a cell complex (Collins's decomposition is a cell complex).

It is easily seen that the complexity of the vertical decomposition of a cell in the plane is proportional to the number of edges in the cell. However, this is no longer the case in higher dimensions: Already for the case of a convex polytope with n facets in $\mathbb{R}^3$, the vertical decomposition may have complexity $\Omega(n^2)$, whereas the undecomposed polytope has only linear complexity. Nevertheless, for the entire arrangement, in dimensions $d = 2, 3, 4$, the vertical decomposition $\mathcal{A}^{\|}(\Gamma)$ behaves well: For $d = 2$, the complexity of $\mathcal{A}^{\|}(\Gamma)$ is $O(n^2)$ (trivial). For $d = 3$, it is $O(n^2 \lambda_q(n))$, for some q that depends on the maximum degree b in (A2); see Chazelle et al. [**215, 216**]. A fairly recent remarkable result of Koltun [**492**] shows that in $d = 4$ dimensions, $\mathcal{A}^{\|}(\Gamma)$ consists of $O(n^{4+\varepsilon})$ subcells, for any $\varepsilon > 0$, where the constant of proportionality depends on ε and b. In general, one can recurse on the dimension, and derive a bound which increases by a factor of $O(n^2)$ for each added dimension. That is, we have:

THEOREM 10.1. *The number of cells in the vertical decomposition $\mathcal{A}^{\|}(\Gamma)$ of the arrangement $\mathcal{A}(\Gamma)$, for a set Γ of n surface patches in $\mathbb{R}^d$ satisfying (A1)–(A2), is $O(n^2)$ for $d = 2$, $O(n^2 \lambda_q(n))$ for $d = 3$, where q is a constant depending on b, and $O(n^{2d-4+\varepsilon})$, for any $\varepsilon > 0$, for $d \geq 4$.*

The only known lower bound on the size of $\mathcal{A}^{\|}(\Gamma)$ is the trivial $\Omega(n^d)$, leaving a considerable gap for $d > 4$; for $d \leq 4$ the two bounds (nearly) coincide. At the moment, we do not offer any conjecture as to whether the gap is real or can be closed.

PROBLEM 10.2. *What is the complexity of the vertical decomposition of the arrangement of n surfaces in $\mathbb{R}^5$ (or in higher dimensions) satisfying assumptions (A1)–(A2)?*

The bound stated above applies to the vertical decomposition of an entire arrangement of surfaces. In many applications, however, one is interested in the vertical decomposition of only a portion of the arrangement, e.g., a single cell, the lower envelope, the sandwich region between two envelopes, the zone of some surface, a specific collection of cells of the arrangement, etc. Since, in general, the complexity of such a portion is known to be smaller than the complexity of the entire arrangement, one would like to conjecture that a similar phenomenon applies to vertical decompositions. For example, it was shown by Schwarzkopf and Sharir [**664**] that the complexity of the vertical decomposition of a single cell in an arrangement of n surface patches in $\mathbb{R}^3$, as above, is $O(n^{2+\varepsilon})$, for any $\varepsilon > 0$. A similar near-quadratic bound has been obtained by Agarwal et al. [**20**] for the vertical decomposition of the region enclosed between the lower envelope and the upper envelope of two sets of bivariate surface patches. Another result by Agarwal et al. [**29**] gives a bound on the complexity of the vertical decomposition of $\mathcal{A}_{\leq k}(\Gamma)$ for a set Γ of surfaces in $\mathbb{R}^3$, which is only slightly larger that the worst-case complexity of $\mathcal{A}_{\leq k}(\Gamma)$.

PROBLEM 10.3. *What is the complexity of the vertical decomposition of the minimization diagram of n surfaces in $\mathbb{R}^4$ satisfying assumptions (A1)–(A2)?*

Agarwal and Sharir [**48**] proved a near-cubic upper bound on the complexity of the vertical decomposition in the special case when the surfaces are graphs of trivariate polynomials and the intersection surface of any pair of surfaces is xy-monotone. In fact, their bound holds for a more general setting; see the original paper for details.

An interesting special case of vertical decomposition is that of hyperplanes. For such arrangements, the vertical decomposition is a too cumbersome construct, because, as described above, one can use the bottom-vertex triangulation (or any other triangulation) to decompose the arrangement into $\Theta(n^d)$ simplices. Still, it is probably a useful exercise to understand the complexity of the vertical decomposition of an arrangement of n hyperplanes in $\mathbb{R}^d$. Guibas *et al.* [**398**] give an almost tight bound of $O(n^4 \log n)$ for this quantity in $\mathbb{R}^4$, which has recently been tightened to $\Theta(n^4)$ by Koltun [**493**] (who has also obtained the sharp bound $O(n^4\alpha(n)\log^2 n)$ on the complexity of the vertical decomposition of n 3-simplices in $\mathbb{R}^4$), but nothing significantly better than the general bound is known for $d > 4$. Another interesting special case is that of triangles in 3-space. This has been studied in [**142, 714**], where almost tight bounds were obtained for the case of a single cell ($O(n^2 \log^2 n)$), and for the entire arrangement ($O(n^2\alpha(n)\log n + K)$, where K is the complexity of the undecomposed arrangement). The first bound is slightly better than the general bound of [**664**] mentioned above. Tagansky [**714**] also derives sharp complexity bounds for the vertical decomposition of many cells in an arrangement of simplices, including the case of all nonconvex cells.

These results are summarized in Table 3.

TABLE 3. Combinatorial bounds on the maximum complexity of the vertical decomposition of an arrangement of n surfaces. In the fourth row, K is the combinatorial complexity of the arrangement.

Objects	Bound	Source
Surfaces in $\mathbb{R}^2$	$O(n^2)$	easy
Surfaces in $\mathbb{R}^3$	$O(n^2\lambda_q(n))$	[**215, 681**]
Surfaces in $\mathbb{R}^d$, $d \geq 4$	$O(n^{2d-4+\varepsilon})$	[**492**]
Triangles in $\mathbb{R}^3$	$O(n^2\alpha(n)\log n + K)$	[**142, 714**]
Surfaces in $\mathbb{R}^3$, single cell	$O(n^{2+\varepsilon})$	[**664**]
Triangles in $\mathbb{R}^3$, zone w.r.t. an algebraic surface	$\Theta(n^2 \log^2 n)$	[**714**]
Surfaces in $\mathbb{R}^3$, ($\leq k$)-level	$O(n^{2+\varepsilon}k)$	[**29**]
Hyperplanes in $\mathbb{R}^4$	$\Theta(n^4)$	[**493**]
3-Simplices in $\mathbb{R}^4$	$O(n^4\alpha(n)\log^2 n)$	[**493**]

10.3. Other decomposition schemes. Sometimes one can devise other decomposition schemes. For example, cells in generalized Voronoi diagrams in $\mathbb{R}^3$ can be decomposed into subcells of constant complexity by first decomposing the boundary of each cell $V(s)$, using, say, a modified variant of planar vertical decomposition, and then, using the star-shapedness of Voronoi cells, extend each resulting cell τ to a 3-dimensional cell τ^*, which is the union of all segments that connect points $x \in \tau$ to their closest points on s. Thus, the overall number of resulting cells is proportional to the complexity of the entire Voronoi diagram. This also applies to the union of Minkowski sums, which, as discussed above, can be regarded as a substructure of an appropriate generalized Voronoi diagram. For example, this scheme yields a decomposition of the union of n congruent cylinders in $\mathbb{R}^3$ into $O(n^{2+\varepsilon})$ cells of constant complexity each.

It is not known how to extend these schemes to higher dimensions. Somewhat surprisingly, the scheme fails (or at least is not known to exist) for the *complement* of the union of Minkowski sums, already for complements of unions of cylinders:

PROBLEM 10.4.

(i) *What is the smallest asymptotic complexity of a decomposition of the cells of a generalized Voronoi diagram in $d \geq 4$ dimensions?*

(ii) *What is the smallest asymptotic complexity of a decomposition of the complement of the union of n congruent cylinders in $\mathbb{R}^3$?*

Linearization, defined in Section 3, can also be used to decompose the cells of the arrangement $\mathcal{A}(\Gamma)$ into cells of constant description complexity as follows. Suppose Γ admits a linearization of dimension k, i.e., there is a transformation $\varphi : \mathbb{R}^d \mapsto \mathbb{R}^k$ that maps each point $\mathbf{x} \in \mathbb{R}^d$ to a point $\varphi(\mathbf{x}) \in \mathbb{R}^k$, each surface $\gamma_i \in \Gamma$ to a hyperplane $h_i \subset \mathbb{R}^k$, and $\mathbb{R}^d$ to a d-dimensional surface $\Sigma \subseteq \mathbb{R}^k$. Let $H = \{h_i \mid 1 \leq i \leq n\}$. We compute the bottom-vertex triangulation $\mathcal{A}^\nabla(H)$ of $\mathcal{A}(H)$. For each simplex $\Delta \in \mathcal{A}^\nabla(H)$, let $\overline{\Delta} = \Delta \cap \Sigma$, and let $\Delta^* = \varphi^{-1}(\overline{\Delta})$ be the back projection of $\overline{\Delta}$ onto $\mathbb{R}^d$; Δ^* is a semialgebraic cell of constant description complexity. Set $\Xi = \{\Delta^* \mid \Delta \in \mathcal{A}^\nabla(H)\}$. Ξ is a decomposition of $\mathcal{A}(\Gamma)$ into cells of constant description complexity. If a simplex $\Delta \in \mathcal{A}^\nabla(H)$ intersects Σ, then Δ lies in the triangulation of a cell in $zone(\Sigma; H)$. Therefore, by Theorem 5.3, $|\Xi| = O(n^{\lfloor (d+k)/2 \rfloor} \log^\gamma n)$, where $\gamma = (d+k) \pmod 2$. Hence, we obtain:

THEOREM 10.5. *Let Γ be a set of hypersurfaces in $\mathbb{R}^d$ of degree at most b. If Γ admits a linearization of dimension k, then $\mathcal{A}(\Gamma)$ can be decomposed into $O(n^{\lfloor (d+k)/2 \rfloor} \log^\gamma n)$ cells of constant description complexity, where $\gamma = d + k$ (mod 2).*

As shown in Section 3, spheres in $\mathbb{R}^d$ admit a linearization of dimension $d+1$; therefore, the arrangement of n spheres in $\mathbb{R}^d$ can be decomposed into $O(n^d \log n)$ cells of constant description complexity.

Aronov and Sharir [**102**] proposed another scheme for decomposing arrangements of triangles in $\mathbb{R}^3$ by combining vertical decomposition and triangulation. They first decompose each three-dimensional cell of the arrangement into convex polyhedra, using an incremental procedure, and then compute a bottom-vertex triangulation of each polyhedron. Other specialized decomposition schemes in $\mathbb{R}^3$ have been proposed in [**422, 538**].

10.4. Cuttings. All the decomposition schemes described in this section decompose $\mathbb{R}^d$ into cells of constant description complexity, so that each cell lies entirely in a single face of $\mathcal{A}(\Gamma)$. In many applications, however, one wants to decompose $\mathbb{R}^d$ into cells of constant description complexity so that each cell intersects only a few surfaces of Γ. Such a decomposition lies at the heart of divide-and-conquer algorithms for numerous geometric problems.

Let Γ be a set of n surfaces in $\mathbb{R}^d$ satisfying assumptions (A1)–(A2). For a parameter $r \leq n$, a family $\Xi = \{\Delta_1, \ldots, \Delta_s\}$ of cells of constant description complexity with pairwise disjoint interiors is called a $(1/r)$-*cutting* of $\mathcal{A}(\Gamma)$ if the interior of each cell in Ξ is crossed by at most n/r surfaces of Γ and Ξ covers $\mathbb{R}^d$.

In this definition we only consider full-dimensional cells; there are refined versions in which we consider a decomposition of $\mathbb{R}^d$ into pairwise disjoint relatively open cells of dimensions $0, 1, \ldots, d$, and require that each cell is *crossed* by at most n/r surfaces, where, for a lower-dimensional cell, a surface crosses the cell if it intersects the cell but does not fully contain it.

Cuttings have led to efficient algorithms for a wide range of geometric problems and to improved bounds for several combinatorial problems. For example, the proof by Clarkson et al. [**245**] on the complexity of m distinct cells in arrangements of lines uses cuttings; see the survey papers [**10, 531**] for a sample of applications of cuttings.

Clarkson [**241**] proved that a $(1/r)$-cutting of size $O(r^d \log^d r)$ exists for a set of hyperplanes in $\mathbb{R}^d$. The bound was improved by Chazelle and Friedman [**221**] to $O(r^d)$; see also [**8, 525, 529, 536**]. An easy counting argument shows that this bound is optimal for any nondegenerate arrangement. There has been considerable work on computing optimal $(1/r)$-cuttings efficiently [**8, 210, 426, 525, 529**]. For example, Chazelle [**210**] showed that a $(1/r)$-cutting for a set of n hyperplanes in $\mathbb{R}^d$ can be computed in time $O(nr^{d-1})$.

Using Haussler and Welzl's result on ε-nets [**431**], one can show that if, for any subset $R \subseteq \Gamma$, there exists a canonical decomposition of $\mathcal{A}(R)$ into at most $g(|R|)$ cells of constant description complexity, then there exists a $(1/r)$-cutting of $\mathcal{A}(\Gamma)$ of size $O(g(r \log r))$. Specifically, one obtains such a cutting by taking a random sample R of $O(r \log r)$ surfaces of Γ, and by constructing the canonical decomposition of $\mathcal{A}(R)$, arguing that, with high probability, this is indeed a $(1/r)$-cutting. By the results of Koltun [**492**] and of Chazelle et al. [**215**] on vertical decompositions, there exists a $(1/r)$-cutting of size $O((r \log r)^{2d-4+\varepsilon})$ of $\mathcal{A}(\Gamma)$ in $d \geq 4$ dimensions, with corresponding sharper bounds for $d = 2, 3$ (see above). On the other hand, if Γ admits a linearization of dimension k, then there exists a $(1/r)$-cutting of size $O((r \log r)^{\lfloor (d+k)/2 \rfloor} \log r)$.

11. Representation of Arrangements

Before we begin to present algorithms for computing arrangements and their substructures, we need to describe how we represent arrangements and their substructures. Planar arrangements of lines can be represented using any standard data structure for representing planar maps, such as quad-edge, winged-edge, and half-edge data structures [**403, 478, 742**]. However, representation of arrangements in higher dimensions is challenging because the topology of cells may be rather complex. Exactly how an arrangement is represented largely depends on the specific application for which one needs to compute it. For example, representations may range from simply computing a representative point within each cell, or the vertices of the arrangement, to storing various spatial relationships between cells. We first review representations of hyperplane arrangements and then discuss surface arrangements.

Hyperplane arrangements. A simple way to represent a hyperplane arrangement $\mathcal{A}(\Gamma)$ is by storing its *1-skeleton* [**287**]. That is, we construct a graph (V, E) whose nodes are the vertices of the arrangement. There is an edge between two nodes v_i, v_j if they are endpoints of an edge of the arrangement. Using the 1-skeleton of $\mathcal{A}(\Gamma)$, we can traverse the entire arrangement in a systematic way. The incidence relationship of various cells in $\mathcal{A}(\Gamma)$ can be represented using a data structure called *incidence graph*. A k-dimensional cell C is called a *subcell* of a $(k+1)$-dimensional cell C' if C lies on the boundary of C'; C' is called the *supercell* of C. We assume that the empty set is a (-1)-dimensional cell of $\mathcal{A}(\Gamma)$, which is a subcell of all vertices of $\mathcal{A}(\Gamma)$; and $\mathbb{R}^d$ is a $(d+1)$-dimensional cell, which is the supercell of all d-dimensional cells of $\mathcal{A}(\Gamma)$. The incidence graph of $\mathcal{A}(\Gamma)$ has a node for each cell of $\mathcal{A}(\Gamma)$, including the (-1)-dimensional and $(d+1)$-dimensional cells. There is a (directed) arc from a node C to another node C' if C is a subcell of C'; see Figure 14. Note that the incidence graph forms a lattice. Many algorithms for computing the arrangement construct the incidence graph of the arrangement.

A disadvantage of 1-skeletons and incidence graphs is that they do not encode ordering information of cells. For examples, in planar arrangements of lines or segments, there is a natural ordering of edges incident to a vertex or of the edges incident to a two-dimensional face. The quad-edge (or doubly-connected edge list) data structure encodes this information for planar arrangements [**146**]. Dobkin and Laszlo [**273**] extended the quad-edge data structure to $\mathbb{R}^3$, which was later extended to higher dimensions [**177, 513, 514**]. Dobkin *et al.* [**269**] described an algorithm for representing a simple polygon as a short Boolean formula, which can be used to store faces of segment arrangements to answer various queries efficiently.

Surface arrangements. Representing arrangements of surface patches is considerably more challenging than representing hyperplane arrangements because of the complex topology that cells in such an arrangement can have. A very simple representation of $\mathcal{A}(\Gamma)$ is to store a representative point from each cell of $\mathcal{A}(\Gamma)$, or to store the vertices of $\mathcal{A}(\Gamma)$. An even coarser representation of arrangements of graphs of polynomials is to store all realizable sign sequences. It turns out that this simple representation is sufficient for some applications [**70, 157**]. The notion of 1-skeleton can be generalized to arrangements of surfaces. However, the full connectivity information cannot be encoded by simply storing vertices and edges of the arrangement. Instead, we need a finer one-dimensional structure, known as the *roadmap*. Roadmaps were originally introduced by Canny [**192, 194**] to determine whether two points lie in the same connected component of a semialgebraic set; see also [**390, 392, 434**]. They were subsequently used for computing a semialgebraic description of connected components of a semialgebraic set [**134, 195, 433**]. We can extend the notion of roadmaps to entire arrangements. Roughly speaking, a roadmap $\mathcal{R}(\Gamma)$ of $\mathcal{A}(\Gamma)$ is a one-dimensional semialgebraic set that satisfies the following two conditions.

(R1) For every cell C in $\mathcal{A}(\Gamma)$, $C \cap \mathcal{R}(\Gamma)$ is nonempty and connected.

(R2) Let C_w be the cross-section of a cell $C \in \mathcal{A}(\Gamma)$ at the hyperplane $x_1 = w$. For any $w \in \mathbb{R}$ and for a cell $C \in \mathcal{A}(\Gamma)$, if $C_w \neq \emptyset$ then every connected component of C_w intersects $\mathcal{R}(\Gamma)$.

We can also define the roadmap of various substructures of arrangements. See [**133, 135, 192**] for more details on roadmaps.

A roadmap does not represent "ordering" of cells in the arrangement or adjacency relationship among various cells. If we want to encode the adjacency relationship among higher dimensional cells of $\mathcal{A}(\Gamma)$, we can compute the vertical decomposition or the cylindrical algebraic decomposition of $\mathcal{A}(\Gamma)$ and compute the adjacency relationship between cells in the decomposition [**88, 660**]. (This adjacency information comes as an integral part of the cylindrical algebraic decomposition, because it is a cell complex; it is harder to extract this information from the vertical decomposition.) Brisson [**177**] describes the *cell-tuple* data structure that encodes topological structures, ordering among cells, the boundaries of cells, and other information for cells of surface arrangements.

Many query-type applications (e.g., point location and ray shooting) call for preprocessing $\mathcal{A}(\Gamma)$ into a data structure so that various queries can be answered efficiently. In these cases, instead of storing various cells of an arrangement explicitly, we can store the arrangement implicitly, e.g., using cuttings. Chazelle *et al.* [**216**] have described how to preprocess arrangements of surfaces for point-location queries; Agarwal *et al.* [**20**] have described data structures for storing lower envelopes in $\mathbb{R}^4$ for point-location queries.

12. Computing Arrangements

We now review algorithms to compute the arrangement $\mathcal{A}(\Gamma)$ of a set Γ of n surface patches satisfying assumptions (A1)–(A2). We need to assume here an appropriate model of computation, in which various primitive operations on a constant number of surfaces can be performed in constant time. We will assume an infinite-precision real arithmetic model in which the roots of any polynomial of constant degree can be computed exactly in constant time. Such a model is available (in theory and in practice) using standard tools from computational real algebraic geometry; see [**135**] and below.

Constructing arrangements of hyperplanes and simplices. Edelsbrunner *et al.* [**301**] describe an incremental algorithm that computes, in time $O(n^d)$, the incidence graph of $\mathcal{A}(\Gamma)$, for a set Γ of n hyperplanes in $\mathbb{R}^d$. Roughly speaking, their algorithm adds the hyperplanes of Γ one by one and maintains the incidence graph of the arrangement of the hyperplanes added so far. Let Γ_i be the set of hyperplanes added in the first i stages, and let γ_{i+1} be the next hyperplane to be added. In the $(i+1)$st stage, the algorithm traces γ_{i+1} through $\mathcal{A}(\Gamma_i)$. If a k-face f of $\mathcal{A}(\Gamma_i)$ does not intersect γ_i, then f remains a face of $\mathcal{A}(\Gamma_{i+1})$. If f intersects γ_{i+1}, then $f \in zone(\gamma_{i+1}; \Gamma_i)$ and f is split into two k-faces f^+, f^-, lying in the two open halfspaces bounded by γ_{i+1}, and a $(k-1)$-face $f' = f \cap \gamma_{i+1}$. The algorithm therefore checks the faces of $zone(\gamma_{i+1}; \Gamma_i)$ to determine which of them intersect γ_{i+1}. For each such intersecting face, it adds corresponding nodes in the incidence graph and updates the edges of that graph. The $(i+1)$st stage can be completed in time proportional to the complexity of $zone(\gamma_{i+1}; \Gamma_i)$, which is $O(i^{d-1})$; see [**288, 301**]. Hence, the overall running time of the algorithm is $O(n^d)$.

A drawback of the algorithm just described is that it requires $O(n^d)$ "working" storage because it has to maintain the entire arrangement constructed so far in order to determine which of the cells intersect the new hyperplane. An interesting question is whether $\mathcal{A}(\Gamma)$ can be computed using only $O(n)$ working storage. Edelsbrunner and Guibas [**293**] proposed the *topological sweep* algorithm, which can construct the arrangement of n lines in the plane in $O(n^2)$ time, using $O(n)$ working storage. Their algorithm, which is a generalization of the sweep-line algorithm of Bentley and Ottmann [**137**], sweeps the plane by a pseudo-line. See [**596**] for analysis of the working storage of the standard straight-line sweeping algorithm

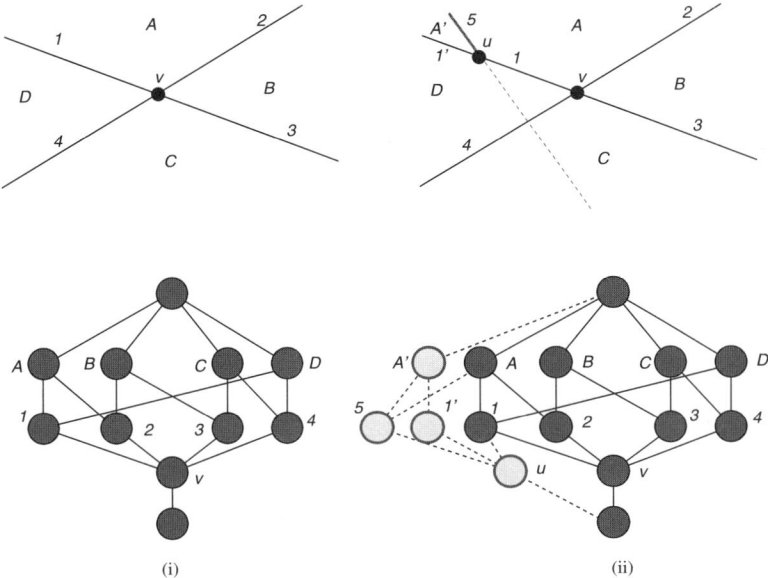

FIGURE 14. (i) Incidence graph of the arrangement of 2 lines. (ii) Adding a new line; incremental changes in the incidence graph as the vertex v, the edge 5, and the face A' are added.

of Bentley and Ottmann. The algorithm of Edelsbrunner and Guibas can be extended to enumerate all vertices in an arrangement of n hyperplanes in $\mathbb{R}^d$ in $O(n^d)$ time using $O(n)$ space. See [**82, 111, 306**] for other topological-sweep algorithms. Avis and Fukuda [**117**] developed an algorithm that can enumerate in $O(n^2k)$ time, using $O(n)$ space, all k vertices of the arrangement of a set Γ of n hyperplanes in $\mathbb{R}^d$, in which every vertex is incident to d hyperplanes. Their algorithm is useful when there are many parallel hyperplanes in Γ. See also [**118, 370**] for some related results.

Using the random-sampling technique, Clarkson and Shor [**246**] developed an $O(n \log n + k)$-expected time algorithm for constructing the arrangement of a set Γ of n line segments in the plane; here k is the number of vertices in $\mathcal{A}(\Gamma)$; see also [**561, 562**]. Chazelle and Edelsbrunner [**212**] developed a deterministic algorithm that can construct $\mathcal{A}(\Gamma)$ in $O(n \log n + k)$ time, using $O(n + k)$ storage. The space complexity was improved to $O(n)$, without affecting the asymptotic running time, by Balaban [**121**]. If Γ is a set of n triangles in $\mathbb{R}^3$, $\mathcal{A}^{\|}(\Gamma)$ can be constructed in $O(n^2 \log n + k)$ expected time, using a randomized incremental algorithm [**219, 681**], where again k is the number of vertices of the arrangement. de Berg et al. [**142**] proposed a deterministic algorithm with $O(n^2 \alpha(n) \log n + k \log n)$ running time for computing $\mathcal{A}^{\|}(\Gamma)$.

Chazelle and Friedman [**222**] described an algorithm that can preprocess a set Γ of n hyperplanes into a data structure of size $O(n^d/\log^d n)$ so that a point-location query can be answered in $O(\log n)$ time. Their algorithm was later simplified by Matoušek [**533**] and Chazelle [**210**]. Mulmuley and Sen [**565**] developed a randomized dynamic data structure of size $O(n^d)$ for point location in arrangements of hyperplanes, which can answer a point-location query in $O(\log n)$ expected time, and can insert or delete a hyperplane in $O(n^{d-1} \log n)$ expected time. Hagerup *et al.* [**410**] described a randomized parallel algorithm for constructing the arrangement of n hyperplanes in $\mathbb{R}^d$ under the CRCW model. Their algorithm runs in $O(\log n)$ time using $O(n^d/\log n)$ expected number of processors. A deterministic algorithm under the CREW model with the same worst-case performance was proposed by Goodrich [**388**].

Constructing arrangements of surfaces. The algorithm by Edelsbrunner *et al.* [**301**] for computing hyperplane arrangements can be extended to computing the vertical decomposition $\mathcal{A}^{\|}(\Gamma)$, for a set Γ of n arcs in the plane. Using the preceding notation, the algorithm traces, in the $(i+1)$st step, the arc γ_{i+1} through $zone(\gamma_{i+1}; \Gamma_i)$, and updates the trapezoids of $\mathcal{A}^{\|}(\Gamma_i)$ that intersect γ_{i+1}. The running time of the $(i+1)$st stage is $O(\lambda_{s+2}(i))$, where s is the maximum number of intersection points between a pair of arcs in Γ. Hence, the overall running time of the algorithm is $O(n\lambda_{s+2}(n))$ [**295**]. If the arcs in Γ are added in a *random* order and a "history dag" (see Chapter 3) is used to efficiently find the trapezoids of $\mathcal{A}^{\|}(\Gamma_i)$ that intersect γ_{i+1}, the expected running time of the algorithm can be improved to $O(n \log n + k)$, where k is the number of vertices in $\mathcal{A}(\Gamma)$ [**219, 681**].

Very little is known about computing the arrangement of a set Γ of surfaces in higher dimensions. Chazelle *et al.* [**215**] gave a randomized algorithm for constructing $\mathcal{A}^{\|}(\Gamma)$ in $d \geq 4$ dimensions, using the random-sampling technique. Combined with Koltun's improved bounds [**492**], the expected running time of the algorithm is $O(n^{2d-4+\varepsilon})$, for any $\varepsilon > 0$. The algorithm can be made deterministic without increasing its asymptotic running time, but the deterministic algorithm is considerably more complex.

13. Computing Substructures in Arrangements

13.1. Lower envelopes. Let Γ be a set of surface patches in $\mathbb{R}^d$ satisfying assumptions (A1)–(A2). We want to compute the minimization diagram $\mathcal{M}(\Gamma)$ of Γ. We will present in Chapter 3 algorithms for computing the minimization diagram of a set of arcs in the plane. In this section we focus on minimization diagrams of sets of surface patches in higher dimensions. There are again several choices, depending on the application, as to what exactly we want to compute. The simplest choice is to compute the vertices or the 1-skeleton of $\mathcal{M}(\Gamma)$. A more difficult task is to compute all the faces of $\mathcal{M}(\Gamma)$ and represent them using any of the mechanisms described in the previous section. Another challenging task, which is required in many applications, is to store Γ into a data structure so that $L_\Gamma(\mathbf{x})$, for any point $\mathbf{x} \in \mathbb{R}^{d-1}$, can be computed efficiently.

For collections Γ of surface patches in $\mathbb{R}^3$, the minimization diagram $\mathcal{M}(\Gamma)$ is a planar subdivision. In this case, the latter two tasks are not significantly harder than the first one, because we can preprocess $\mathcal{M}(\Gamma)$ using any optimal planar point-location algorithm [**146**]. Several algorithms have been developed for computing the minimization diagram of bivariate (partial) surface patches [**43, 141, 158, 159, 678, 681**]. Some of these techniques use randomized algorithms, and their expected running time is $O(n^{2+\varepsilon})$, which is comparable with the maximum complexity of the minimization diagram of bivariate surface patches. The simplest algorithm is probably the deterministic divide-and-conquer algorithm presented by Agarwal *et al.* [**43**]. It partitions Γ into two subsets Γ_1, Γ_2 of roughly equal size, and computes recursively the minimization diagrams $\mathcal{M}_1$, $\mathcal{M}_2$ of Γ_1 and Γ_2, respectively. It then computes the overlay $\mathcal{M}^*$ of $\mathcal{M}_1$ and $\mathcal{M}_2$. Over each face f of $\mathcal{M}^*$ there are only (at most) two surface patches that can attain the final envelope (the one attaining $L(\Gamma_1)$ over f and the one attaining $L(\Gamma_2)$ over f), so we compute the minimization diagram of these two surface patches over f, replace f by this refined diagram, and repeat this step for all faces of $\mathcal{M}^*$. We finally merge any two adjacent faces f, f' of the resulting subdivision if the same surface patches attain $L(\Gamma)$ over both f and f'. The cost of this step is proportional to the number of faces of $\mathcal{M}^*$. By the result of Agarwal et al. [43] (see also [495]), $\mathcal{M}^*$ has $O(n^{2+\varepsilon})$ faces. This implies that the complexity of the above divide-and-conquer algorithm is $O(n^{2+\varepsilon})$, for any $\varepsilon > 0$. If Γ is a set of triangles in $\mathbb{R}^3$, the running time of the algorithm is $O(n^2 \alpha(n))$ [296]. This divide-and-conquer algorithm can also be used to compute the *sandwich region* $S(\Gamma, \Gamma')$, namely, the region lying above all surface patches of one collection Γ' and below all surface patches of another collection Γ, in time $O(n^{2+\varepsilon})$, where $n = |\Gamma| + |\Gamma'|$ [43].

A more difficult problem is to devise *output-sensitive* algorithms for computing $\mathcal{M}(\Gamma)$, whose complexity depends on the actual combinatorial complexity of the envelope. A rather complex algorithm is presented by de Berg [139] for the case of triangles in $\mathbb{R}^3$, whose running time is $O(n^{4/3+\varepsilon} + n^{4/5+\varepsilon} k^{4/5})$, where k is the number of vertices in $\mathcal{M}(\Gamma)$. If the triangles in Γ are pairwise disjoint, the running time can be improved to $O(n^{1+\varepsilon} + n^{2/3+\varepsilon} k^{2/3})$ [37, 139].

This issue of output-sensitivity is particularly important when constructing planar Voronoi diagrams (which, as we recall, are minimization diagrams of a corresponding collection of bivariate "distance functions"). Since most Voronoi diagrams in the plane have linear complexity, it is desirable to implement the above divide-and-conquer algorithm so that it runs in nearly-linear time. It was recently shown by Setter et al. [672] that if the divide step is done at random, the expected complexity of the overlay $\mathcal{M}^*$ is linear, so the algorithm does indeed take only nearly-linear time. (For arbitrary partitions, the overlay $\mathcal{M}^*$ can have quadratic complexity.)

The algorithm by Edelsbrunner et al. [296] can be extended to compute, in $O(n^{d-1}\alpha(n))$ time, all faces of the minimization diagram of $(d-1)$-simplices in $\mathbb{R}^d$, for $d \geq 4$. However, little is known about computing the minimization diagram of more general surface patches in $d \geq 4$ dimensions. Let Γ be a set of surface patches in $\mathbb{R}^d$ satisfying assumptions (A1)–(A2). Agarwal et al. [20] showed that all vertices, edges, and 2-faces of $\mathcal{M}(\Gamma)$ can be computed in randomized expected time $O(n^{d-1+\varepsilon})$. We sketch their algorithm below.

Assume that Γ satisfies assumptions (A1)–(A5). Fix a $(d-2)$-tuple of surface patches, say $\gamma_1, \ldots, \gamma_{d-2}$, and decompose their common intersection $\bigcap_{i=1}^{d-2} \gamma_i$ into smooth, $x_1 x_2$-monotone, connected patches, using a stratification algorithm. Let Π be one such piece. Each surface γ_i, for $i \geq d-1$, intersects Π at a curve ξ_i, which partitions Π into two regions. If we regard each γ_i as the graph of a partially defined $(d-1)$-variate function, then we can define $K_i \subseteq \Pi$ to be the region whose projection on the hyperplane $H : x_d = 0$ consists of points $\mathbf{x}$ at which $\gamma_i(\mathbf{x}) \geq \gamma_1(\mathbf{x}) = \cdots = \gamma_{d-2}(\mathbf{x})$. The intersection $Q = \bigcap_{i \geq d-1} K_i$ is equal to the portion of Π that appears along the lower envelope $L(\Gamma)$. We repeat this procedure for all patches of the intersection $\bigcap_{i=1}^{d-2} \gamma_i$ and for all $(d-2)$-tuples of surface patches. This will give all the vertices, edges and 2-faces of $L(\Gamma)$.

Since Π is an x_1x_2-monotone 2-manifold, computing Q is essentially the same as computing the intersection of $n-d+2$ planar regions. Q can thus be computed using an appropriate variant of the randomized incremental approach [141, 219]. It adds the intersection curves $\xi_i = \gamma_i \cap \Pi$ one by one, in a random order (each ξ may consist of $O(1)$ arcs), and maintains the intersection of the regions K_i for the arcs added so far. Let Q_r denote this intersection after r arcs have been added. We maintain the "vertical decomposition" of Q_r (within Π), and represent Q_r as a collection of *pseudo-trapezoids*. We maintain additional data structures, including a *history dag* and a union-find structure, and proceed exactly as in [141, 219] (see Chapter 3). We omit here the details.

We define the *weight* of a pseudo-trapezoid τ to be the number of surface patches γ_i, for $i \geq d-1$, whose graphs either cross τ or hide τ completely from the lower envelope (excluding the up to four function graphs whose intersections with Π *define* τ). The cost of the above procedure, summed over all $(d-2)$-tuples of Γ, is proportional to the number of pseudo-trapezoids that are created during the execution of the algorithm, plus the sum of their weights, plus an overhead term of $O(n^{d-1})$ needed to prepare the collections of arcs ξ_i over all two-dimensional patches Π. Modifying the analysis in the papers cited above, Agarwal *et al.* prove the following.

THEOREM 13.1 (Agarwal *et al.* [20]). *Let Γ be a set of n surface patches in $\mathbb{R}^d$ satisfying assumptions (A1)–(A2). The vertices, edges, and 2-faces of $\mathcal{M}(\Gamma)$ can be computed in randomized expected time $O(n^{d-1+\varepsilon})$, for any $\varepsilon > 0$.*

For $d = 4$, the above algorithm can be extended to compute the incidence graph (or cell-tuple structure) of $\mathcal{M}(\Gamma)$. The approach of [20], however, falls short of computing such representations for $d > 4$. Agarwal *et al.* also show that the three-dimensional point-location algorithm of Preparata and Tamassia [638] can be extended to preprocess a set of trivariate surface patches, in time $O(n^{3+\varepsilon})$, into a data structure of size $O(n^{3+\varepsilon})$, so that $L_\Gamma(\mathbf{x})$, for any point $\mathbf{x} \in \mathbb{R}^3$, can be computed in $O(\log^2 n)$ time.

PROBLEM 13.2. *Let Γ be a set of n surface patches in $\mathbb{R}^d$, for $d > 4$, satisfying assumptions (A1)–(A3). How fast can Γ be preprocessed, so that $L_\Gamma(\mathbf{x})$, for a query point $\mathbf{x} \in \mathbb{R}^{d-1}$, can be computed efficiently?*

13.2. Single cells. Computing a single cell in an arrangement of n hyperplanes in $\mathbb{R}^d$ is equivalent, by duality, to computing the convex hull of a set of n points in $\mathbb{R}^d$, which is a widely studied problem; see, e.g., [288, 670] for a summary of known results. For $d \geq 4$, an $O(n^{\lfloor d/2 \rfloor})$ expected-time algorithm for this problem was proposed by Clarkson and Shor [246] (see also [668]), which is optimal in the worst case. By derandomizing this algorithm, Chazelle [211] developed an $O(n^{\lfloor d/2 \rfloor})$-time deterministic algorithm. A somewhat simpler algorithm with the same running time was later proposed by Brönnimann *et al.* [179]. These results imply that the Euclidean Voronoi diagram of a set of n points in $\mathbb{R}^d$ can be computed in time $O(n^{\lceil d/2 \rceil})$.

Since the complexity of a cell may vary between $O(1)$ and $O(n^{\lfloor d/2 \rfloor})$, it pays off to seek output-sensitive algorithms for computing a single cell in hyperplane arrangements; see [225, 480, 666]. For $d \leq 3$, Clarkson and Shor [246] gave randomized algorithms with expected time $O(n \log h)$, where h is the complexity of the cell, provided that the planes are in general position. Simple deterministic algorithms with the same worst-case bound were developed by Chan [**199**]. Seidel [**666**] proposed an algorithm whose running time is $O(n^2 + h \log n)$; the first term can be improved to $O(n^{2-2/(\lfloor d/2 \rfloor+1)} \log^c n)$ [**532**] or to $O((nh)^{1-1/(\lfloor d/2 \rfloor+1)} \log^c n)$ [**200**], for some absolute constant c. Chan et al. [**208**] described another output-sensitive algorithm whose running time is $O((n + (nf)^{1-1/\lceil d/2 \rceil} + fn^{1-2/\lceil d/2 \rceil}) \log^c n)$, where f is the number of vertices of the cell. Avis et al. [**116**] described an algorithm that can compute in $O(nf)$ time, using $O(n)$ space, all f vertices of a cell in an arrangement of n hyperplanes in $\mathbb{R}^d$; see also [**176, 369**]. All these output-sensitive bounds hold only for simple arrangements. Although many of these algorithms can be extended to nonsimple arrangements, the running time increases.

As will be mentioned in Chapter 3, Guibas et al. [**402**] give an $O(\lambda_{s+2}(n) \log^2 n)$-time algorithm for computing a single face in an arrangement of n arcs in the plane, each pair of which intersect in at most s points. Later, a randomized algorithm with expected time $O(\lambda_{s+2}(n) \log n)$ was developed by Chazelle et al. [**219**]. Since the complexity of the vertical decomposition of a single cell in an arrangement of n surface patches in $\mathbb{R}^3$ is $O(n^{2+\varepsilon})$ [**664**], an application of the random-sampling technique yields an algorithm for computing a single cell in randomized expected time $O(n^{2+\varepsilon})$ in an arrangement of n surface patches in $\mathbb{R}^3$ [**664**]. If Γ is a set of triangles, the running time can be improved to $O(n^2 \log^3 n)$ [**141**]. Halperin [**412, 413**] developed faster algorithms for computing a single cell in arrangements of "special" classes of bivariate surfaces that arise in motion-planning applications.

13.3. Levels.

Constructing the $(\leq k)$-level. Let Γ be a set of n arcs in the plane, each pair of which intersect in at most s points. $\mathcal{A}_{\leq k}(\Gamma)$ can be computed by a simple divide-and-conquer algorithm, as follows [**676**]. Partition Γ into two subsets Γ_1, Γ_2, each of size at most $\lceil n/2 \rceil$, compute recursively $\mathcal{A}_{\leq k}(\Gamma_1), \mathcal{A}_{\leq k}(\Gamma_2)$, and then use a sweep-line algorithm to compute $\mathcal{A}_{\leq k}(\Gamma)$ from $\mathcal{A}_{\leq k}(\Gamma_1)$ and $\mathcal{A}_{\leq k}(\Gamma_2)$. The time spent in the merge step is proportional to the number of vertices in $\mathcal{A}_{\leq k}(\Gamma_1), \mathcal{A}_{\leq k}(\Gamma_2)$ and the number of intersections points between the edges of two subdivisions, each of which is a vertex of $\mathcal{A}(\Gamma)$ whose level is at most $2k$. Using Theorem 6.1, the total time spent in the merge step is $O(\lambda_{s+2}(n) k \log n)$. Hence, the overall running time of the algorithm is $O(\lambda_{s+2}(n) k \log^2 n)$. We can also use a randomized incremental algorithm, which adds arcs one by one in a random order and maintains $\mathcal{A}_{\leq k}(\Gamma_i)$, where Γ_i is the set of arcs added so far. The expected running time of the algorithm is $O(\lambda_{s+2}(n) k \log(n/k))$; see, e.g., [**563**]. Everett et al. [**341**] showed that if Γ is a set of n lines, the expected running time can be improved to $O(n \log n + nk)$. Agarwal et al. [**26**] gave another randomized incremental algorithm that can compute $\mathcal{A}_{\leq k}(\Gamma)$ in expected time $O(\lambda_{s+2}(n)(k + \log n))$.

In higher dimensions, little is known about computing $\mathcal{A}_{\leq k}(\Gamma)$, for collections Γ of surface patches. For $d = 3$, Mulmuley [**563**] gave a randomized incremental algorithm for computing the $(\leq k)$-level in an arrangement of n planes whose expected running time is $O(nk^2 \log(n/k))$. The expected running time can be improved to $O(n \log^3 n + nk^2)$ using the algorithm by Agarwal et al. [**26**]. There are, however, several technical difficulties in extending this approach to computing levels in arrangements of surface patches. Using the random-sampling technique, Agarwal et al. [**29**] developed an $O(n^{2+\varepsilon} k)$ expected-time algorithm for computing $\mathcal{A}_{\leq k}(\Gamma)$, for a collection Γ of n surface patches in $\mathbb{R}^3$. Their algorithm can be derandomized without affecting the asymptotic running time bound. For $d \geq 4$, Agarwal et al.'s and Mulmuley's algorithms can compute the ($\leq k$)-level in arrangements of n hyperplanes in expected time $O(n^{\lfloor d/2 \rfloor} k^{\lceil d/2 \rceil})$. These algorithms do not extend to efficient algorithms for computing the ($\leq k$)-level in arrangements of more general surfaces, because no nontrivial bound is known for the complexity of a triangulation (or, more concretely, the vertical decomposition) of $\mathcal{A}_{\leq k}(\Gamma)$ in four and higher dimensions.

Constructing a single level. Edelsbrunner and Welzl [**309**] gave an $O(n \log n + b \log^2 n)$-time algorithm to construct the k-level in an arrangement of n lines in the plane, where b is the number of vertices of the k-level. This bound was slightly improved by Cole et al. [**249**] to $O(n \log n + b \log^2 k)$. However, these algorithms do not extend to computing the k-level in arrangements of curves. The approach by Agarwal et al. [**26**] can compute the k-level in an arrangement of lines in randomized expected time $O(n \log^2 n + nk^{1/3} \log^{2/3} n)$, and it extends to arrangements of curves and to arrangements of hyperplanes. In an unpublished note, Chan [**202**] slightly improves these bounds; in particular, his technique computes a single level in a line arrangement in randomized expected time $O(n \log n + nk^{1/3})$. Agarwal and Matoušek [**39**] describe an output-sensitive algorithm for computing the k-level in an arrangement of planes. The running time of their algorithm, after a slight improvement by Chan [**200**], is $O(n \log b + b^{1+\varepsilon})$, for any $\varepsilon > 0$, where b is the number of vertices of the k-level. A variant of this technique was recently given by Kapelushnik [**456**] for computing the k-level in an arrangement of n triangles in $\mathbb{R}^3$; the expected running time is $O(n^{2+\varepsilon} k^{2/3})$, for any $\varepsilon > 0$. The algorithm of Agarwal and Matoušek can compute the k-level in an arrangement of hyperplanes in $\mathbb{R}^d$ in time $O(n \log b + (nb)^{1-1/(\lfloor d/2 \rfloor+1)+\varepsilon} + bn^{1-2/(\lfloor d/2 \rfloor+1)+\varepsilon})$. As in the case of single cells, all the output-sensitive algorithms assume that the hyperplanes are in general position.

13.4. Marked cells and incidences. Let Γ be a set of n lines in the plane and P a set of m points in the plane, none lying on any line of Γ. Edelsbrunner et al. [**297**] presented a randomized algorithm, based on the random-sampling technique, for computing $\mathcal{C}(P, \Gamma)$, the set of cells in $\mathcal{A}(\Gamma)$ that contain at least one point of P, whose expected running time is $O(m^{2/3-\varepsilon} n^{2/3+2\varepsilon} + m \log n + n \log n \log m)$, for any $\varepsilon > 0$. A deterministic algorithm with running time $O(m^{2/3} n^{2/3} \log^c n + n \log^3 n + m \log n)$ was developed by Agarwal [**9**]. However, both algorithms are rather complicated. A simple randomized divide-and-conquer algorithm, with $O((m\sqrt{n} + n) \log n)$ expected running time, was recently proposed by Agarwal et al. [**40**]. Using random sampling, they improved the expected running time to $O(m^{2/3} n^{2/3} \log^{2/3}(n/\sqrt{m}) + (m+n) \log n)$. If we let the points of P lie on lines of Γ, and are interested in computing the incidences between Γ and P (see Chapter 4 for a comprehensive survey on incidences), the best known algorithm is by Matoušek; its expected running time is $O(m^{2/3} n^{2/3} 2^{O(\log^*(m+n))} + (m+n) \log(m+n))$ [**534**]. An $\Omega(m^{2/3} n^{2/3} + (m+n) \log(m+n))$ lower bound for this problem is proved by Erickson [**338**] under a restricted model of computation. Matoušek's algorithm can be extended to higher dimensions, to count the number of incidences between m points and n hyperplanes in $\mathbb{R}^d$, in $O((mn)^{1-1/(d+1)} 2^{O(\log^*(m+n))} + (m+n) \log(m+n))$ time.

The above algorithms can be modified to compute marked cells in arrangements of segments in the plane. The best known solution is a randomized algorithm, presented by Agarwal et al. [**40**], whose running time is $O(m^{2/3}n^{2/3}\log^2(n/\sqrt{m}) + (m + n\log m + n\alpha(n))\log n)$. Little is known about computing marked cells in arrangements of arcs in the plane. Using a randomized incremental algorithm, $\mathcal{C}(P,\Gamma)$ can be computed in expected time $O(\lambda_{s+2}(n)\sqrt{m}\log n)$, where s is the maximum number of intersection points between a pair of arcs in Γ [**681**]. If Γ is a set of n unit-radius circles and P is a set of m points in the plane, the incidences between Γ and P can be computed using Matoušek's algorithm [**534**]. A recent result of Agarwal and Sharir [**55**] computes all incidences between m points and n circles in time close to $O(m^{2/3}n^{2/3} + m^{6/11}n^{9/11} + m + n)$; see also Chapter 4.

Randomized incremental algorithms can be used to construct marked cells in arrangements of hyperplanes in higher dimensions, in time close to their worst-case complexity. For example, if Γ is a set of n planes and P is a set of m points in $\mathbb{R}^3$, then the incidence graph of cells in $\mathcal{C}(P,\Gamma)$ can be computed in expected time $O(nm^{2/3}\log n)$ [**141**]. For $d \geq 4$, the expected running time is $O(m^{1/2}n^{d/2}\log^\gamma n)$, where $\gamma = (\lfloor d/2 \rfloor - 1)/2$. de Berg et al. [**147**] describe an efficient point-location algorithm in the zone of a k-flat in an arrangement of hyperplanes in $\mathbb{R}^d$. Their algorithm can answer a query in $O(\log n)$ time using $O(n^{\lfloor(d+k)/2\rfloor}\log^\gamma n)$ space, where $\gamma = d + k \pmod{2}$.

13.5. Union of objects. Let Γ be a set of n semialgebraic simply connected regions in the plane, each of constant description complexity. The union of Γ can be computed in $O(f(n)\log^2 n)$ time by a divide-and-conquer technique, similar to that described in Section 13.3 for computing $\mathcal{A}_{\leq k}(\Gamma)$. Here $f(m)$ is the maximum complexity of the union of a subset of Γ of size m. Alternatively, $\bigcup \Gamma$ can be computed in $O(f(n)\log n)$ expected time using the *lazy* randomized incremental algorithm by de Berg et al. [**141**]. As a consequence, the union of n convex fat objects, each of constant description complexity, can be computed in $O(n^{1+\varepsilon})$ time, for any $\varepsilon > 0$. The same holds for κ-round or (δ, β)-covered objects; see Section 9.

While the union of triangles can have quadratic complexity in the worst case, there have been several attempts to compute the union efficiently in certain favorable situations; see [**345, 347**].

Aronov et al. [**109**] modified the approach of Agarwal et al. [**20**] so that the union of n convex polytopes in $\mathbb{R}^3$ with a total of s vertices can be computed in expected time $O(sn\log n\log s + n^3)$. The same approach can be used to compute the union of n congruent cylinders in $\mathbb{R}^3$ in time $O(n^{2+\varepsilon})$, for any $\varepsilon > 0$. (Again, consult Section 9 for the corresponding bounds on the complexity of the unions.)

Many applications call for computing the volume or surface area of $\bigcup \Gamma$ instead of its combinatorial structure. Overmars and Yap [**578**] showed that the volume of the union of n axis-parallel boxes in $\mathbb{R}^d$ can be computed in $O(n^{d/2}\log n)$ time. The running time $O(n^{3/2}\log n)$, for $d = 3$, can be improved to $O(n^{4/3}\log n)$ in the special case where all the boxes are cubes [**35**]. Edelsbrunner [**290**] gave an elegant formula for the volume and the surface area of the union of n balls in $\mathbb{R}^d$, which can be used to compute the volume efficiently.

14. Applications

In this section we present a sample of applications of arrangements. We discuss a few specific problems that can be reduced to bounding the complexity of various substructures of arrangements of surfaces or to computing these substructures. We also mention a few general areas that have motivated the study of several basic problems involving arrangements, and in which arrangements have played an important role.

14.1. Range searching. A typical range-searching problem can be stated as follows: *Preprocess a set P of n points in $\mathbb{R}^d$, so that all points of P lying in a query region can be reported (or counted) quickly.* A special case of range searching is *halfspace range searching*, in which the query regions are halfspaces. Because of numerous applications, range searching has received much attention during the last twenty years. See [**13, 31, 535**] for surveys on range searching and its applications.

If we define the dual of a point $p = (a_1, \ldots, a_d)$ to be the hyperplane p^* : $x_d = -a_1 x_1 - \cdots - a_{d-1} x_{d-1} + a_d$, and the dual of a hyperplane h : $x_d = b_1 x_1 + \cdots + b_{d-1} x_{d-1} + b_d$ to be the point $h^* = (b_1, \ldots b_d)$, then p lies above (resp., below, on) h if and only if the hyperplane p^* passes above (resp., below, through) the point h^*. Hence, halfspace range searching has the following equivalent "dual" formulation: Preprocess a set Γ of n hyperplanes in $\mathbb{R}^d$ so that the hyperplanes of H lying below (or above) a query point can be reported quickly, or the level of a query point can be computed quickly. Using the point-location data structure for hyperplane arrangements given in [**210**], the level of a query point can be computed in $O(\log n)$ time using $O(n^d / \log^d n)$ space. This data structure can be modified to report all t hyperplanes lying below a query point in time $O(\log n + t)$. Chazelle *et al.* [**224**] showed, using results on arrangements, that a two-dimensional halfspace *range-reporting* query can be answered in $O(\log n + t)$ time using only $O(n)$ space. In higher dimensions, by constructing $(1/r)$-cuttings for $\mathcal{A}_{\leq k}(\Gamma)$, for shallow values of k, Matoušek [**530**] developed a data structure that can answer a halfspace range-reporting query in time $O(\log n + t)$, using $O(n^{\lfloor d/2 \rfloor} \log^c n)$ space, for some constant c. He also developed a data structure that can answer a query in time $O(n^{1-1/\lfloor d/2 \rfloor} \log^c n + t)$, using $O(n \log \log n)$ space [**530**]. See also [**16, 226**]. Using linearization, a semialgebraic range-searching query, where one wants to count or report all points of P lying inside a semialgebraic set of constant description complexity, can be answered efficiently using halfspace range-searching data structures [**38, 752**].

Point location in hyperplane arrangements can be used for simplex range searching [**227**], ray shooting [**37, 38, 538**], and several other geometric searching problems [**36**].

14.2. Transversals. Let S be a set of n compact convex sets in $\mathbb{R}^d$. A hyperplane h is called a *transversal* of S if h intersects every member of S. Let $\mathcal{T}(S)$ denote the space of all hyperplane transversals of S. We wish to study the structure of $\mathcal{T}(S)$. To facilitate this study, we apply the dual transform described in Section 14.1. Let $h : x_d = a_1 x_1 + \cdots + a_{d-1} x_{d-1} + a_d$ be a hyperplane which intersects a set $s \in S$. Translate h up and down until it becomes tangent to s. Denote the resulting upper and lower tangent hyperplanes by

$$x_d = a_1 x_1 + \cdots + a_{d-1} x_{d-1} + U_s(a_1, \ldots, a_{d-1})$$

and

$$x_d = a_1 x_1 + \cdots + a_{d-1} x_{d-1} + L_s(a_1, \ldots, a_{d-1}),$$

respectively. Then we have
$$L_s(a_1,\ldots,a_{d-1}) \leq a_d \leq U_s(a_1,\ldots,a_{d-1}).$$

Now if h is a transversal of S, we must have
$$\max_{s \in S} L_s(a_1,\ldots,a_{d-1}) \leq a_d \leq \min_{s \in S} U_s(a_1,\ldots,a_{d-1}).$$

In other words, if we define $\Gamma = \{U_s \mid s \in S\}$ and $\Gamma' = \{L_s \mid s \in S\}$, then $\mathcal{T}(S)$ is $\mathcal{S}(\Gamma,\Gamma')$, the sandwich region lying below the lower envelope of Γ and above the upper envelope of Γ'. The results of Agarwal *et al.* [**43**] (resp., of Koltun and Sharir [**495**]) imply that if each set in S has constant description complexity, then so do the functions L_s, U_s, for $s \in S$, Thus the complexity of $\mathcal{T}(S)$ is $O(n^{2+\varepsilon})$, for any $\varepsilon > 0$, in $\mathbb{R}^3$, and $O(n^{3+\varepsilon})$, for any $\varepsilon > 0$, in $\mathbb{R}^4$. The results in [**43**] concerning the complexity of the vertical decomposition of $\mathcal{S}(\Gamma,\Gamma')$ imply that, in three dimensions, $\mathcal{T}(S)$ can be constructed in $O(n^{2+\varepsilon})$ time, for any $\varepsilon > 0$. In four dimensions, the results of [**495**] do not yield a near-cubic bound on the complexity of the vertical decomposition of the sandwich region. However, using the algorithm by Agarwal *et al.* [**20**] for point location in the minimization diagram of trivariate functions, we can preprocess S into a data structure of size $O(n^{3+\varepsilon})$, so that we can determine, in $O(\log n)$ time, whether a hyperplane h is a transversal of S. No sharp bounds are known on the complexity of $\mathcal{T}(S)$ in higher dimensions.

The problem can be generalized by considering lower-dimensional transversals. For example, in $\mathbb{R}^3$ we can also consider the space of all line transversals of S (lines that meet every member of S), where a similar reduction to the sandwich region between two envelopes can be established. This will be discussed in detail in Chapter 7.

14.3. Geometric optimization. In the past couple of decades, many problems in geometric optimization have been successfully attacked by techniques that reduce the problem to constructing and searching various substructures of surface arrangements. Hence, the area of geometric optimization is a natural extension, and a good application area, of the study of arrangements. See [**44, 50, 553**] for surveys on geometric optimization.

One of the basic techniques for geometric optimization is the *parametric searching* technique, originally proposed by Megiddo [**545**]. This technique reduces the optimization problem to a *decision problem*, where one needs to compare the optimal value to a given parameter. In most cases, the decision problem is easier to solve than the optimization problem. The parametric searching technique proceeds by a parallel simulation of a generic version of the decision procedure with the (unknown) optimum value as an input parameter. In most applications, careful implementation of this technique leads to a solution of the optimization problem whose running time is larger than that of the decision algorithm only by a polylogarithmic factor. See [**50**] for a more detailed survey of parametric searching and its applications.

Several alternatives to parametric searching have been developed during the past couple of decades. They use randomization [**48, 201, 528**], expander graphs [**467**], and searching in monotone matrices [**366**]. Like parametric searching, all these techniques are based on the availability of an efficient procedure for the decision problem. When applicable, they lead to algorithms with running times that are similar to, and sometimes slightly better than, those yielded by parametric searching.

These methods have been used to solve a wide range of geometric optimization problems, many of which involve arrangements. We mention a sample of such results.

Slope selection. Given a set S of n points in $\mathbb{R}^2$ and an integer $k \leq \binom{n}{2}$, find the line with the kth smallest slope among the lines connecting pairs of points of S. If we dualize the points in S to a set Γ of lines in $\mathbb{R}^2$, the problem becomes that of computing the kth leftmost vertex of $\mathcal{A}(\Gamma)$. Cole et al. [**247**] developed a rather sophisticated $O(n \log n)$-time algorithm for this problem, which is based on parametric searching. (Here the decision problem is to determine whether at most k vertices of $\mathcal{A}(\Gamma)$ lie to the left of a given vertical line, which is easy to solve in $O(n \log n)$ time, by counting inversions in an appropriate permutation of the lines.) A considerably simpler algorithm, based on $(1/r)$-cuttings, was later proposed by Brönnimann and Chazelle [**178**]. See also [**466, 528**].

Distance selection. Given a set S of n points in $\mathbb{R}^2$ and a parameter $k \leq \binom{n}{2}$, find the kth smallest distance among the points of S [**24, 467**]. The corresponding decision problem reduces to repeated point location in an arrangement of congruent disks in $\mathbb{R}^2$. Specifically, given a set Γ of m congruent disks in the plane, we wish to count efficiently the number of containments between disks of Γ and points of S. This is a simple extension of the problem of computing the cells of $\mathcal{A}(\Gamma)$ marked by the points of S; see Section 13.4. This problem can be solved using parametric searching [**24**], expander graphs [**467**], or randomization [**528**]. The deterministic algorithm of Katz and Sharir [**467**], runs in $O(n^{4/3} \log^{3+\varepsilon} n)$ time.

Segment center. Given a set S of n points in $\mathbb{R}^2$ and a line segment e, find a placement of e (allowing translation and rotation) which minimizes the largest distance from the points of S to e [**30, 316**]. The decision problem reduces to determining whether, for two given families Γ and Γ' of bivariate surfaces, the sandwich region $\mathcal{S}(\Gamma, \Gamma')$, lying between L_Γ and $U_{\Gamma'}$, is empty. Exploiting the special properties of the resulting collections Γ and Γ', Efrat and Sharir [**316**] showed that the complexity of $\mathcal{S}(\Gamma, \Gamma')$ is $O(n \log n)$, and used this property to design an $O(n^{1+\varepsilon})$-time algorithm to determine whether $S(\Gamma, \Gamma')$ is empty, which leads to an $O(n^{1+\varepsilon})$-time algorithm for the segment-center problem.

Extremal polygon placement. Given a convex m-gon P and a closed polygonal environment Q with n vertices, find the largest similar copy of P that is fully contained in Q [**686**]. Here the decision problem is to determine whether P, with a fixed scaling factor, can be placed inside Q; this is a variant of the corresponding motion-planning problem for P inside Q, and is solved by constructing an appropriate representation of the 3-dimensional free configuration space, as a collection of cells in a corresponding 3-dimensional arrangement of surfaces. The running time of the whole algorithm is only slightly larger than the time needed to solve the fixed-size placement problem (i.e., the decision problem). The best running time is $O(mn\lambda_6(mn) \log^3 mn \log^2 n)$ [**22**]; see also [**473, 686**]. If Q is a convex n-gon, the largest similar copy of P that can be placed inside Q can be computed in $O(mn^2 \log n)$ time [**14**]. See Agarwal et al. [**28**] for some improvements and a study of related questions.

Diameter in 3D. Given a set S of n points in $\mathbb{R}^3$, determine the maximum distance between a pair of points in S. The corresponding decision problem can be reduced to determining whether S is fully contained in the intersection of a set Γ of n congruent balls, centered at the points of S. A randomized algorithm for the decision problem, with $O(n \log n)$ expected time, was proposed by Clarkson and Shor [**246**]. A series of papers [**149, 217, 539, 640, 641**] describe near-linear-time deterministic algorithms, for the optimization problem, culminating in an optimal $O(n \log n)$ algorithm by Ramos [**642**].

Width in 3D. Given a set S of n points in $\mathbb{R}^3$, determine the smallest distance between two parallel planes enclosing S between them. This problem has been studied in a series of papers [**20, 48, 217**], and the currently best known exact (randomized) algorithm computes the width in $O(n^{3/2+\varepsilon})$ expected time [**48**]. The technique used in attacking the decision problems for this and the two following problems reduce them to point location in the region above the lower envelope of a collection of trivariate functions in $\mathbb{R}^4$. Given the basic nature of this problem, it is an intriguing challenge to further improve the running time.

Biggest stick in a simple polygon. The problem is to compute the longest line segment that can fit inside a given simple polygon with n edges. The current best solution runs in $O(n^{3/2+\varepsilon})$ time [**48**] (see also [**20, 57**]).

Minimum-width annulus. Here we wish to compute the annulus of smallest width that encloses a given set of n points in the plane. This problem arises in fitting a circle through a set of points in the plane, using the so-called L_∞ measure of resemblance. Again, the current best solution runs in $O(n^{3/2+\varepsilon})$ time [**48**] (see also [**20, 57**]).

Approximations. Since the last three solutions are fairly complicated, it is desirable to seek simpler approximation algorithms. Several recent works successfully address this issue, such as Chan's elegant algorithms [**203**], and the coreset approach of Agarwal *et al.* [**33, 34**].

Geometric matching. Consider the problem where we are given two sets S_1, S_2 of n points each in the plane, and we wish to compute a minimum-weight matching in the complete bipartite graph $S_1 \times S_2$, where the weight of an edge (p,q) is the Euclidean distance between p and q. One can also consider the analogous nonbipartite version of the problem, which involves just one set S of $2n$ points, and the complete graph on S. The goal is to explore the underlying geometric structure of these graphs, to obtain faster algorithms than those available for general abstract graphs. Vaidya [**730**] had shown that both the bipartite and the nonbipartite versions of the problem can be solved in time close to $O(n^{5/2})$. A fairly sophisticated application of vertical decomposition in three-dimensional arrangements, given in [**29**], has improved the running time for the bipartite case to $O(n^{2+\varepsilon})$. A better improvement, for the nonbipartite case, was given by Varadarajan [**734**]; his algorithm runs in $O(n^{3/2} \log^c n)$ time, for some absolute constant c.

Center point. A *center point* of a set S of n points in the plane is a point $\pi \in \mathbb{R}^2$ so that each line ℓ passing through π has the property that at least $\lfloor n/3 \rfloor$ points lie in each (closed) halfplane bounded by ℓ. It is well known that such a center point always exists [**288**]. If we dualize S to a set Γ of n lines in the plane, then π^*, the line dual to π, lies between $\mathcal{A}_{\lfloor n/3 \rfloor}(\Gamma)$ and $\mathcal{A}_{\lceil 2n/3 \rceil}(\Gamma)$. The existence of a center point implies that the convex hulls of these two levels are disjoint, and the goal is to find a line π^* which separates them. Cole *et al.* [**249**] described an $O(n \log^3 n)$-time algorithm for computing a center point of S, using parametric searching. A simple $O(n)$-time algorithm was later given by Jadhav and Mukhopadhyay [**451**]. Following the observation just made, the problem of computing the set of all center points reduces to computing the convex hull of $\mathcal{A}_k(\Gamma)$ for a given k. Matoušek [**526**] described an $O(n \log^4 n)$-time algorithm for computing the convex hull of $\mathcal{A}_k(\Gamma)$ for any $k \leq n$; recall, in contrast, that the best known upper bound for the number of vertices of $\mathcal{A}_k(\Gamma)$ is $O(n(k+1)^{1/3})$. In three dimensions, Agarwal *et al.* [**58**] show that the complexity of the center region of a set of n points is $O(n^2)$, and that it can be computed in $O(n^{2+\varepsilon})$ time. A much earlier work by Naor and Sharir [**567**] gives an algorithm for finding a single center point in three dimensions in nearly quadratic time.

Ham sandwich cuts. Let $S_1, S_2, \ldots, S_d$ be d sets of points in $\mathbb{R}^d$, each containing n points. Suppose n is even. A *ham sandwich cut* is a hyperplane h so that each open halfspace bounded by h contains at most $n/2$ points of S_i, for $i = 1, \ldots, d$. It is known [**288, 748**] that such a cut always exists. Let Γ_i be the set of hyperplanes dual to S_i. Then the problem reduces to computing a vertex of the intersection of $\mathcal{A}_{n/2}(\Gamma_1)$ and $\mathcal{A}_{n/2}(\Gamma_2)$. Megiddo [**546**] developed a linear-time algorithm for computing a ham sandwich cut in the plane if S_1 and S_2 can be separated by a line. For arbitrary point sets in the plane, a linear-time algorithms was later developed by Lo *et al.* [**517**], who also described an algorithm for computing a ham sandwich cut in $\mathbb{R}^3$, whose running time is $O(\psi_{n/2}(n) \log^2 n)$, where $\psi_k(n)$ is the maximum complexity of the k-level in an arrangement of n lines in the plane. By Dey's result on k-levels [**264**], the running time of their algorithm is $O(n^{4/3} \log^2 n)$. A recent paper by Agarwal *et al.* [**27**] gives an algorithm for computing a center-transversal line to two sets in $\mathbb{R}^3$, a notion which "interpolates" between center points and ham sandwich cuts.

14.4. Robotics. As mentioned in the introduction, *motion planning* for a robot system has been a major motivation for the study of arrangements. Let B be a robot system with d degrees of freedom, which is allowed to move freely within a given two- or three-dimensional environment cluttered with obstacles. Given two placements I and F of B, determining whether there exists a collision-free path between these placements reduces to the problem of determining whether I and F lie in the same cell of the arrangement of the family Γ of "contact surfaces" in $\mathbb{R}^d$, which define the configuration space of B (see the introduction for more details). If I and F lie in the same cell, then a path between I and F in $\mathbb{R}^d$ that does not intersect any surface of Γ corresponds to a collision-free path of B in the physical environment from I to F. Schwartz and Sharir [**660**] developed an $n^{2^{O(d)}}$-time algorithm for this problem. If d is part of the input, the problem was later proved to be PSPACE-complete [**193, 645**]. Canny [**192, 194**] gave an $n^{O(d)}$-time algorithm for computing the roadmap of a single cell in an arrangement $\mathcal{A}(\Gamma)$ of a set Γ of n surfaces in $\mathbb{R}^d$, provided that the cells in $\mathcal{A}(\Gamma)$ form a Whitney regular stratification of $\mathbb{R}^d$ (see [**389**] for the definition of Whitney stratification). Using a perturbation argument, Canny showed that his approach can be extended to obtain a Monte Carlo algorithm for determining whether two points lie in the same cell of $\mathcal{A}(\Gamma)$. The algorithm was subsequently extended and improved by many researchers; see [**133, 390, 434**]. The best known algorithm, due to Basu *et al.* [**133**], can compute the roadmap in time $n^{d+1}b^{O(d^2)}$. Much work has been done on developing efficient algorithms for robots with a small number of degrees of freedom, say, two or three [**412, 422, 472**]. The result by Schwarzkopf and Sharir [**664**] gives an efficient algorithm for computing a collision-free path between two given placements for fairly general robot systems with three degrees of freedom. See [**415, 662, 680**] for surveys on motion-planning algorithms.

It is impractical to compute the roadmap, or any other explicit representation, of a single cell in $\mathcal{A}(\Gamma)$ if d is large. A general Monte Carlo algorithm for computing a *probabilistic roadmap* of a cell in $\mathcal{A}(\Gamma)$ is described by Kavraki *et al.* [**470**]. This approach avoids computing the cell explicitly. Instead, it samples a large number of random points in the configuration space and only those configurations that lie in the free configuration space (FP) are retained (they are called *milestones*); we also add I and F as milestones. The algorithm then builds a "connectivity graph" whose nodes are these milestones, and whose edges connect pairs of milestones if the line segment joining them in configuration space lies in FP (or if they satisfy some other "local reachability" rule). Various strategies have been proposed for choosing random configurations [**80, 130, 446, 469**]. The algorithm returns a path from I to F if they lie in the same connected component of the resulting network. Note that this algorithm may fail to return a collision-free path from I to F even if there exists one. This technique nevertheless has been successful in several real-world applications.

Assembly planning is another area in which the theory of arrangements has led to efficient algorithms. An *assembly* is a collection of objects (called *parts*) placed rigidly in some specified relative positions so that no two objects overlap (i.e., penetrate into one another). A *subassembly* of an assembly A is a subset of objects in A in their relative placements in A. An assembly operation is a motion that merges some subassemblies of A into a new and larger subassembly. An *assembly sequence* for A is a sequence of assembly operations that starts with the individual parts separated from each other and ends up with the full assembly A. The goal of assembly planning is to compute an assembly sequence for a given assembly. A classical approach to assembly sequencing is *disassembly sequencing*, which separates an assembly into its individual parts [**444**]. The reverse order of a sequence of disassemblying operations yields an assembly sequence. Several kinds of motion have been considered in separating parts of an assembly, including translating a subassembly along a straight line, arbitrary translational motion, rigid motion, etc. A common approach to generate a disassembly sequence is the so-called *nondirectional blocking graph* approach. It partitions the space of all allowable motions of separation into a finite number of cells so that within each cell the set of "blocking relations" between all pairs of parts remains fixed. The problem is then reduced to computing representative points in the cells of the arrangement of an appropriate family of surfaces. This approach has been successful in many instances, including polyhedral assembly with infinitesimal rigid motion [**397**]; see also [**415, 416**].

Other problems in robotics that have exploited arrangements include fixturing [**643**], MEMS (micro electronic mechanical systems) [**157**], path planning with uncertainty [**143**], and manufacturing [**59**].

14.5. Molecular modeling. In the introduction of this chapter, we described the Van der Waals model, in which a molecule M is represented as a collection Γ of balls in $\mathbb{R}^3$. (See [**251, 291, 548**] for other geometric models of molecules.) Let $\Sigma = \partial(\bigcup \Gamma)$. Σ is called the "surface" of M. Many problems in molecular biology, especially those that study the interaction of a protein with another molecule, involve computing the molecular surface, a portion of the surface (e.g., the so-called *active site* of a protein), or various features of the molecular surface [**292, 417, 511, 735**]. We briefly describe two problems in molecular modeling that can be formulated in terms of arrangements.

The chemical behavior of solute molecules in a solution is strongly dependent on the interactions between the solute and solvent molecules. These interactions are critically dependent on those molecular fragments that are accessible to the solvent molecules. Suppose we use the Van der Waals model for the solute molecule and model the solvent by a sphere S. By rolling S on the molecular surface Σ, we obtain a new surface Σ', traced by the center of the rolling sphere. If we enlarge each sphere of Γ by the radius of S, Σ' is the boundary of the union of the enlarged spheres.

As mentioned above, several methods have been proposed to model the surface of a molecule. The best choice of the model depends on the chemical problem the molecular surface is supposed to represent. For example, the Van der Waals model represents the space requirement of molecular conformations, while isodensity contours and molecular electrostatic potential contour surfaces [**548**] are useful in studying molecular interactions. An important problem in molecular modeling is to study the interrelations among various molecular surfaces of the same molecule. For example, let $\mathbf{\Sigma} = \{\Sigma_1, \ldots, \Sigma_m\}$ be a family of molecular surfaces of the same molecule. We may want to compute the arrangement $\mathcal{A}(\mathbf{\Sigma})$, or we may want to compute the subdivision of Σ_i induced by $\{\Sigma_j \cap \Sigma_i \mid 1 \leq j \neq i \leq m\}$.

Researchers have also been interested in computing "connectivity" of a molecule, e.g., computing voids, tunnels, and pockets of Σ. A *void* of Σ is a bounded component of $\mathbb{R}^3 \setminus (\bigcup \Gamma)$; a *tunnel* is a hole through $\bigcup \Gamma$ that is accessible from the outside, i.e., an "inner" part of a non-contractible loop in $\mathbb{R}^3 \setminus \bigcup \Gamma$; and a *pocket* is a depression or cavity on Σ. Pockets are not holes in the topological sense and are not well defined; see [**251, 292**] for some of the definitions proposed so far. Pockets and tunnels are interesting because they are good candidates to be *binding sites* for other molecules.

Efficient algorithms have been developed for computing Σ, the connectivity of Σ, and the arrangement $\mathcal{A}(\Gamma)$ [**291, 417, 735**]. Halperin and Shelton [**423**] describe an efficient perturbation scheme to handle degeneracies while constructing $\mathcal{A}(\Gamma)$ or Σ. Some applications require computing the measure of different substructures of $\mathcal{A}(\Gamma)$, including the volume of Σ, the surface area of Σ, or the volume of a void of Σ. Edelsbrunner *et al.* [**292**] describe an efficient algorithm for computing these measures; see also [**290, 291**].

15. Conclusions

In this chapter we have reviewed a wide range of topics on arrangements of surfaces. We mentioned a few old results, but the emphasis of the presentation was on the tremendous progress made in this area during the past 20 years, mostly in three and higher dimensions, and for arrangements of nonlinear surfaces. We discussed the combinatorial complexity of arrangements and of their various substructures, representation of arrangements, algorithms for computing arrangements and their substructures, and several geometric problems in which arrangements play pivotal roles. Although the chapter does cover a broad spectrum of results, many topics on arrangements were either not included or very briefly touched upon. For example, we did not discuss arrangements of pseudo-lines and oriented matroids, we discussed algebraic and topological issues only very briefly, and we mentioned a rather short list of applications that exploit arrangements. We have also not discussed the topic of robust computation of arrangements, a subject of extensive contemporary research; see [**161**]. There are numerous other sources where more details on arrangements and their applications can be found; see, e.g., the books [**152, 536, 585, 681**] and the survey papers [**51, 385, 414, 552, 584**].

CHAPTER 3

Davenport–Schinzel Sequences: The Inverse Ackermann Function in Geometry

1. Introduction

Davenport–Schinzel sequences, introduced by H. Davenport and A. Schinzel in the 1960s, are interesting and powerful combinatorial structures that arise in the analysis and construction of the lower (or upper) envelope of collections of univariate functions, and therefore have applications in a variety of geometric problems that can be reduced to computing such an envelope. In addition, Davenport–Schinzel sequences play a central role in many related geometric problems involving arrangements of curves and surfaces. For these reasons, they have become one of the major tools in the analysis of combinatorial and algorithmic problems in geometry.

DEFINITION 1.1. *Let n and s be positive integers. A sequence $U = \langle u_1, \ldots, u_m \rangle$ of integers is an (n, s) Davenport–Schinzel sequence (a $DS(n, s)$-sequence for short) if it satisfies the following conditions:*

(i) $1 \leq u_i \leq n$ *for each* $i \leq m$,
(ii) $u_i \neq u_{i+1}$ *for each* $i < m$, *and*
(iii) *there do not exist $s + 2$ indices $1 \leq i_1 < i_2 < \cdots < i_{s+2} \leq m$ such that*

$$u_{i_1} = u_{i_3} = u_{i_5} = \cdots = a, \quad u_{i_2} = u_{i_4} = u_{i_6} = \cdots = b,$$

and $a \neq b$.

In other words, the third condition forbids the presence of long alternations of any pair of distinct symbols in a Davenport–Schinzel sequence. We refer to s as the *order* of U, to n as the *number of symbols* composing U, and to $|U| = m$ as the *length* of the sequence U. Define

$$\lambda_s(n) = \max\{\,|U| \mid U \text{ is a } DS(n, s)\text{-sequence}\,\}.$$

Curiously, the original papers by Davenport and Schinzel [**259, 260**] were entitled *On a combinatorial problem connected with differential equations*, because they were motivated by a particular application that involved the pointwise maximum of a collection of independent solutions of a linear differential equation. This, however, is only a special case of more general lower or upper envelopes. Davenport and Schinzel did establish in [**259, 260**] the connection between envelopes and these sequences, and obtained several non-trivial bounds on the length of the sequences, which were later strengthened by Szemerédi [**710**]. The potential of *DS*-sequences for geometric problems, however, remained unnoticed until Atallah rediscovered and applied them to several problems in dynamic computational geometry [**113**]. It is easy to show that $\lambda_1(n) = n$ and $\lambda_2(n) = 2n - 1$ (see Theorem 3.1). Hart and Sharir [**430**] proved that $\lambda_3(n) = \Theta(n\alpha(n))$, where $\alpha(n)$ is the inverse Ackermann function (see below for details), and later Agarwal et al. [56] (see also Sharir [**673**, **674**]) proved sharp bounds on $\lambda_s(n)$ for $s > 3$. These bounds were recently slightly tightened by Nivasch [**570**]. These bounds are of the form $\lambda_s(n) \leq n\beta_s(n)$, where $\beta_s(n)$ is some elementary function of $\alpha(n)$. Given how slowly $\alpha(n)$ increases, it follows that $\lambda_s(n)$ is nearly linear in n for any fixed s. Davenport–Schinzel sequences have become a useful and powerful tool for solving numerous problems in discrete and computational geometry, usually by showing that the geometric structure being analyzed has smaller combinatorial complexity than what more naive methods would have implied. Many such geometric applications have been obtained in the past twenty years, and we will review some of them below. The book by Sharir and Agarwal [**681**] gives a more detailed description of the theory of DS-sequences and of their geometric applications. See also the survey of Agarwal and Sharir [**52**] and the book of Matoušek [**536**].

As noted above, and will be shown in more detail below, Davenport–Schinzel sequences provide a complete combinatorial characterization of the lower envelope of a collection of *univariate* functions. In many geometric problems, though, one faces the more difficult problem of calculating or analyzing the envelope of a collection of multivariate functions. Even for bivariate functions this problem appears to be considerably harder than the univariate case. Nevertheless, as extensively reviewed in Chapter 2, considerable progress has been made on the multivariate case, leading to almost-tight bounds on the complexity of envelopes in higher dimensions (specifically, see [**419**, **678**]). Alas, most of these higher-dimensional bounds do not involve Davenport–Schinzel sequences.

The material reviewed in this chapter mostly covers the basic combinatorial analysis of Davenport–Schinzel sequences and some of their basic geometric applications, both combinatorial and algorithmic. Section 2 shows the connection between DS-sequences and lower envelopes. Sections 3–5 discuss the analysis of the maximum length of (n, s) Davenport–Schinzel sequences. Section 6 presents basic combinatorial geometric applications of Davenport–Schinzel sequences to two-dimensional arrangements of lines, segments, and arcs, and studies the role that these sequences play in various structures in such arrangements, including envelopes, individual faces, zones, and levels.

2. Davenport–Schinzel Sequences and Lower Envelopes

2.1. Lower envelopes of totally defined functions. Let $\mathcal{F} = \{f_1, \ldots, f_n\}$ be a collection of n real-valued, continuous totally defined functions, so that the graphs of every pair of distinct functions intersect in at most s points (this is the case for polynomials of fixed degree, Chebychev systems, etc.). The *lower envelope* of $\mathcal{F}$ is defined as

$$E_{\mathcal{F}}(x) = \min_{1 \leq i \leq n} f_i(x),$$

i.e., $E_{\mathcal{F}}$ is the *pointwise minimum* of the functions f_i; see Figure 1. Let $I_1, \ldots, I_m$ be the partition of the x-axis into maximal connected intervals, with pairwise disjoint interiors, so that, for each $k \leq m$, the same function f_{u_k} appears on $E_{\mathcal{F}}$ for all points in I_k (i.e., $E_{\mathcal{F}}(x) = f_{u_k}(x)$ for all $x \in I_k$). In other words, m is the number of (maximal) connected portions of the graphs of the f_i's that constitute the graph of $E_{\mathcal{F}}$. The endpoints of the intervals I_k are called the *breakpoints* of $E_{\mathcal{F}}$. Assuming

that $I_1, \ldots, I_m$ are sorted from left to right, put
$$U(\mathcal{F}) = \langle u_1, \ldots, u_m \rangle.$$
$U(\mathcal{F})$ is called the *lower-envelope sequence* of $\mathcal{F}$; see Figure 1 again. The *minimization diagram* of $\mathcal{F}$, denoted by $M_{\mathcal{F}}$, is the partition of the x-axis induced by the intervals $I_1, \ldots, I_m$. The endpoints of these intervals are called the *breakpoints* of $M_{\mathcal{F}}$. For convenience, we add $-\infty, +\infty$ as extreme breakpoints of $M_{\mathcal{F}}$.

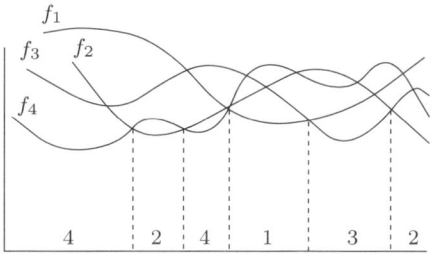

FIGURE 1. A lower-envelope sequence.

The *upper envelope* of $\mathcal{F}$ is defined, in a fully symmetric manner, to be
$$E^*_{\mathcal{F}}(x) = \max_{1 \leq i \leq n} f_i(x),$$
and the *maximization diagram* $M^*_{\mathcal{F}}$ is defined as the corresponding partition of the real line, as in the case of lower envelopes. In this chapter we mostly consider lower envelopes; this choice is arbitrary, and all the results, of course, apply equally well to upper envelopes.

THEOREM 2.1 ([**113, 260**]). *$U(\mathcal{F})$ is a $DS(n,s)$-sequence. Conversely, for any given $DS(n,s)$-sequence U, one can construct a set $\mathcal{F} = \{f_1, \ldots, f_n\}$ of continuous, totally defined, univariate functions, each pair of whose graphs intersect in at most s points, such that $U(\mathcal{F}) = U$.*

PROOF (Sketch). For the first part, note that, by definition, the lower-envelope sequence $U = U(\mathcal{F})$ does not contain a pair of adjacent equal elements. (For simplicity, assume that the graphs of functions in $\mathcal{F}$ intersect transversally at each intersection point. The proof can easily be extended to the case when the graphs of two functions touch each other.) Suppose U contains $s+2$ indices $i_1 < i_2 < \cdots < i_{s+2}$ so that $u_{i_1} = u_{i_3} = \cdots = a$ and $u_{i_2} = u_{i_4} = \cdots = b$ for $a \neq b$. By definition of the lower-envelope sequence, we must have $f_a(x) < f_b(x)$ for $x \in (int(I_{i_1}) \cup int(I_{i_3}) \cup \cdots)$ and $f_a(x) > f_b(x)$ for $x \in (int(I_{i_2}) \cup int(I_{i_4}) \cup \cdots)$, where $int(J)$ denotes the interior of the interval J. Since f_a and f_b are continuous, there must exist $s+1$ distinct points $x_1, \ldots, x_{s+1}$ so that x_r lies between the intervals I_{i_r} and $I_{i_{r+1}}$ and $f_a(x_r) = f_b(x_r)$, for $r = 1, \ldots, s+1$. This, however, contradicts the assumption that the graphs of f_a and f_b intersect in at most s points.

For the converse statement, let $U = \langle u_1, \ldots, u_m \rangle$ be a given $DS(n,s)$-sequence. Without loss of generality, suppose the symbols $1, 2, \ldots, n$, of which U is composed, are ordered so that the leftmost appearance of symbol i in U precedes the leftmost appearance of symbol j in U if and only if $i < j$. We now define the required collection of functions $\mathcal{F} = \{f_1, \ldots, f_n\}$ as follows. We choose $m - 1$ distinct "transition points" $x_2 < x_3 < \ldots < x_m$ on the x-axis, and $n + m - 1$ distinct horizontal "levels," say, at $y = 1, 2, \ldots, n, -1, -2, \ldots, -(m - 1)$. For each symbol $1 \leq a \leq n$, the graph of the corresponding function f_a is always horizontal at one of these levels, except at short intervals near some of the transition points, where it can drop very steeply from one level to a lower one. At each transition point exactly one function changes its level. More specifically:

  (i) Before x_2, the function f_a is at the level $y = a$, for $a = 1, \ldots, n$.
  (ii) At the transition point x_i, let $a = u_i$; then f_a drops down from its current level to the highest still "unused" level. See Figure 2 for an illustration.

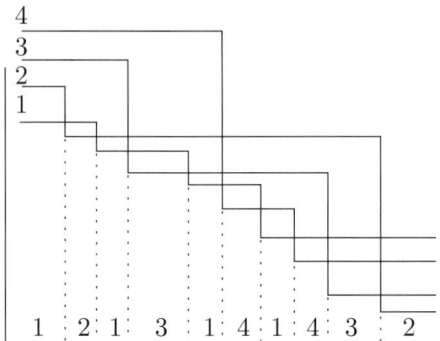

FIGURE 2. Realization of the $DS(4, 3)$-sequence $\langle 1, 2, 1, 3, 1, 4, 1, 4, 3, 2 \rangle$.

It is clear from this construction that $U(\mathcal{F}) = U$, and it can be shown that each pair of functions intersect in at most s points. This completes the proof of the theorem. $\square$

COROLLARY 2.2. *For any collection $\mathcal{F} = \{f_1, \ldots, f_n\}$ of n continuous, totally defined, univariate functions, each pair of whose graphs intersect in at most s points, the length of the lower-envelope sequence $U(\mathcal{F})$ is at most $\lambda_s(n)$, and this bound can be attained by such a collection $\mathcal{F}$.*

COROLLARY 2.3. *Let $\mathcal{F} = \{f_1, \ldots, f_n\}$ and $\mathcal{G} = \{g_1, \ldots, g_n\}$ be two collections of n continuous, totally defined, univariate functions, such that the graphs of any pair of functions of $\mathcal{F}$, or of any pair of functions of $\mathcal{G}$, intersect in at most s points, and the graphs of any pair of functions in $\mathcal{F} \times \mathcal{G}$ intersect in a (possibly larger) constant number of points. Let $\mathcal{S}_{\mathcal{F},\mathcal{G}}$ denote the "sandwich" region lying between the upper envelope of $\mathcal{G}$ and the lower envelope of $\mathcal{F}$ (see Figure 3), i.e., the region*
$$\mathcal{S}_{\mathcal{F},\mathcal{G}} = \{(x, y) \mid E_{\mathcal{G}}^*(x) \leq y \leq E_{\mathcal{F}}(x)\}.$$
Then the number of intersection points of graphs of functions in $\mathcal{F} \cup \mathcal{G}$ that lie on the boundary of $\mathcal{S}_{\mathcal{F},\mathcal{G}}$ is $O(\lambda_s(n))$.

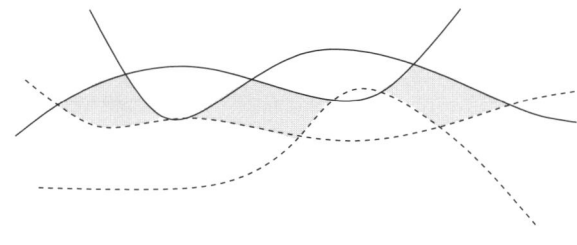

FIGURE 3. The sandwich region between $E_{\mathcal{G}}^*$ and $E_{\mathcal{F}}$ is shown shaded; the graphs of the functions in $\mathcal{F}$ (resp., in $\mathcal{G}$) are drawn solid (resp., dashed).

PROOF. Let $L = (b_1, \ldots, b_t)$ be the sequence of breakpoints of $M_{\mathcal{F}}$ and $M_{\mathcal{G}}^*$, sorted from left to right. By definition, $t \leq 2\lambda_s(n)$. In each interval (b_i, b_{i+1}), the envelopes $E_{\mathcal{F}}$, $E_{\mathcal{G}}^*$ are attained by a unique pair of functions $f^{(i)} \in \mathcal{F}$, $g^{(i)} \in \mathcal{G}$. Hence, there are $O(1)$ intersection points on the boundary of $\mathcal{S}_{\mathcal{F},\mathcal{G}}$ whose x-coordinates lie in (b_i, b_{i+1}), and the corollary follows. □

2.2. Lower envelopes of partially defined functions. It is useful to note that a similar equivalence exists between Davenport–Schinzel sequences and lower envelopes of *partially defined* functions. Specifically, let $f_1, \ldots, f_n$ be a collection of partially defined and continuous functions, so that the domain of definition of each function f_i is an interval I_i, and suppose further that the graphs of each pair of these functions intersect in at most s points. The lower envelope of $\mathcal{F}$ is now defined as
$$E_{\mathcal{F}}(x) = \min f_i(x),$$
where the minimum is taken over those functions that are defined at x. One can then define the minimization diagram $M_{\mathcal{F}}$ and the lower-envelope sequence $U(\mathcal{F})$ in much the same way as for totally defined functions; see Figure 4. In this case the following theorem holds.

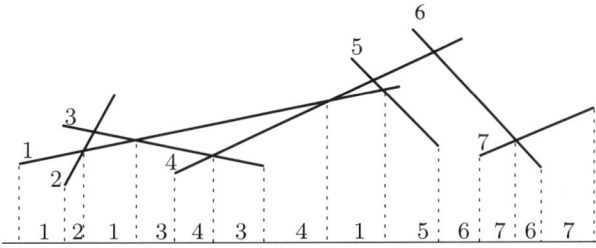

FIGURE 4. The lower envelope of a collection of (nonvertical) segments.

THEOREM 2.4 ([**430**]). *The lower-envelope sequence $U(\mathcal{F})$ is a $DS(n, s + 2)$-sequence. Conversely, for any $DS(n, s + 2)$-sequence U one can construct a collection $\mathcal{F} = \{f_1, \ldots, f_n\}$ of partially-defined, continuous functions, each defined over an interval, and each pair of which intersect in at most s points, such that $U(\mathcal{F}) = U$.*

Hence, we can conclude:

THEOREM 2.5 ([**430**]). *Let $\mathcal{F}$ be a collection of n partially-defined, continuous, univariate functions, with at most s intersection points between the graphs of any pair. Then the length of the lower-envelope sequence $U(\mathcal{F})$ is at most $\lambda_{s+2}(n)$.*

The functions constructed in Theorems 2.1 and 2.4, to realize arbitrary $DS(n, s)$-sequences, have fairly irregular structure (cf. Figure 2). A problem that arises naturally in this context is whether any $DS(n, s)$-sequence can be realized as the lower envelope sequence of a collection of n partial or total functions of some canonical form. For example, can any $(n, 3)$-Davenport–Schinzel sequence be realized as the lower envelope sequence of a collection of n line segments (see Figure 4)? Some partially affirmative results on geometric realization of $DS(n, s)$-sequences will be mentioned below, although the problem is still wide open.

2.3. Constructing lower envelopes. We conclude this section by presenting a simple, efficient divide-and-conquer algorithm for computing the minimization diagram of a set $\mathcal{F}$ of n continuous, totally defined, univariate functions, each pair of whose graphs intersect at most s times, for some constant parameter s. Here we assume a model of computation that allows us to compute the intersections between any pair of functions in $\mathcal{F}$ in $O(1)$ time.

We partition $\mathcal{F}$ into two subsets $\mathcal{F}_1, \mathcal{F}_2$, each of size at most $\lceil n/2 \rceil$, compute the minimization diagrams $M_{\mathcal{F}_1}, M_{\mathcal{F}_2}$ recursively, and merge the two diagrams to obtain $M_{\mathcal{F}}$. We merge the lists of breakpoints of $M_{\mathcal{F}_1}$ and of $M_{\mathcal{F}_2}$ into a single list $V = (v_1 = -\infty, v_2, \ldots, v_t = +\infty)$, sorted from left to right. Notice that, for any $1 \leq i < t$, there is a unique pair of functions $f_i^{(1)} \in \mathcal{F}_1$, $f_i^{(2)} \in \mathcal{F}_2$, that attain the respective envelopes $E_{\mathcal{F}_1}, E_{\mathcal{F}_2}$ over (v_i, v_{i+1}). We compute the real roots $r_1, \ldots, r_k$ ($k \leq s$) of the function $f_i^{(1)} - f_i^{(2)}$ that lie in the interval (v_i, v_{i+1}), and add them to the list V. Let $V' = (v'_1, \ldots, v'_{t'})$ denote the new list of points. It is clear that, for each i, a unique function $f_i \in \mathcal{F}$ attains $E_{\mathcal{F}}$ over the interval (v'_i, v'_{i+1}). We associate f_i with this interval. If the same function is associated with two adjacent intervals (v'_{i-1}, v'_i) and (v'_i, v'_{i+1}), we delete the breakpoint v'_i from V'. The resulting list represents the minimization diagram $M_{\mathcal{F}}$ of $\mathcal{F}$. The total time spent in the merge step is

$$O(|V'|) = O(|V|) = O(|M_{\mathcal{F}_1}| + |M_{\mathcal{F}_2}|) = O(\lambda_s(n)).$$

Hence, the overall running time of the algorithm is $O(\lambda_s(n) \log n)$. (We anticipate here the near-linear bounds on $\lambda_s(n)$, which will be established later in the chapter.)

If the functions in $\mathcal{F}$ are partially defined, an easy modification of the above algorithm constructs $M_{\mathcal{F}}$ in time $O(\lambda_{s+2}(n) \log n)$. In this case, however, $M_{\mathcal{F}}$ can be computed in time $O(\lambda_{s+1}(n) \log n)$, using a more clever algorithm due to Hershberger [**437**].

THEOREM 2.6 ([**113, 437**]). *The lower envelope of a set $\mathcal{F}$ of n continuous, totally defined, univariate functions, each pair of whose graphs intersect in* at most s points, can be constructed, in an appropriate model of computation, in $O(\lambda_s(n)\log n)$ time. If the functions in $\mathcal{F}$ are partially defined, then $E_{\mathcal{F}}$ can be computed in $O(\lambda_{s+1}(n)\log n)$ time. In particular, the lower envelope of a set of n segments in the plane can be computed in optimal $O(n\log n)$ time.

3. Simple Bounds and Variants

One of the main goals in the analysis of DS-sequences is to estimate the value of $\lambda_s(n)$. In this section we review some of the earlier results that established nontrivial bounds on $\lambda_s(n)$. These bounds are somewhat weaker than the best known bounds, but have simpler proofs. We begin our analysis by disposing of the two simple cases $s = 1$ and $s = 2$.

THEOREM 3.1 ([**260**]). (a) $\lambda_1(n) = n$. (b) $\lambda_2(n) = 2n - 1$.

PROOF (Sketch). (a) Let U be a $DS(n,1)$-sequence. U cannot contain any subsequence of the form $\langle a \cdots b \cdots a \rangle$, for $a \neq b$, and any two adjacent elements of U are distinct, therefore all elements of U are distinct, which implies that $|U| \leq n$. The bound is tight, because $U = \langle 1\ 2\ 3 \cdots n \rangle$ is a $DS(n,1)$-sequence.

(b) The proof proceeds by induction on n. The case $n = 1$ is obvious. Suppose the claim holds for $n - 1$, and let U be any $DS(n,2)$-sequence. Without loss of generality, we can assume that the leftmost occurrence of i in U is before the leftmost occurrence of j if and only if $i < j$. It can then be shown that there is only one occurrence of n in U, or else a forbidden subsequence of the form $\langle x \cdots n \cdots x \cdots n \rangle$ would arise. Remove this single appearance of n from U, and if the two symbols adjacent to n are equal, remove also one of them from U. The resulting sequence is clearly a $DS(n-1,2)$-sequence, and is one or two elements shorter than U. The induction hypothesis then implies $|U| \leq 2n - 3 + 2 = 2n - 1$. Since the sequence $\langle 1\ 2\ 3 \cdots n-1\ n\ n-1 \cdots 3\ 2\ 1 \rangle$ is clearly a $DS(n,2)$-sequence of length $2n - 1$, the bound is tight. □

A cyclic sequence U is called a $DS(n,2)$-cycle if no two adjacent symbols are equal and if U does not contain a subcycle of the form $\langle a \cdots b \cdots a \cdots b \rangle$, for any $a \neq b$. Notice that the maximum length of a $DS(2,2)$-cycle is 2. The same argument as in Theorem 3.1(b) can be used to prove the following.

COROLLARY 3.2. *The maximum length of a $DS(n,2)$-cycle consisting of n symbols is $2n - 2$.*

As we will see later, obtaining sharp bounds on the maximum length of a $DS(n,s)$-sequence, for $s \geq 3$, is not as simple. Let us first give a simple proof of the following bound:

THEOREM 3.3 ([**260**]). $\lambda_3(n) = O(n\log n)$.

PROOF (Sketch). Let U be a $DS(n,3)$-sequence of length $\lambda_3(n)$. There must exist a symbol x that appears in U at most $\lambda_3(n)/n$ times. For any appearance of x which is neither the leftmost nor the rightmost, the symbols immediately preceding and succeeding x must be different, or else we would have obtained a forbidden subsequence of the form $\langle x \cdots yxy \cdots x \rangle$. Hence, if we erase from U all appearances of x, and, if necessary, at most two other elements, near the first and last appearances of x, we obtain a $DS(n-1,3)$-sequence, so this analysis implies the recurrence

$$\lambda_3(n) \leq \lambda_3(n-1) + \frac{\lambda_3(n)}{n} + 2,$$

or

$$\frac{\lambda_3(n)}{n} \leq \frac{\lambda_3(n-1)}{n-1} + \frac{2}{n-1},$$

from which the claim follows easily. □

This bound was later improved by Davenport [259] to $O(n \log n / \log \log n)$. For any given n and s, a trivial upper bound on $\lambda_s(n)$ is $sn(n-1)/2 + 1$ (use Corollary 2.2 and the observation that the total number of intersections between the functions in $\mathcal{F}$ is at most $s\binom{n}{2}$). Roselle and Stanton [654] proved that, for $s > n$, one has

$$\lambda_s(n) \geq sn(n-1)/2 - cn^3,$$

where $c < 1$ is a constant. Davenport and Schinzel [260] proved that, for any fixed s, there is a constant C_s depending on s, such that $\lambda_s(n) \leq n \cdot 2^{C_s \sqrt{\log n}}$. The problem was also studied in several other early papers [274, 549, 560, 627, 653, 655, 700, 701], but the next significant improvement on the bound of $\lambda_s(n)$ was made by Szemerédi [710], who proved that $\lambda_s(n) \leq A_s n \log^* n$, for each $s \geq 3$ and for appropriate positive constants A_s (doubly exponential in s); recall that $\log^* n$ is the smallest number of iterated applications of the $\log_2(x)$ function, starting at n, until we get down to 1. The currently best known bounds on $\lambda_s(n)$ for $s \geq 3$, stated below, are by Hart and Sharir [430], Agarwal *et al.* [56], and Nivasch [570].

$$\begin{aligned}
\lambda_3(n) &= \Theta(n\alpha(n)), \\
\lambda_4(n) &= \Theta(n \cdot 2^{\alpha(n)}), \\
\lambda_{2s+2}(n) &= n \cdot 2^{\frac{1}{s!}\alpha^s(n) \pm O(\alpha^{s-1}(n))}, \quad \text{for } s \geq 2, \\
\lambda_{2s+3}(n) &\leq n \cdot 2^{\frac{1}{s!}\alpha^s(n)\log\alpha(n) + O(\alpha^s(n))}, \quad \text{for } s \geq 1;
\end{aligned}$$

See Theorems 4.3, 4.5, 5.1, and 5.2 below. The best known lower bound for $\lambda_{2s+3}(n)$ is the one for $\lambda_{2s+2}(n)$; tightening this (rather small) gap is the main remaining open problem in the analysis of the quantities $\lambda_s(n)$.

To conclude this section we mention some generalizations of $DS(n,s)$-sequences. Let $U = \langle u_1, u_2, \ldots, u_m \rangle$ be a $DS(n,s)$-sequence. For $1 \leq j \leq m$, let $\mu(j)$ denote the number of symbols whose leftmost occurrences in U occur at an index $\leq j$ and whose rightmost occurrences occur at an index $> j$. We define the *depth* of U to be the maximum value of $\mu(j)$, for $j \leq m$. Define a $DS(n,s,t)$-sequence to be a $DS(n,s)$-sequence whose depth is at most t, and let $\lambda_{s,t}(n)$ denote the maximum length of a $DS(n,s,t)$-sequence. Huttenlocher *et al.* [449] proved that $\lambda_{s,t}(n) \leq \lceil n/t \rceil \lambda_s(2t)$ (see also Har-Peled [425]). This result has the following interesting consequence:

COROLLARY 3.4 ([449]). *Let $\mathcal{F} = \{f_1, \ldots, f_t\}$ be a collection of t continuous, real-valued, piecewise-linear functions (i.e., the graph of each f_i is an x-monotone polygonal chain). Let n be the total number of edges in the graphs of the functions of $\mathcal{F}$. Then the lower envelope of $\mathcal{F}$ has at most $\lambda_{3,t}(n) \leq \lceil n/t \rceil \lambda_3(2t) = O(n\alpha(t))$ breakpoints.*

Similar corollaries can be obtained for lower envelopes of piecewise algebraic functions.

Adamec et al. [**7**] have studied some generalizations of Davenport–Schinzel sequences. In particular, they bound the length of sequences not containing more general forbidden subsequences, for example, subsequences consisting of more than two symbols. They also showed that the maximum length of a sequence not containing any forbidden (not necessarily contiguous) subsequence $\langle a^{i_1} b^{i_2} a^{i_3} b^{i_4} \rangle$, where i_1, i_2, i_3, i_4 are some fixed positive constants, is linear, extending the bound on $\lambda_2(n)$. See also [**483, 484, 485, 486, 488, 570, 630**] for related results.

These generalized Davenport–Schinzel sequences have found several interesting applications, such as in Valtr's proof of a linear bound on the number of edges in a geometric graph with no k pairwise disjoint edges [**732**], and in Pettie's analysis of the so-called deque conjecture for splay trees [**629**].

4. Sharp Upper Bounds on $\lambda_s(n)$

In the previous section we mentioned some weak upper bounds on $\lambda_s(n)$. The problem of bounding $\lambda_s(n)$ lay dormant for about 10 years after Szemerédi's result [**710**], until Hart and Sharir [**430**] proved a tight bound of $\Theta(n\alpha(n))$ on $\lambda_3(n)$; here $\alpha(n)$ is the inverse Ackermann function, defined below. Later, Sharir [**673**] extended the analysis of Hart and Sharir to prove that $\lambda_s(n) = n \cdot \alpha(n)^{O(\alpha(n)^{s-3})}$, for $s > 3$. Applying a more careful analysis, Agarwal et al. [**56**] improved the bounds further, and obtained sharp, nearly tight bounds on $\lambda_s(n)$, for any fixed s. Recently, Nivasch [**570**] has slightly tightened these bounds. The best known upper bounds on $\lambda_s(n)$ are summarized in Theorem 4.3 (for $s = 3$) and Theorem 4.5 (for larger values of s). Since the proofs of these theorems are quite technical, we will sketch the proof of Theorem 4.3, and only briefly mention how the proof extends to the case $s > 3$.

4.1. Ackermann's function—A review. In this subsection we recall the definition of *Ackermann's function* and its functional inverse, which appears in the upper and lower bounds for $\lambda_s(n)$. Ackermann's function (also called "generalized exponentials") is an extremely fast growing function, defined over the integers in the following recursive manner [**6**].

Let $\mathbb{N}$ denote the set of positive integers. Given a function g from a set into itself, denote by $g^{(s)}$ the composition $g \circ g \circ \ldots \circ g$ of g with itself s times, for $s \in \mathbb{N}$. Define inductively a sequence $\{A_k\}_{k=1}^{\infty}$ of functions from $\mathbb{N}$ into itself as follows:

$$\begin{aligned} A_1(n) &= 2n, & n &\geq 1, \\ A_k(1) &= 2, & k &\geq 2, \\ A_k(n) &= A_{k-1}(A_k(n-1)), & n &\geq 2, k \geq 2. \end{aligned}$$

Finally, define *Ackermann's function* itself as $A(n) = A_n(n)$. The function A grows very quickly; its first few values are: $A(1) = 2$, $A(2) = 4$, $A(3) = 16$, and $A(4)$ is an exponential "tower" of 65536 2s. See [**477, 618, 652**] for a discussion on Ackermann's function and other rapidly growing functions.

Let α_k and α denote the functional inverses of A_k and A, respectively. That is,

$$\alpha_k(n) = \min\{s \geq 1 \mid A_k(s) \geq n\} \quad \text{and} \quad \alpha(n) = \min\{s \geq 1 \mid A(s) \geq n\}.$$

The functions α_k are easily seen to satisfy the following recursive formula:

(4.5) $$\alpha_k(n) = \min\{s \geq 1 \mid \alpha_{k-1}^{(s)}(n) = 1\};$$

that is, $\alpha_k(n)$ is the number of iterations of α_{k-1} needed to go from n to 1. In particular, (4.5) implies that, for $n \in \mathbb{N}$,

$$\alpha_1(n) = \lceil n/2 \rceil, \quad \alpha_2(n) = \lceil \log n \rceil, \quad \text{and} \quad \alpha_3(n) = \log^* n.$$

For each k, the function α_k is nondecreasing and unbounded. The same holds for α too, which grows more slowly than any of the α_k. Note that $\alpha(n) \leq 4$ for all $n \leq A(4)$, which is an exponential tower with 65536 2s, thus $\alpha(n) \leq 4$ for all "practical" values of n. We will need the following two easily established properties of $\alpha_k(n)$:

(4.6) $\qquad \alpha_{\alpha(n)}(n) = \alpha(n) \quad$ and, for $n > 4, \quad \alpha_{\alpha(n)+1}(n) \leq 4.$

We note that there are several alternative ways of defining Ackermann's function and its inverse in the literature. The particular choice of definition has little effect on the analysis—at worst, it can change $\alpha(n)$ by a (small) additive constant.

4.2. The upper bound for $\lambda_3(n)$. Let U be a $DS(n,3)$-sequence. A *chain* in U is a contiguous subsequence in which each symbol appears at most once. One can show that any such U can be decomposed into at most $2n-1$ pairwise disjoint chains, by splitting U just before the leftmost and rightmost appearances of each symbol. Let $\Psi(m,n)$ denote the maximum length of a $DS(n,3)$-sequence that can be decomposed into at most m chains.

LEMMA 4.1. *Let $m, n \geq 1$, and let $b > 1$ be a divisor of m. Then there exist integers $n^\star, n_1, n_2, \ldots, n_b \geq 0$ such that*

$$n^\star + \sum_{i=1}^{b} n_i = n,$$

and

(4.7) $\qquad \Psi(m,n) \leq 4m + 4n^\star + \Psi(b, n^\star) + \sum_{i=1}^{b} \Psi\left(\frac{m}{b}, n_i\right).$

PROOF. Let U be a $DS(n,3)$-sequence, consisting of at most m chains $c_1, \ldots, c_m$, of length $\Psi(m,n)$, and let $b > 1$ be a divisor of m. Partition the sequence U into b *blocks* (contiguous subsequences) $L_1, \ldots, L_b$, so that the block L_i consists of $p = m/b$ chains $c_{(i-1)p+1}, c_{(i-1)p+2}, \ldots, c_{ip}$. Call a symbol a *internal* to block L_i if all the occurrences of a in U are within L_i. A symbol is called *external* if it is not internal to any block. Suppose that there are n_i internal symbols in block L_i and $n^\star$ external symbols; thus $n^\star + \sum_{i=1}^{b} n_i = n$.

We estimate the total number of occurrences in U of symbols that are internal to L_i, as follows. Erase all external symbols from L_i. Next scan L_i from left to right and erase each element that has become equal to the element immediately preceding it. This leaves us with a sequence $L_i^\star$, which is clearly a $DS(n_i, 3)$-sequence consisting of at most m/b chains, and thus its length is at most $\Psi(m/b, n_i)$. Moreover, if two equal internal elements in L_i have become adjacent after erasing the external symbols, then these two elements must have belonged to two distinct chains, thus the total number of deletions of internal symbols is at most $(m/b) - 1$. Hence, summing over all blocks, we conclude that the total contribution of internal symbols to $|U|$ is at most

$$m - b + \sum_{i=1}^{b} \Psi\left(\frac{m}{b}, n_i\right).$$

Next, to estimate the contribution of external symbols to $|U|$, we argue as follows. For each L_i, call an external symbol a a *middle* symbol if none of its occurrences in L_i is the first or the last occurrence of a in U. Otherwise we call a a *non-middle* symbol. We will consider the contribution of middle and non-middle external symbols separately.

Consider first the middle symbols. To estimate their contribution to the length of L_i, we erase all internal and non-middle symbols from L_i, and also erase a middle symbol if it has become equal to the symbol immediately preceding it. As above, at most $(m/b) - 1$ deletions of external middle symbols will be performed. Let $L_i^\star$ be the resulting subsequence, and suppose that it is composed of p_i distinct symbols. It is easily seen that $L_i^\star$ is a $DS(p_i, 1)$-sequence, so its length is at most p_i. Hence, summing over all blocks, the total contribution of external middle symbols is at most $m - b + \sum_{i=1}^{b} p_i$. But $\sum_{i=1}^{b} p_i$ is the length of the sequence obtained by concatenating all the subsequences $L_i^\star$. This concatenation can contain at most b pairs of adjacent equal elements, and if we erase each element that is equal to its predecessor, we obtain a sequence $U^\star$ which is clearly a $DS(n^\star, 3)$-sequence composed of b chains (namely the subsequences $L_i^\star$). The length of $U^\star$ is thus at most $\Psi(b, n^\star)$. Hence, the contribution of middle external elements to the length of U is at most $m + \Psi(b, n^\star)$.

Consider next the contribution of non-middle symbols. A symbol is called *starting* (resp., *ending*) in block L_i if it does not occur in any block before (resp., after) L_i. To estimate the contribution of starting symbols to the length of L_i, we erase from L_i all symbols occurring there except for starting symbols, and, if necessary, also erase each occurrence of a *starting* symbol that has become equal to the element immediately preceding it. As above, at most $(m/b) - 1$ deletions of external starting symbols will be performed. Let $L_i^\#$ be the resulting subsequence, and suppose that it is composed of p_i distinct symbols.

Note first that each external symbol can appear as a starting symbol in exactly one block, thus $\sum_{i=1}^{b} p_i = n^\star$. It is easily seen that $L_i^\#$ is a $DS(p_i, 2)$-sequence, so the length of $L_i^\#$ is at most $2p_i - 1$, and, summing over all blocks, we conclude that the contribution of all external starting symbols to the length of U is at most

$$m - b + \sum_{i=1}^{b}(2p_i - 1) = m - 2b + 2n^\star.$$

In a completely symmetric manner, the contribution of external *ending* symbols to the length of U is also at most $m - 2b + 2n^\star$. Summing up all these contributions, we finally obtain the asserted inequality (4.7). $\square$

Next, we solve the recurrence derived in the previous lemma.

LEMMA 4.2. *For all $m, n \geq 1$, and for $k \geq 2$,*

(4.8) $$\Psi(m, n) \leq (8k - 8)m\alpha_k(m) + (4k - 2)n.$$

PROOF (Sketch). For the sake of simplicity, we will only show that, for $n, s \geq 1$, $k \geq 2$, and m dividing $A_k(s)$, one has

(4.9) $$\Psi(m,n) \leq (4k-4)ms + (4k-2)n.$$

If $m = A_k(s)$, then $s = \alpha_k(m)$, and (4.9) implies the assertion of the lemma for these values of m also. The case of an arbitrary m is then easy to handle; see [**430, 681**] for details.

We will use (4.7) repeatedly to obtain the series of upper bounds on Ψ, stated in (4.9) for $k = 2, 3, \ldots$. At each step we choose b in an appropriate manner, and estimate $\Psi(b, n^\star)$ using the bound obtained in the preceding step. This yields a new recurrence relation on Ψ, which we solve to obtain a better upper bound on Ψ.

Specifically, we proceed by double induction on k and s. For $k = 2$, m divides $A_2(s) = 2^s$, so m is a power of 2. Choose $b = 2$ in (4.7); it is easily checked that $\Psi(b, n^\star) = \Psi(2, n^\star) = 2n^\star$ for all $n^\star$, so (4.7) becomes

$$\Psi(m,n) \leq 4m + 6n^\star + \Psi\left(\frac{m}{2}, n_1\right) + \Psi\left(\frac{m}{2}, n_2\right).$$

The solution to this recurrence relation, for m a power of 2 and $n = n^\star + n_1 + n_2$ arbitrary, is easily verified to be

$$\Psi(m,n) \leq 4m \log m + 6n.$$

The case $k > 2$ and $s = 1$ is now a consequence of this bound (because m divides $A_k(1) = 2$ in this case).

Suppose next that $k > 2$ and $s > 1$, and that the induction hypothesis is true for all $k' < k$ and $s' \geq 1$, and for $k' = k$ and all $s' < s$. Let $m = A_k(s)$, and $t = A_k(s-1)$, and choose $b = m/t$, which is an integer dividing $m = A_k(s) = A_{k-1}(t)$. Hence, by the induction hypothesis for $k-1$ and t, we have

$$\Psi(b, n^\star) \leq (4k-8)bt + (4k-6)n^\star = (4k-8)m + (4k-6)n^\star.$$

Then (4.7) becomes

$$\Psi(m,n) \leq (4k-8)m + (4k-6)n^\star + 4m + 4n^\star + \sum_{i=1}^{b} \Psi(t, n_i).$$

Using the induction hypothesis once more (for k and $s-1$), we obtain

$$\begin{aligned}\Psi(m,n) &\leq (4k-4)m + (4k-2)n^\star + \sum_{i=1}^{b}((4k-4)t(s-1) + (4k-2)n_i) \\ &= (4k-4)ms + (4k-2)n,\end{aligned}$$

because $n^\star + \sum_{i=1}^{b} n_i = n$.

The case where m only divides $A_k(s)$ is handled by taking a concatenation of $p = A_k(s)/m$ copies of a sequence whose length is $\Psi(m,n)$, using pairwise-disjoint sets of symbols for the copies. The concatenated sequence is composed of pn symbols and has at most pm chains, so

$$p\Psi(m,n) \leq \Psi(pm, pn) \leq (4k-4)pms + (4k-2)pn,$$

from which (4.9) follows.

This completes the proof of the asserted bound. $\square$

THEOREM 4.3 ([**430**]). $\lambda_3(n) = O(n\alpha(n))$.

PROOF. By putting $k = \alpha(m) + 1$ in (4.8) and using (4.6), we obtain

$$\Psi(m,n) \leq 32m\alpha(m) + (4\alpha(m) + 2)n.$$

As noted in the beginning of this subsection, $\lambda_3(n) \leq \Psi(2n-1,n)$. Since $\alpha(2n-1) \leq \alpha(n) + 1$, the theorem follows. □

Applying a more careful analysis, Klazar [**487**] has shown that

(4.10) $$\lambda_3(n) \leq 2n\alpha(n) + O(n\sqrt{\alpha(n)}).$$

An immediate corollary of Theorem 4.3 is that the lower envelope of n segments in the plane has $O(n\alpha(n))$ breakpoints.

4.3. Upper bounds on $\lambda_s(n)$. We now briefly mention how the upper bounds on $\lambda_s(n)$, for $s > 3$, are derived in [**56**]. Let $\Psi_s^t(m,n)$ denote the maximum length of a $DS(n,s)$-sequence composed of at most m contiguous subsequences, each of which is a $DS(n,t)$-sequence. As above, Agarwal et al. [**56**] obtain a recurrence relation for $\Psi_s(m,n)$, the length of a $DS(n,s)$-sequences composed of at most m chains, but the recurrence is now written in terms of Ψ_s and Ψ_s^{s-2}. Let S be a given $DS(n,s)$-sequence composed of at most m chains. The analysis in [**56**] divides S into b blocks and counts the contributions of internal, middle, and non-middle symbols separately, in a manner similar to that given above. This leads to the following lemma.

LEMMA 4.4. *Let $m, n \geq 1$ and $1 < b < m$ be integers. For any partitioning $m = \sum_{i=1}^b m_i$, with $m_1, \ldots, m_b \geq 1$, there exist integers $n^\star, n_1, n_2, \ldots, n_b \geq 0$ such that*

$$n^\star + \sum_{i=1}^b n_i = n$$

and

(4.11) $$\Psi_s(m,n) \leq \Psi_s^{s-2}(b, n^\star) + 2\Psi_{s-1}(m, n^\star) + 4m + \sum_{i=1}^b \Psi_s(m_i, n_i).$$

If we choose $b = 2$, the solution of the recurrence is $O(n \log^{s-2} n)$. However, extending the proof of Lemma 4.2, but using a rather involved analysis, one can obtain much sharper bounds. This analysis has recently been refined by Nivasch [**570**], who has managed to slightly tighten the older bounds of [**56**]. All in all, these are the best known upper bounds on $\lambda_s(n)$.

THEOREM 4.5 ([**56, 570**]).
  (i) $\lambda_4(n) = O(n \cdot 2^{\alpha(n)})$.
  (ii) *For $s > 1$, there exists a polynomial $C_s(q)$ of degree at most $s - 1$, such that*

$$\lambda_{2s+1}(n) \leq n \cdot 2^{\frac{1}{(s-1)!}\alpha^{s-1}(n)\log\alpha(n) + C_s(\alpha(n))},$$
$$\lambda_{2s+2}(n) \leq n \cdot 2^{\frac{1}{s!}\alpha^s(n) + C_s(\alpha(n))}.$$

Recently, Nivasch [**570**] introduced an alternative, and arguably simpler, technique for deriving upper bounds for $\lambda_s(n)$. They re-derived Theorem 4.3 and Theorem 4.5 using this new technique.

5. Lower Bounds on $\lambda_s(n)$

An even more surprising result in the theory of Davenport–Schinzel sequences is that the bounds stated in Theorems 4.3 and 4.5 are optimal for $s = 3$ and 4, and are very close to optimal for even $s > 4$. The first superlinear lower bound on $\lambda_s(n)$ was obtained by Hart and Sharir [430], who proved that $\lambda_3(n) = \Omega(n\alpha(n))$. Their original proof transforms $DS(n, 3)$-sequences into certain path compression schemes on rooted trees. A more direct proof for the lower bound on $\lambda_3(n)$ was given by Wiernik and Sharir [747] — they describe an explicit recursive scheme for constructing a $DS(n, 3)$-sequence of length $\Omega(n\alpha(n))$. See also [499] for another proof of the same lower bound. We sketch Wiernik and Sharir's construction, omitting many details, which can be found in [681, 747].

Let $\{C_k(m)\}_{k \geq 1}$ be a sequence of functions from $\mathbb{N}$ to itself, defined by

$$\begin{aligned} C_1(m) &= 1, & m &\geq 1, \\ C_k(1) &= 2C_{k-1}(2), & k &\geq 2, \\ C_k(m) &= C_k(m-1) \cdot C_{k-1}(C_k(m-1)), & k &\geq 2, m \geq 2. \end{aligned}$$

It can be shown that, for all $k \geq 4, m \geq 1$,

(5.12) $$A_{k-1}(m) \leq C_k(m) \leq A_k(m+3).$$

In what follows, let $\mu = C_k(m-1)$, $\nu = C_{k-1}(C_k(m-1))$, and $\gamma = \mu \cdot \nu$.

For each $k, m \geq 1$, we construct a sequence $S_k(m)$ that satisfies the following two properties:

- (P1) $S_k(m)$ is composed of $N_k(m) = m \cdot C_k(m)$ distinct symbols. These symbols are named (d, l), for $d = 1, \ldots, m$, $l = 1, \ldots, \gamma$, and are ordered in lexicographical order, so that $(d, l) < (d', l')$ if $l < l'$ or $l = l'$ and $d < d'$.
- (P2) $S_k(m)$ contains γ *fans* of size m, where each fan is a contiguous subsequence of the form $\langle (1, l) (2, l) \cdots (m, l) \rangle$, for $l = 1, \ldots, \gamma$.

Since fans are pairwise disjoint, by definition, the naming scheme of the symbols of $S_k(m)$ can be interpreted as assigning to each symbol the index l of the fan in which it appears, and its index d within that fan.

The construction of $S_k(m)$ proceeds by double induction on k and m, as follows.

(1) $k = 1$: The sequence is a single fan of size m: $S_1(m) = \langle (1,1) (2,1) \cdots (m,1) \rangle$. Properties (P1) and (P2) clearly hold here ($C_1(m) = 1$).

(2) $k = 2$: The sequence contains a pair of disjoint fans of size m, with a block of elements following each of these fans. Specifically,

$$\begin{aligned} S_2(m) =\ & \langle (1,1) (2,1) \cdots (m-1,1) (m,1) (m-1,1) \cdots (1,1) \\ & (1,2) (2,2) \cdots (m-1,2) (m,2) (m-1,2) \cdots (1,2) \rangle. \end{aligned}$$

Indeed, $S_2(m)$ contains $C_2(m) = 2$ fans and is composed of $2m$ distinct symbols.

(3) $k \geq 3, m = 1$: The sequence is identical to the sequence for $k' = k - 1$ and $m' = 2$, except for renaming of its symbols and fans: $S_{k-1}(2)$ contains $C_{k-1}(2) = \frac{1}{2}C_k(1)$ fans, each of which consists of two symbols; the symbol renaming in $S_k(1)$ causes each of these two elements to become a 1-element fan. Properties (P1) and (P2) clearly hold.

(4) The general case $k \geq 3, m > 1$:

5. LOWER BOUNDS ON $\lambda_s(n)$

(i) Generate inductively the sequence $S' = S_k(m-1)$; by induction, it contains μ fans of size $m-1$ each and is composed of $(m-1) \cdot \mu$ symbols.

(ii) Create ν copies of S' whose sets of symbols are pairwise disjoint. For each $j \leq \nu$, rename the symbols in the jth copy S'_j of S' as (d, i, j), where $1 \leq d \leq m-1$ is the index of the symbol in the fan of S'_j containing it, and $1 \leq i \leq \mu$ is the index of this fan in S'_j.

(iii) Generate inductively the sequence $S^\star = S_{k-1}(\mu)$ whose set of symbols is disjoint from that of any S'_j; by induction, it contains ν fans of size μ each. Rename the symbols of $S^\star$ as (m, i, j) (where i is the index of that symbol within its fan, and j is the index of that fan in $S^\star$). Duplicate the last element (m, μ, j) in each of the ν fans of $S^\star$.

(iv) For each $1 \leq i \leq \mu$, $1 \leq j \leq \nu$, extend the ith fan of S'_j by duplicating its last element $(m-1, i, j)$, and by inserting the corresponding symbol (m, i, j) of $S^\star$ between these duplicated appearances of $(m-1, i, j)$. This process extends the $(m-1)$-fans of S'_j into m-fans and adds a new element after each extended fan.

(v) Finally construct the desired sequence $S_k(m)$ by merging the ν copies S'_j of S' with the sequence $S^\star$. This is done by replacing, for each $1 \leq j \leq \nu$, the jth fan of $S^\star$ by the corresponding copy S'_j of S', as modified in (iv) above. Note that the duplicated copy of the last element in each fan of $S^\star$ (formed in step (iii) above) appears now after the copy S'_j that replaces this fan; see Figure 5 for an illustration of this process.

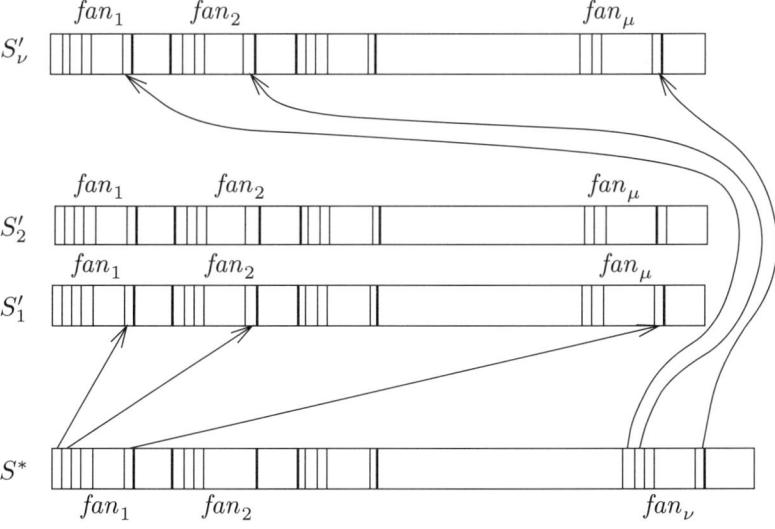

FIGURE 5. Lower bound construction: merging the subsequences.

It is easily checked that $S_k(m)$ consists of
$$N_k(m) = \nu(m-1)\mu + \mu C_{k-1}(\mu) = mC_k(m)$$
symbols, and it can also be shown that $S_k(m)$ is a $DS(N_k(m), 3)$-sequence satisfying properties (P1) and (P2). If we let $\sigma_k(m)$ denote the length of $S_k(m)$, then
$$\begin{aligned} \sigma_1(m) &= m, \\ \sigma_2(m) &= 4m - 2, \\ \sigma_k(1) &= \sigma_{k-1}(2), \\ \sigma_k(m) &= \nu\sigma_k(m-1) + \sigma_{k-1}(\mu) + \nu(\mu+1). \end{aligned}$$

The third term in the last equation is due to the duplication of the rightmost symbol of each fan of S^* and of each S'_j (see Steps 4 (iii)–(iv)). Using a double induction on k and m, one can prove that
$$\sigma_k(m) > (km - 2)C_k(m) + 1.$$

THEOREM 5.1 ([**430, 747**]). $\lambda_3(n) = \Omega(n\alpha(n))$.

PROOF. Choose $m_k = C_{k+1}(k-3)$. Then
$$n_k = N_k(m_k) = C_{k+1}(k-2) \leq A_{k+1}(k+1)$$
where the last inequality follows from (5.12). Therefore $\alpha(n_k) \leq k+1$, and hence
$$\lambda_3(n_k) \geq \sigma_k(m_k) \geq kn_k - 2C_k(m_k) \geq (k-2)n_k \geq n_k(\alpha(n_k) - 3).$$

As shown in [**681**], this bound can be extended to any integer n, to prove that $\lambda_3(n) = \Omega(n\alpha(n))$. □

Nivasch [**570**] recently proved that, in fact, $\lambda_3(n) \geq 2n\alpha(n) - O(n)$. Together with the above-mentioned upper bound for $\lambda_3(n)$ of Klazar (4.10), this implies that $\lim_{n\to\infty} \lambda_3(n)/(n\alpha(n)) = 2$.

Generalizing the above construction and using induction on s — basically replacing each chain of the sequence $S_k(m)$ by a $DS(n, s-2)$-sequence, which, in turn, is constructed recursively — Sharir [**674**] proved that $\lambda_{2s+1}(n) = \Omega(n\alpha(n)^s)$. Later Agarwal et al. [**56**] proved that the upper bounds stated in Theorem 4.5 are almost optimal. In particular, using a rather involved doubly-inductive scheme, they constructed a $DS(n, 4)$-sequence of length $\Omega(n2^{\alpha(n)})$. Then, by recursing on s, they generalized their construction of $DS(n, 4)$-sequences to higher-order sequences. Nivasch [**570**] has recently simplified this construction.

The following theorem summarizes these lower bounds.

THEOREM 5.2 ([**56**]).
(i) $\lambda_4(n) = \Omega(n \cdot 2^{\alpha(n)})$.
(ii) For $s > 1$, there exists a polynomial $Q_s(q)$ of degree at most $s - 1$, such that
$$\lambda_{2s+2}(n) \geq n \cdot 2^{\frac{1}{s!}\alpha^s(n) + Q_s(\alpha(n))}.$$

PROBLEM 5.3. *Obtain tight bounds on $\lambda_s(n)$ for $s > 4$ for odd values of s.*

Wiernik and Sharir [**747**] proved that the $DS(n, 3)$-sequence $S_k(m)$ constructed above can be realized as the lower envelope sequence of a set of n segments, which leads to the following fairly surprising result:

THEOREM 5.4 ([**747**]). *The lower envelope of n segments can have $\Omega(n\alpha(n))$ breakpoints in the worst case.*

Shor [**689**] gave a simpler construction of n segments whose lower envelope also has $\Omega(n\alpha(n))$ breakpoints. These results also yield an $\Omega(n\alpha(n))$ lower bound on many other unrelated problems, including searching in totally monotone matrices [**481**] and counting the number of distinct edges in the convex hull of a planar point set as the points are being updated dynamically [**717**]. Shor has also shown that there exists a set of n degree-4 polynomials whose lower envelope has $\Omega(n\alpha(n))$ breakpoints [**688**] (which is somewhat weak, because the upper bound for this quantity is $\lambda_4(n) = O(n \cdot 2^{\alpha(n)})$). We conclude this section by mentioning another open problem, which we believe is one of the most challenging and interesting problems related to Davenport–Schinzel sequences.

PROBLEM 5.5. *Is there a natural geometric realization of higher order sequences? For example, can the lower envelope of n conic sections, or of n circular arcs, have $\Omega(n 2^{\alpha(n)})$ breakpoints?*

6. Davenport–Schinzel Sequences and Arrangements

In this section we consider certain geometric and topological structures induced by a family of arcs in the plane, where Davenport–Schinzel sequences play a major role in their analysis.

This section somewhat overlaps the material given in Chapter 2, except that here the discussion is specialized to planar arrangements. Readers unfamiliar with this basic theory might want to read this section first.

Specifically, let $\Gamma = \{\gamma_1, \ldots, \gamma_n\}$ be a collection of n Jordan arcs in the plane,[1] each pair of which intersect in at most s points, for some fixed constant s. Specializing the general definition in Chapter 2, we have:

DEFINITION 6.1. *The* arrangement $\mathcal{A}(\Gamma)$ *of* Γ *is the planar subdivision induced by the arcs of Γ; that is, $\mathcal{A}(\Gamma)$ is a planar map whose* vertices *are the endpoints of the arcs of Γ and their pairwise intersection points, whose* edges *are maximal (relatively open) connected portions of the γ_i's that do not contain a vertex, and whose* faces *are the connected components of $\mathbb{R}^2 \setminus \bigcup \Gamma$. The* combinatorial complexity *of a face is the number of vertices (or edges) on its boundary, and the* combinatorial complexity *of $\mathcal{A}(\Gamma)$ is the total complexity of all of its faces.*

The maximum combinatorial complexity of $\mathcal{A}(\Gamma)$ is clearly $\Theta(sn^2) = \Theta(n^2)$, and $\mathcal{A}(\Gamma)$ can be computed in time $O(n^2 \log n)$, in an appropriate model of computation, using the sweep-line algorithm of Bentley and Ottmann [**137**]. A slightly faster algorithm, with running time $O(n\lambda_{s+2}(n))$, is mentioned in Section 6.3. Many applications, however, need to compute only a small portion of the arrangement, such as a single face, a few faces, or some other substructures that we will consider shortly. Using *DS*-sequences, one can show that the combinatorial complexity of these substructures is substantially smaller than that of the entire arrangement. This fact is then exploited in the design of efficient algorithms, whose running time

---

[1]A *Jordan arc* is an image of the closed unit interval under a continuous bijective mapping. Similarly, a *closed Jordan curve* is an image of the unit circle under a similar mapping, and an *unbounded Jordan curve* is an image of the open unit interval (or of the entire real line) that separates the plane.

is close to the bound on the complexity of the substructures that these algorithms aim to construct. In this section we review combinatorial and algorithmic results related to these substructures, in which *DS*-sequences play a crucial role.

6.1. Complexity of a single face. It is well known that the complexity of a single face in an arrangement of n lines is at most n [**637**], and a linear bound on the complexity of a face in an arrangement of rays is also known (see Alevizos *et al.* [**64, 65**]). The lower bound of Wiernik and Sharir [**747**] on the complexity of lower envelopes of segments implies that the unbounded face in an arrangement of n line segments can have $\Omega(n\alpha(n))$ vertices in the worst case. A matching upper bound was proved by Pollack *et al.* [**635**], which was later extended by Guibas *et al.* [**402**] to general Jordan arcs. The case of closed or unbounded Jordan curves was treated in [**663**].

THEOREM 6.2 ([**402, 663**]). *Let Γ be a set of n Jordan arcs in the plane, each pair of which intersect in at most s points, for some fixed constant s. Then the combinatorial complexity of any single face in $\mathcal{A}(\Gamma)$ is $O(\lambda_{s+2}(n))$. If each arc in Γ is a Jordan curve (closed or unbounded), then the complexity of a single face is at most $\lambda_s(n)$.*

PROOF (Sketch). We only consider the first part of the theorem; the proof of the second part is simpler, and can be found in [**663, 681**]. Let f be a given face in $\mathcal{A}(\Gamma)$, and let C be a connected component of its boundary. We can assume that C is the only connected component of ∂f. Otherwise, we repeat the following analysis for each connected component and sum their complexities. Since each arc appears in at most one connected component, the bound follows. For each arc γ_i, let u_i and v_i denote its endpoints, and let γ_i^+ (resp., γ_i^-) be the directed arc γ_i oriented from u_i to v_i (resp., from v_i to u_i).

Without loss of generality, assume that C is the exterior boundary component of f. Traverse C in counterclockwise direction (so that f lies to our left) and let $S = \langle s_1, s_2, \ldots, s_t \rangle$ be the circular sequence of oriented arcs in Γ in the order in which they appear along C (if C is unbounded, S is a linear, rather than circular, sequence). More precisely, if during our traversal of C we encounter an arc γ_i and follow it in the direction from u_i to v_i (resp., from v_i to u_i) then we add γ_i^+ (resp., γ_i^-) to S. See Figure 6 for an illustration. Note that in this example *both* sides of an arc γ_i belong to the outer connected component.

Let $\xi_1, \ldots, \xi_{2n}$ denote the oriented arcs of Γ. For each ξ_i we denote by $|\xi_i|$ the nonoriented arc γ_j coinciding with ξ_i. For the purpose of the proof, we transform each arc γ_i into a very thin closed Jordan curve $\gamma_i^\star$ by taking two nonintersecting copies of γ_i lying very close to one another, and by joining them at their endpoints. This will perturb the face f slightly but can always be done in such a way that the combinatorial complexity of C does not decrease. Note that this transformation allows a natural identification of one of the two "sides" of $\gamma_i^\star$ with γ_i^+ and the other side with γ_i^-.

A crucial topological property (see [**402, 681**]) is that the portions of each arc ξ_i appear in S in a circular order that is consistent with their order along the oriented ξ_i. In particular, there exists a starting point in S (which depends on ξ_i) so that if we read S in circular order starting from that point, we encounter these portions of ξ_i in their order along ξ_i. For each directed arc ξ_i, consider the linear sequence V_i of all appearances of ξ_i in S, arranged in the order they appear along

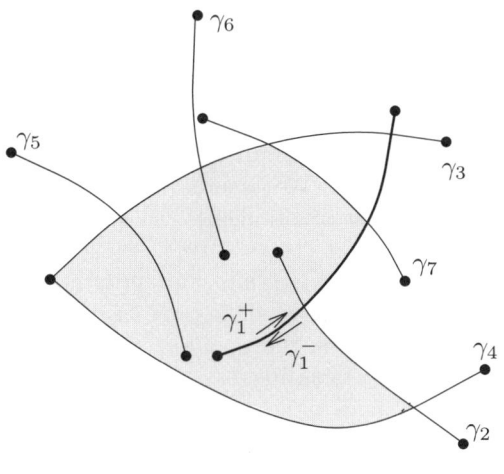

$$S = \langle \gamma_1^+ \gamma_2^- \gamma_2^+ \gamma_1^+ \gamma_7^- \gamma_3^- \gamma_6^+ \gamma_6^- \gamma_3^- \gamma_5^+ \gamma_5^- \gamma_3^- \gamma_4^+ \gamma_2^- \gamma_1^- \rangle$$

FIGURE 6. A single face and its associated boundary sequence; all arcs are positively oriented from left to right.

ξ_i. Let μ_i and ν_i denote, respectively, the index in S of the first and of the last element of V_i. Consider $S = \langle s_1, \ldots, s_t \rangle$ as a linear, rather than a circular, sequence (this change is not needed if C is unbounded). For each arc ξ_i, if $\mu_i > \nu_i$ we split the symbol ξ_i into two distinct symbols ξ_{i1}, ξ_{i2}, and replace all appearances of ξ_i in S between the places μ_i and t (resp., between 1 and ν_i) by ξ_{i1} (resp., by ξ_{i2}). Note that the above claim implies that we can actually split the arc ξ_i into two connected subarcs, so that all appearances of ξ_{i1} in the resulting sequence represent portions of the first subarc, whereas all appearances of ξ_{i2} represent portions of the second subarc. This splitting produces a sequence $S^\star$, of the same length as S, composed of at most $4n$ symbols.

With all these modifications, one can then prove that $S^\star$ is a $DS(4n, s+2)$-sequence. This is done by showing that each quadruple of the form $\langle a \cdots b \cdots a \cdots b \rangle$ in S^* corresponds, in a unique manner, to an intersection point between the two arcs of Γ that a and b represent. Thus, an alternation in $S^\star$ of two symbols a, b of length $s + 4$ would have forced the two arcs a, b to intersect in at least $s + 1$ points, which is impossible. See [**402, 681**] for more details. This completes the proof of the first part of the theorem. □

Theorem 6.2 has the following interesting consequence. Let $\Gamma = \{\gamma_1, \ldots, \gamma_n\}$ be a set of n closed Jordan curves, each pair of which intersects in at most s points. Let $K = conv(\Gamma)$ be the convex hull of the curves in Γ. Partition the boundary of K into a minimum number of subarcs, $\alpha_1, \alpha_2, \ldots, \alpha_m$, such that the relative interior of each α_i has a nonempty intersection with exactly one of the curves γ_j. Then the number m of such arcs is at most $\lambda_s(n)$; see [**663**] for a proof.

Arkin *et al.* [**87**] showed that the complexity of a single face in an arrangement of line segments with h distinct endpoints is only $O(h \log h)$ (even though the number of segments can be $\Theta(h^2)$). A matching lower bound was proved by Matoušek and Valtr [**541**]. The upper bound by Arkin *et al.* does not extend to general arcs. Har-Peled [**425**] has also obtained improved bounds on the complexity of a single face in many special cases.

6.2. Computing a single face. Let Γ be a collection of n Jordan arcs, as above, and let x be a point that does not lie on any arc of Γ. We wish to compute the face of $\mathcal{A}(\Gamma)$ that contains x. We assume that each arc in Γ has at most a constant number of points of vertical tangency, so that we can break it into $O(1)$ x-monotone Jordan arcs.

We assume a model of computation allowing infinite-precision real arithmetic, in which certain primitive operations involving one or two arcs (e.g., computing the intersection points of a pair of arcs, the points of vertical tangency of an arc, the intersections of an arc with a vertical line, etc.) are assumed to take constant time. Such models exist, in theory and, progressively, also in practice, for many classes of arcs.

If Γ is a set of n lines, or a set of n rays, then a single face can be computed in time $O(n \log n)$. In the case of lines, this is done by dualizing the lines to points and using any optimal convex hull algorithm [**637**]; the case of rays is somewhat more involved, and is described in [**64, 65**]. However, these techniques do not extend to arrangements of more general Jordan arcs. Pollack *et al.* [**635**] presented an $O(n\alpha(n) \log^2 n)$-time algorithm for computing the unbounded face in certain arrangements of line segments, but the first algorithm that works for general arcs (satisfying the above mild assumptions) was given by Guibas *et al.* [**402**]. Later, several other efficient algorithms—both randomized and deterministic—have been proposed. We first present randomized (Las Vegas) algorithms[2] for computing a single face, and then review the deterministic solution of [**402**], and mention some other related results.

Randomized algorithms. The randomized algorithms that we will describe actually compute the so-called *vertical decomposition* of f. (This notion has been discussed in detail and in more generality in Chapter 2, but we specialize here its simple definition for planar arrangements.) The vertical decomposition, which we denote by $f^{\parallel}$, is obtained by drawing a vertical segment from each vertex and from each point of vertical tangency of the boundary of f in both directions, and extend it until it meets another edge of f, or else all the way to $\pm\infty$. The vertical decomposition partitions f into 'pseudo-trapezoidal' cells, each bounded by at most two arcs of Γ and at most two vertical segments. To simplify the presentation, we will refer to these cells simply as trapezoids; see Figure 7 for an illustration.

We first present a rather simple randomized divide-and-conquer algorithm due to Clarkson [**243**] (see also [**681**]). The basic idea of the algorithm is as follows: Randomly choose a subset $\Gamma_1 \subseteq \Gamma$ of $\lfloor n/2 \rfloor$ arcs. Recursively compute the vertical decompositions $f_1^{\parallel}, f_2^{\parallel}$ of the faces f_1, f_2 containing x in $\mathcal{A}(\Gamma_1)$ and in $\mathcal{A}(\Gamma \setminus \Gamma_1)$, respectively. Then merge $f_1^{\parallel}$ and $f_2^{\parallel}$ to obtain the vertical decomposition of the face f of $\mathcal{A}(\Gamma)$ that contains x. The merge step essentially performs a simultaneous depth-first search over the trapezoids of $f_1^{\parallel}$ and of $f_2^{\parallel}$, in which it computes the

---
[2]A *Las Vegas* algorithm always terminates with the correct output, but its running time is a random variable (over the internal randomizations it performs).

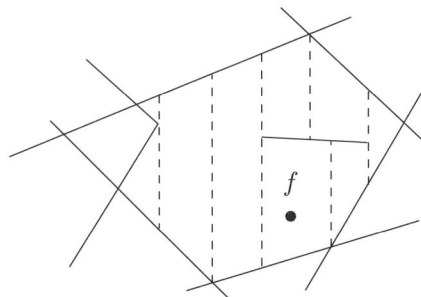

FIGURE 7. Vertical decomposition of a face in an arrangement of line segments; here each cell is indeed a trapezoid or a triangle.

intersection cells $\tau_1 \cap \tau_2$, for $\tau_1 \in f_1^{||}$, $\tau_2 \in f_2^{||}$, that lie in $f^{||}$. After having computed all these intersection cells, $f^{||}$ can be computed in additional $O(|f^{||}|)$ time; see [**681**] for details. Although the merge step is quite naive, and can take quadratic time in the worst case, one can nevertheless show that the randomization makes this step fast—its expected running time is only $O(\lambda_{s+2}(n))$. Hence, the expected running time of the whole algorithm is $O(\lambda_{s+2}(n) \log n)$.

The second randomized algorithm, due to Chazelle *et al.* [**219**], constructs the vertical decomposition $f^{||}$ of the face containing x incrementally, by adding the arcs of Γ one by one in a random order (the choice of the insertion order is the only randomized step in the algorithm), where each permutation of Γ is chosen with equal probability. While the worst-case running time of this algorithm is also quadratic, the expected running time is only $O(\lambda_{s+2}(n) \log n)$, as for the preceding algorithm.

The basic idea of this algorithm is as follows. Let $\langle \gamma_1, \gamma_2, \ldots, \gamma_n \rangle$ denote the (random) insertion sequence, let $\Gamma_i = \{\gamma_1, \ldots, \gamma_i\}$, and let $f_i^{||}$ be the vertical decomposition of the face containing x in $\mathcal{A}(\Gamma_i)$, for $i = 1, \ldots, n$. When γ_{i+1} is inserted, it may chop off a portion of f_i by separating it from the point x, so some of the trapezoids of $f_i^{||}$ may not appear in $f_{i+1}^{||}$, and some of them, which are crossed by γ_{i+1}, will have to be replaced by new trapezoids that have γ_{i+1} on their boundary. Thus, adding γ_{i+1} requires the following steps: (i) Compute the set of trapezoids in $f_i^{||}$ that γ_{i+1} intersects; (ii) determine the set of new cells that appear in $f_{i+1}^{||}$, and (iii) find the portion of f_i that is chopped off by γ_{i+1}, if any; (iv) finally, discard the trapezoids of $f_i^{||}$ that do not appear in $f_{i+1}^{||}$.

To facilitate the execution of these steps, the algorithm stores $f_i^{||}$ as a vertical adjacency graph, whose edges connect pairs of trapezoids sharing a vertical edge (more precisely, having overlapping vertical edges); for each trapezoid in $f_i^{||}$, we store the list of trapezoids that are its neighbors in the vertical adjacency graph. The algorithm also maintains a directed acyclic graph (dag) G, referred to as the *history dag*. The nodes of G, after the ith insertion stage, correspond to the trapezoids that appeared in at least one $f_j^{||}$, for $j \leq i$. The root of the dag corresponds to the entire plane. There is a directed edge from a node v to a node w if the corresponding trapezoids τ_v and τ_w intersect and if τ_v (resp., τ_w) appeared in $f_j^{||}$ (resp., $f_k^{||}$) for some $j < k$. If τ_v is a trapezoid of $f_i^{||}$, then v is an *active* leaf (in the version of G after the ith insertion), and if τ_v was a trapezoid of $f_{i-1}^{||}$ but is not in $f_i^{||}$, and γ_i does not cross τ_v, then v is an *inactive* leaf, in the sense that no successor of τ_v will ever be created. All other nodes of G are *inner* nodes, and represent trapezoids that existed in some $f_j^{||}$, but were crossed by some arc γ_k, for $j < k \leq i$. The purpose of the history dag G is to facilitate, through a top-down traversal of it, a simple and efficient technique for finding all active trapezoids that the newly inserted arc intersects.

How exactly the above steps are executed and how the data structures are updated is somewhat involved, and is described in detail in [**219, 681**]. As mentioned above, the expected running time of the algorithm is $O(\lambda_{s+2}(n) \log n)$. Moreover, the expected size and depth of G are $O(\lambda_{s+2}(n))$ and $O(\log n)$, respectively, so we also obtain a point-location data structure that can determine, in $O(\log n)$ expected time, whether a query point lies in f. A somewhat simpler variant of the randomized incremental algorithm is given in [**141**].

THEOREM 6.3 ([**219, 243, 141**]). *Given a collection Γ of n Jordan arcs, each pair of which intersect in at most s points, and a point x not lying on any arc, the face of $\mathcal{A}(\Gamma)$ containing x can be computed by a randomized algorithm in $O(\lambda_{s+2}(n) \log n)$ expected running time (in an appropriate model of computation).*

Deterministic algorithms. We now sketch a deterministic, divide-and-conquer algorithm, due to Guibas *et al.* [**402**], for computing the face f containing the input point x. The high-level description of the algorithm is quite simple, and is similar to the first randomized algorithm described above. That is, we partition Γ into two subsets Γ_1, Γ_2, of roughly $n/2$ arcs each, recursively compute the faces, f_1, f_2, of $\mathcal{A}(\Gamma_1)$, $\mathcal{A}(\Gamma_2)$, respectively, that contain x, and then "merge" these two faces to obtain the desired face f. Note that f is the connected component of $f_1 \cap f_2$ containing x. However, as already noted, it is generally too expensive to compute this intersection in its entirety, and then select the component containing x, because the boundaries of f_1 and f_2 might have $\Omega(n^2)$ points of intersection. We therefore need a more careful way of performing the merge.

The setup for the merge step is as follows. We are given two connected (but not necessarily simply connected) regions in the plane, which we denote, respectively, as the red region R and the blue region B. Both regions contain the point x in their interior, and our task is to calculate the connected component f of $R \cap B$ that contains x. The boundaries of R and B are composed of (maximal connected x-monotone) portions of the given curves in Γ, each of which will be denoted in what follows as an *arc segment* (or "subarc").

For technical reasons, we extend this task as follows. Let P be the set of points containing x and all endpoints of the arcs of Γ that lie on the boundary of either R or B. Clearly, $|P| \leq 2n + 1$. For each $w \in P$, let f_w denote the connected component of $R \cap B$ that contains w (these components are not necessarily distinct, and some may be empty). Our task is now to calculate all these components (but produce each distinct component just once, even if it contains several points of P). We refer to this task as the *red–blue merge*. We call the resulting components f_w

purple regions, as each of them is covered by both the red and the blue regions. An illustration of this merge is shown below in Figure 8.

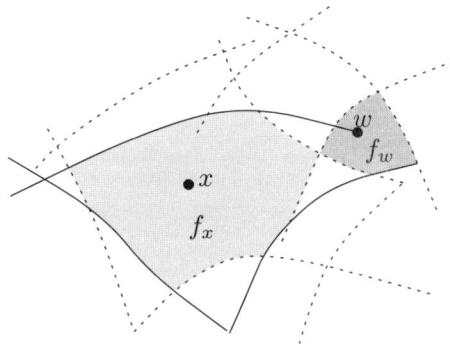

FIGURE 8. The red–blue merge; the solid arcs are the blue arcs, and the dashed arcs are red.

The algorithm relies heavily on the following technical result, called the *combination lemma*, which is interesting in its own right. We first introduce a few notations. Let $R_1, \ldots, R_m$ be a collection of distinct faces in an arrangement of a set Γ_r of "red" Jordan arcs, and let $B_1, \ldots, B_n$ be a similar collection of faces in an arrangement of a set Γ_b of "blue" Jordan arcs (where, again, each pair of arcs from $\Gamma_r \cup \Gamma_b$ are assumed to intersect in at most s points). Let $P = \{p_1, \ldots, p_k\}$ be a collection of points, so that each $p_i \in P$ belongs to one red face R_{m_i} and to one blue face B_{n_i}. Let E_i be the connected component of $R_{m_i} \cap B_{n_i}$ containing p_i (i.e., E_i is the "purple" face of the combined arrangement $\mathcal{A}(\Gamma_r \cup \Gamma_b)$ containing p_i). Then we have the following result.

LEMMA 6.4 (Combination Lemma, [**402**]). *The total combinatorial complexity of all the regions E_i is at most $O(r + b + k)$, where r and b are the total number of arc segments composing the boundaries of the red faces and of the blue faces, respectively.*

REMARK 6.5. *A stronger combination lemma was obtained by Edelsbrunner et al.* [**297**] *for the case of line segments. They proved that the total complexity of the purple regions E_i is bounded by $r + b + O(u + v + k)$, where u (resp., v) is the total number of connected components of the boundaries of the red (resp., blue) faces. Har-Peled* [**425**] *generalized the combination lemma to the overlay of more than two arrangements.*

The combination lemma implies that the complexity of all the "purple" regions in the overlay of the faces f_1 and f_2 is $O(r + b) = O(\lambda_{s+2}(n))$. Exploiting this bound, Guibas *et al.* [**402**] describe a somewhat involved sweep-line algorithm, which sweeps over f_1 and f_2, and computes the purple regions in time $O(\lambda_{s+2}(n) \log n)$. The main idea behind this sweep is that it is performed separately, but simultaneously, over the red, blue, and purple arrangements, in a manner that processes only a small number of red-blue intersections. See [**402, 681**] for more details. Hence, the overall running time of the algorithm is $O(\lambda_{s+2}(n) \log^2 n)$.

Amato et al. [**81**] have succeeded in derandomizing the algorithm by Chazelle et al. [**219**], described above, for a set of segments. The worst-case running time of their algorithm is $O(n\alpha^2(n) \log n)$. Hence, we can conclude the following.

THEOREM 6.6 ([**81, 402**]). *Given a collection Γ of n Jordan arcs, each pair of which intersect in at most s points, and a point x not lying on any arc, the face of $\mathcal{A}(\Gamma)$ containing x can be computed by a deterministic algorithm in time $O(\lambda_{s+2}(n) \log^2 n)$, in an appropriate model of computation. The running time improves to $O(\lambda_s(n) \log^2 n)$ for collections of Jordan curves (closed or unbounded), and to $O(n\alpha^2(n) \log n)$ for collections of line segments.*

We conclude this subsection by mentioning two open problems.

PROBLEM 6.7.
  (i) *Given a set Γ of n segments and a point x, can the face in $\mathcal{A}(\Gamma)$ containing x be computed in time $O(n \log n)$? Or even in time $O(n \log h)$, where h is the number of edges in the face?*
  (ii) *Given a set Γ of n Jordan arcs, each pair of which intersect in at most s points, and a point x, can the face in $\mathcal{A}(\Gamma)$ containing x be computed in deterministic time $O(\lambda_{s+2}(n) \log n)$?*

6.3. Zones. The *zone* of a curve γ_0 in the arrangement $\mathcal{A}(\Gamma)$ of a collection Γ of n Jordan arcs is the set of all faces of $\mathcal{A}(\Gamma)$ that γ_0 intersects. The complexity of the zone is the sum of the complexities of all the faces in the zone.

Zones were initially studied for arrangements of lines and hyperplanes [**288, 301, 303**], but they are also easy to analyze in the context of general arcs. The following theorem demonstrates a close relationship between zones and faces in an arrangement.

THEOREM 6.8 ([**295**]). *The complexity of the zone of a curve γ_0 in an arrangement $\mathcal{A}(\Gamma)$ of n Jordan arcs, each pair of which intersect in at most s points, is $O(\lambda_{s+2}(n))$, assuming that γ_0 intersects every arc of Γ in at most some constant number (possibly larger than s) of points.*

PROOF. Split every arc $\gamma \in \Gamma$ into two subarcs at each intersection point of γ and γ_0, and leave sufficiently small gaps between these pieces. In this manner all faces in the zone of γ_0 are merged into one face, at the cost of increasing the number of arcs from n to $O(n)$. Now apply Theorem 6.2 to conclude the proof. □

If Γ is a set of n lines and γ_0 is also a line, then, after splitting each line of Γ at its intersection point with γ_0, we obtain a collection of $2n$ rays, and therefore the complexity of the zone, which now becomes one of the unbounded faces of the new arrangement, is $O(n)$. In fact, in this case one can show that the edges of the zone form a $DS(4n, 2)$ sequence, thereby obtaining an upper bound of $8n - 1$ on the complexity of the zone. Applying a more careful analysis, Bern et al. [**148**] established the following sharp bound.

THEOREM 6.9 ([**148**]). *The complexity of the zone of a line in an arrangement of n lines is at most $5.5n$, and this bound is tight, up to an additive constant term, in the worst case.*

See [64, 148, 228, 295] for other results and applications of zones of arcs.

An immediate consequence of Theorem 6.8 is an efficient algorithm for computing the arrangement $\mathcal{A}(\Gamma)$. Suppose we add the arcs of Γ one by one, and maintain the arrangement of the arcs added so far. Let Γ_i be the set of arcs added in the first i stages, and let γ_{i+1} be the next arc to be added. Then in the $(i+1)$st stage one has to update only those faces of $\mathcal{A}(\Gamma_i)$ which lie in the zone of γ_{i+1}, and this can easily be done in time proportional to the complexity of the zone; see Edelsbrunner et al. [295] for details. By Theorem 6.8, the total running time of the algorithm is $O(n\lambda_{s+2}(n))$. This can be slightly strengthened in the case of lines, using Theorem 6.9, resulting in an algorithm for computing the arrangement of n lines in optimal $O(n^2)$ time. If the arcs of Γ are added in a *random* order, then the expected running time of the above algorithm is $O(n\log n + k)$, where k is the number of vertices in $\mathcal{A}(\Gamma)$ [219, 246, 564], which is at most quadratic in n. The latter time bound is worst-case optimal.

Theorem 6.8 can also be used to obtain an upper bound on the complexity of any m faces of $\mathcal{A}(\Gamma)$. Specifically, let $\{f_1, \ldots, f_m\}$ be a subset of m distinct faces in $\mathcal{A}(\Gamma)$, and let n_f denote the number of vertices in a face f of $\mathcal{A}(\Gamma)$. Then, using the Cauchy–Schwarz inequality,

$$\sum_{i=1}^m n_{f_i} \leq m^{1/2}\left(\sum_i n_{f_i}^2\right)^{1/2} \leq m^{1/2}\left(\sum_{f\in\mathcal{A}(\Gamma)} n_f^2\right)^{1/2}$$

$$= O\left[m^{1/2}\left(\sum_{f\in\mathcal{A}(\Gamma)} n_f \lambda_{s+2}(k_f)\right)^{1/2}\right]$$

$$= O\left[m^{1/2}\left(\frac{\lambda_{s+2}(n)}{n}\right)^{1/2}\left(\sum_{f\in\mathcal{A}(\Gamma)} n_f k_f\right)^{1/2}\right],$$

where k_f is the number of arcs in Γ that appear along the boundary of f. It is easily verified that

$$\sum_{f\in\mathcal{A}(\Gamma)} n_f k_f \leq \sum_{\gamma\in\Gamma}\sum_{f\in zone(\gamma,\Gamma\setminus\{\gamma\})} n_f = O(n\lambda_{s+2}(n)).$$

Hence, we obtain the following result.

THEOREM 6.10 ([295, 425]). *Let Γ be a set of n arcs satisfying the conditions stated earlier. The maximum number of edges bounding any m distinct faces of $\mathcal{A}(\Gamma)$ is $O(m^{1/2}\lambda_{s+2}(n))$.*

It should be noted that Theorem 6.10 is weaker than the best bounds known for the complexity of m distinct faces in arrangements of several special types of arcs, such as lines, segments, and circles (see Chapter 4 and [92, 191, 245]), but it applies to arrangements of more general arcs.

6.4. Levels in arrangements. For the convenience of the reader, we repeat here some definitions pertaining to levels in (planar) arrangements. Let Γ be a set of n x-monotone, unbounded Jordan curves, each pair of which intersects in at most s points. The *level* of a point $p \in \mathbb{R}^2$ in $\mathcal{A}(\Gamma)$ is the number of curves of Γ lying strictly below p, and the level of an edge $e \in \mathcal{A}(\Gamma)$ is the common level of all the points lying in the relative interior of e. For a nonnegative integer $k < n$, the

k-*level* (respectively, $(\leq k)$-*level*) of $\mathcal{A}(\Gamma)$ is (the closure of) the union of all edges in $\mathcal{A}(\Gamma)$ whose level is k (respectively, at most k). Note that the graph of the lower envelope E_Γ is the 0-level, so the complexity of the 0-level is at most $\lambda_s(n)$. As reviewed in Chapter 2, there are still significant gaps between the known upper and lower bounds on the complexity of an arbitrary single level, even for arrangements of lines (which, as reviewed earlier, is the dual version of the k-*set* problem for point sets); see [**263, 330, 518, 603, 715**]. However, tight bounds are known for the complexity of $(\leq k)$-levels in arrangements of curves:

THEOREM 6.11 ([**246, 676**]). *Let Γ be a set of n x-monotone unbounded Jordan curves, each pair intersecting in at most s points, and let $0 < k < n$ be an integer. The number of edges in $\mathcal{A}(\Gamma)$ of level at most k is $O(k^2 \lambda_s(\lfloor n/k \rfloor))$, and this bound is tight in the worst case.*

The proof of the theorem is based on the elegant probabilistic analysis technique of Clarkson and Shor [**246**], which has been applied to a variety of other problems as well (see Chapter 2 for some of these applications). An immediate corollary of the above theorem is the following claim (the exact upper bound is in fact $n(k+1)$ [**73, 383**]).

COROLLARY 6.12 ([**73**]). *The number of edges in the $(\leq k)$-level of an arrangement of n lines in the plane is $\Theta(nk)$.*

Corollary 6.12 can be extended to arrangements of hyperplanes in higher dimensions as well, where the number of vertices in the $(\leq k)$-level is $\Theta(n^{\lfloor d/2 \rfloor} k^{\lceil d/2 \rceil})$ [**246**] (see also Chapter 2). Efficient algorithms for computing $(\leq k)$-levels in arrangements are given in [**26, 341, 563**], and reviewed in Chapter 2.

Theorem 6.11 can be extended to a more general setting. Let $\mathcal{K} = \{K_1, \ldots, K_n\}$ be a collection of n regions in $\mathbb{R}^2$ so that the boundaries of any two them intersect in at most s points. Let $f(r)$ denote the expected number of vertices on the boundary of the union of a random subset of r regions of $\mathcal{K}$. For example, if the boundary of each K_i is an x-monotone curve, then $f(r) = O(\lambda_s(r))$ (see Corollary 2.3). Using the same probabilistic technique of [**246**], one can establish the following result.

THEOREM 6.13 ([**676**]). *Let $\mathcal{K} = \{K_1, \ldots, K_n\}$ be a collection of n regions in $\mathbb{R}^2$ so that the boundaries of any two them intersect in at most s points. For any integer $1 \leq k \leq n-2$, the number of intersection points of the boundaries of regions in $\mathcal{K}$ that lie in the interior of at most k regions of $\mathcal{K}$ is $O(k^2 f(\lfloor n/k \rfloor))$. If each ∂K_i is an x-monotone curve, then the number of such vertices is $O(k^2 \lambda_s(\lfloor n/k \rfloor))$.*

CHAPTER 4

Incidences and Their Relatives: From Szemerédi and Trotter to Cutting Lenses

1. Introduction

The problem and its relatives. Let P be a set of m distinct points, and let L be a set of n distinct lines in the plane. Let $I(P, L)$ denote the number of *incidences* between the points of P and the lines of L, i.e.,

$$I(P, L) = |\{(p, \ell) \mid p \in P,\ \ell \in L,\ p \in \ell\}|.$$

How large can $I(P, L)$ be? More precisely, we wish to determine or estimate $\max_{|P|=m, |L|=n} I(P, L)$.

This simplest formulation of the incidence problem, due to Erdős and first settled by Szemerédi and Trotter, has been the starting point of extensive research that has picked up considerable momentum during the past two decades. It is the purpose of this chapter to review the results obtained so far, describe the main techniques used in the analysis of this problem, and discuss many variations and extensions.

The problem can be generalized in many natural directions. One can ask the same question when the set L of lines is replaced by a set C of n curves of some other simple shape; the two cases involving respectively unit circles and arbitrary circles are of particular interest—see below.

A related problem involves the same kind of input—a set P of m points and a set C of n curves, but now we assume that no point of P lies on any curve of C. Let $\mathcal{A}(C)$ denote the *arrangement* of the curves of C (see Chapters 2 and 3). The combinatorial complexity of a single *face* is defined as the number of lower-dimensional cells (i.e., vertices and edges) belonging to its boundary. The points of P then mark certain faces in the arrangement $\mathcal{A}(C)$ of the curves (assume for simplicity that there is at most one point of P in each face), and the goal is to establish an upper bound on $K(P, C)$, the combined combinatorial complexity of the marked faces. This problem is often referred to in the literature as the *Many-Faces Problem*; see Section 7 of Chapter 2.

One can extend the above questions to d-dimensional spaces, for $d > 2$. Here we can either continue to consider incidences between points and *curves*, or incidences between points and *surfaces* of any larger dimension k, $1 < k < d$. In the special case when $k = d - 1$, we may also wish to study the natural generalization of the 'many-faces problem' described in the previous paragraph: to estimate the total combinatorial complexity of m marked (d-dimensional) cells in the arrangement of n given surfaces.

All of the above problems have many algorithmic variants. Perhaps the simplest question of this type is *Hopcroft's problem*: Given m points and n lines in the plane, how fast can one determine whether there exists any point that lies on any line? One can consider more general problems, like counting the number of incidences or reporting all of them, doing the same for a collection of curves rather than lines, computing m marked faces in an arrangement of n curves, and so on. See also Section 13.4.

It turned out that two exciting *metric* problems (involving interpoint distances) proposed by Erdős in 1946 [**323**] are strongly related to problems involving incidences.

(1) *Repeated Distances Problem*: Given a set P of n points in the plane, what is the maximum number of pairs that are at a fixed distance, say 1, from each other? To see the connection, let C be the set of unit circles centered at the points of P. Then two points $p, q \in P$ are at distance 1 apart if and only if the circle centered at p passes through q and vice versa. Hence, $I(P, C)$ is twice the number of unit distances determined by P.

(2) *Distinct Distances Problem*: Given a set P of n points in the plane, at least how many distinct distances must there always exist between its point pairs? Later we will show the connection between this problem and the problem of incidences between P and an appropriate set of circles of different radii.

Some other applications of the incidence problem will be reviewed at the end of this chapter. They include the analysis of the maximum number of isosceles triangles, or triangles with a fixed area or perimeter, whose vertices belong to a planar point set; estimating the maximum number of mutually congruent simplices determined by a point set in higher dimensions; etc.

Historical perspective and overview. The first derivation of the tight upper bound

$$I(P, L) = \Theta(m^{2/3}n^{2/3} + m + n)$$

was given by Szemerédi and Trotter in their 1983 seminal paper [**712**]. They proved Erdős's conjecture, who found the matching lower bound (which was rediscovered many years later by Edelsbrunner and Welzl [**310**]). A somewhat different and simpler lower bound construction was exhibited by Elekes [**318**] (see Section 2).

The original proof of Szemerédi and Trotter is rather involved, and yields a rather astronomical constant of proportionality hidden in the O-notation. A considerably simpler proof was found by Clarkson *et al.* [**245**] in 1990, using extremal graph theory combined with a geometric partitioning technique based on random sampling (see Section 3). Their paper contains many extensions and generalizations of the Szemerédi–Trotter theorem. Many further extensions can be found in subsequent papers by Edelsbrunner *et al.* [**297, 298**], by Agarwal and Aronov [**17**], by Aronov *et al.* [**92**], and by Pach and Sharir [**597**].

The next breakthrough occurred in 1997. In a surprising paper, Székely [**709**] gave an embarrassingly short proof (which we will review in Section 4) of the upper bound on $I(P, L)$ using a simple lower bound of Ajtai *et al.* [**61**] and of Leighton [**509**] on the *crossing number* of a graph G, i.e., the minimum number of edge crossings in the best drawing of G in the plane, where the edges are represented by Jordan arcs. In the literature this result is often referred to as the 'Crossing Lemma.' See also later in the book, most notably in Chapter 5. Székely's method can easily be extended to several other variants of the problem, but appears to be less general than the previous technique of Clarkson *et al.* [**245**].

Székely's paper has triggered an intensive re-examination of the problem. In particular, several attempts were made to improve the existing upper bound on the number of incidences between m points and n circles of arbitrary radii in the plane [**598**]. This was the simplest instance where Székely's proof technique failed. By combining Székely's method with a seemingly unrelated technique of Tamaki and Tokuyama [**715**] for cutting circles into 'pseudo-segments', Aronov and Sharir [**107**] managed to obtain an improved bound for this variant of the problem. Their work has then been followed by Agarwal *et al.* [**23**], who studied the complexity of many faces in arrangements of circles and pseudo-segments, and by Agarwal *et al.* [**41**], who extended this result to arrangements of pseudo-circles (see Section 5). A combinatorial result of Marcus and Tardos [**524**] (see below) has led to slight improvements and generalizations of these results. Aronov *et al.* [**97**] generalized the problem to higher dimensions, while Sharir and Welzl [**687**] studied incidences between points and *lines* in three dimensions (see Section 6).

The related problems involving distances in a point set have had mixed success in recent studies. As for the Repeated Distances problem in the plane, the best known upper bound on the number of times the same distance can occur among n points is $O(n^{4/3})$, which was obtained about 25 years ago by Spencer *et al.* [**699**] (with progressively simpler proofs in [**245, 709**]). This is far from the best known lower bound of Erdős, which is only slightly super-linear (see [**585**]). The best known upper bound for the 3-dimensional case, due to Clarkson *et al.* [**245**], is roughly $O(n^{3/2})$, while the corresponding lower bound of Erdős is $\Omega(n^{4/3}\log\log n)$ (see [**584**]). Many variants of the problem have been studied; see, e.g., [**329**].

While the Repeated Distances problem has been "stuck" for quite some time, more progress has been made on the companion problem of Distinct Distances. In the planar case, L. Moser [**555**], Chung [**239**], and Chung *et al.* [**240**] proved that the number of distinct distances determined by n points in the plane is $\Omega(n^{2/3})$, $\Omega(n^{5/7})$, and $\Omega(n^{4/5}/\mathrm{polylog}(n))$, respectively. Székely [**709**] managed to get rid of the polylogarithmic factor, while Solymosi and Tóth [**695**] improved this bound to $\Omega(n^{6/7})$. This was a real breakthrough. Their analysis was subsequently refined by Tardos [**718**] and then by Katz and Tardos [**468**], who obtained the current record of $\Omega(n^{(48-14e)/(55-16e)-\varepsilon})$, for any $\varepsilon > 0$, which is $\Omega(n^{0.8641})$. In spite of this steady improvement, there is still a considerable gap to the best known upper bound of $O(n/\sqrt{\log n})$, due to Erdős [**323**] (see Section 7). In three dimensions, a recent result of Aronov *et al.* [**99**] yields a lower bound of $\Omega(n^{77/141-\varepsilon})$, for any $\varepsilon > 0$, which is $\Omega(n^{0.546})$. This is still far from the best known upper bound of $O(n^{2/3})$. A better lower bound of $\Omega(n^{0.5794})$ in a special case (involving "homogeneous" point sets) has recently been given by Solymosi and Vu [**696**]. Their analysis also applies to higher-dimensional homogeneous point sets, and yields the bound $\Omega(n^{2/d-1/d^2})$. In a subsequent paper [**697**], the same authors tackled the general case and obtained a lower bound $\Omega(n^{2/d-2/d(d+2)})$.

For other surveys on related subjects, consult [**174**], Chapter 4 of [**536**], [**584**], and [**585**]. See also [**600, 601**] for earlier versions of this chapter.

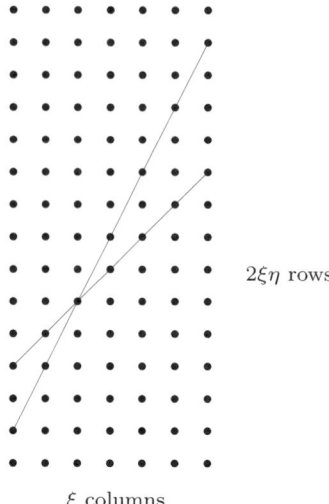

FIGURE 1. Elekes's construction.

2. Lower Bounds

We describe a simple construction due to Elekes [**318**] of a set P of m points and a set L of n lines, so that $I(P, L) = \Omega(m^{2/3}n^{2/3} + m + n)$. We fix two integer parameters ξ, η. We take P to be the set of all lattice points in $\{1, 2, \ldots, \xi\} \times \{1, 2, \ldots, 2\xi\eta\}$. The set L consists of all lines of the form $y = ax + b$, where a is an integer in the range $1, \ldots, \eta$, and b is an integer in the range $1, \ldots, \xi\eta$. Clearly, each line in L passes through exactly ξ points of P. See Figure 1.

We have $m = |P| = 2\xi^2\eta$, $n = |L| = \xi\eta^2$, and

$$I(P, L) = \xi|L| = \xi^2\eta^2 = \Omega(m^{2/3}n^{2/3}).$$

Given any sizes m, n so that $n^{1/2} \leq m \leq n^2$, we can find ξ, η that give rise to sets P, L whose sizes are within a constant factor of m and n, respectively. If m lies outside this range then $m^{2/3}n^{2/3}$ is dominated by $m + n$, and then it is trivial to construct sets P, L of respective sizes m, n so that $I(P, L) = \Omega(m + n)$. We have thus shown that

$$I(P, L) = \Omega(m^{2/3}n^{2/3} + m + n).$$

We note that this construction is easy to generalize to incidences involving other curves. For example, we can take P to be the grid $\{1, 2, \ldots, \xi\} \times \{1, 2, \ldots, 3\xi^2\eta\}$, and define C to be the set of all parabolas of the form $y = ax^2 + bx + c$, where $a \in \{1, \ldots, \eta\}$, $b \in \{1, \ldots, \xi\eta\}$, $c \in \{1, \ldots, \xi^2\eta\}$. Now we have $m = |P| = 3\xi^3\eta$, $n = |C| = \xi^3\eta^3$, and

$$I(P, C) = \xi|C| = \xi^4\eta^3 = \Omega(m^{1/2}n^{5/6}).$$

Note that in the construction we have $m = O(n)$. When m is larger, we use the preceding construction for points and lines, which can be easily transformed into a construction for points and parabolas, to obtain the overall lower bound for points

and parabolas:
$$I(P,C) = \begin{cases} \Omega(m^{2/3}n^{2/3} + m), & \text{if } m \geq n, \\ \Omega(m^{1/2}n^{5/6} + n), & \text{if } m \leq n. \end{cases}$$

These constructions can be generalized to incidences involving graphs of polynomials of higher degrees.

From incidences to many faces. Let P be a set of m points and L a set of n lines in the plane, and put $I = I(P, L)$. Fix a sufficiently small parameter $\varepsilon > 0$, and replace each line $\ell \in L$ by two lines ℓ^+, ℓ^-, obtained by translating ℓ parallel to itself by distance ε in the two possible directions. We obtain a new collection L' of $2n$ lines. If ε is sufficiently small then each point $p \in P$ that is incident to $k \geq 2$ lines of L becomes a point that lies in a small face of $\mathcal{A}(L')$ that has $2k$ edges; note also that the circle of radius ε centered at p is tangent to all these edges. Moreover, these faces are distinct for different points p, when ε is sufficiently small.

We have thus shown that $K(P, L') \geq 2I(P, L) - 2m$ (where the last term accounts for points that lie on just one line of L). In particular, in view of the preceding construction, we have, for $|P| = m$, $|L| = n$,
$$K(P, L) = \Omega(m^{2/3}n^{2/3} + m + n).$$

An interesting consequence of this construction is as follows. Take $m = n$ and sets P, L that satisfy $I(P, L) = \Theta(n^{4/3})$. Let C be the collection of the $2n$ lines of L' and of the n circles of radius ε centered at the points of P. By applying an inversion,[1] we can turn all the curves in C into circles. We thus obtain a set C' of $3n$ circles with $\Theta(n^{4/3})$ tangent pairs. If we replace each of the circles centered at the points of P by circles with a slightly larger radius, we obtain a collection of $3n$ circles with $\Theta(n^{4/3})$ *empty lenses*, namely faces of degree 2 in their arrangement. Empty lenses play an important role in the analysis of incidences between points and circles; see Section 5.

Lower bounds for incidences with unit circles. As noted, this problem is equivalent to the problem of Repeated Distances. Erdős [**323**] has shown that, for the vertices of an $n^{1/2} \times n^{1/2}$ grid, there exists a distance that occurs $\Omega(n^{1+c/\log\log n})$ times, for an appropriate absolute constant $c > 0$. The details of this analysis, based on number-theoretic considerations, can be found in the monographs [**536**] and [**585**].

Lower bounds for incidences with arbitrary circles. As we will see later, we are still far from a sharp bound on the number of incidences between points and circles, especially when the number of points is small relative to the number of circles.

By taking sets P of m points and L of n lines with $I(P, L) = \Theta(m^{2/3}n^{2/3} + m + n)$, and by applying inversion to the plane, we obtain a set C of n circles and a set P' of m points with $I(P', C) = \Theta(m^{2/3}n^{2/3} + m + n)$. Hence the maximum number of incidences between m points and n circles is $\Omega(m^{2/3}n^{2/3} + m + n)$. However, we can slightly increase this lower bound, as follows.

Let P be the set of vertices of the $m^{1/2} \times m^{1/2}$ integer lattice. As shown by Erdős [**323**], there are $t = \Theta(m/\sqrt{\log m})$ distinct distances between pairs of points of P. Draw a set C of mt circles, centered at the points of P and having as radii

---

[1] An inversion about, say, the unit circle centered at the origin, maps each point (x, y) to the point $\left(\frac{x}{x^2+y^2}, \frac{y}{x^2+y^2}\right)$. It maps lines to circles passing through the origin.

the t possible inter-point distances. Clearly, the number of incidences $I(P, C)$ is exactly $m(m-1)$. If the bound on $I(P, C)$ were $O(m^{2/3}n^{2/3} + m + n)$, then we would have

$$m(m-1) = I(P, C) = O(m^{2/3}(mt)^{2/3} + mt) = O(m^2/((\log m)^{1/3}),$$

a contradiction. This shows that, under the most optimistic conjecture, the maximum value of $I(P, C)$ should be larger than the corresponding bound for lines by at least some polylogarithmic factor, at least when m is much smaller than n.

3. Upper Bounds for Incidences via the Partition Technique

The approach presented in this section is due to Clarkson *et al.* [**245**]. It predated Székely's method, but it seems to be more flexible, and suitable for generalizations. It can also be used for the refinement of some proofs based on Székely's method.

We exemplify this technique by establishing an upper bound for the number of point-line incidences. Let P be a set of m points and L a set of n lines in the plane. First, we give a weaker bound on $I(P, L)$, as follows. Consider the bipartite graph $H \subseteq P \times L$ whose edges represent all incident pairs (p, ℓ), for $p \in P$, $\ell \in L$. Clearly, H does not contain $K_{2,2}$ as a subgraph. By the Kővári–Sós–Turán theorem in extremal graph theory (see [**585**]), we have

$$(3.13) \qquad I(P, L) = O(mn^{1/2} + n).$$

To improve this bound, we partition the plane into subregions, apply this bound within each subregion separately, and sum up the bounds. We fix a parameter $1 \leq r \leq n$, whose value will be determined shortly, and construct a so-called $(1/r)$-*cutting* of the arrangement $\mathcal{A}(L)$ of the lines of L. This is a decomposition of the plane into $O(r^2)$ vertical trapezoids with pairwise disjoint interiors, such that each trapezoid is crossed by at most n/r lines of L. The existence of such a cutting has been established by Chazelle and Friedman [**221**], following earlier and somewhat weaker results of Clarkson and Shor [**246**]. See Chapter 2, [**536**] and [**681**] for more details.

For each cell τ of the cutting, let P_τ denote the set of points of P that lie in the interior of τ, and let L_τ denote the set of lines that cross τ. Put $m_\tau = |P_\tau|$ and $n_\tau = |L_\tau| \leq n/r$. Using (3.13), we have

$$I(P_\tau, L_\tau) = O(m_\tau n_\tau^{1/2} + n_\tau) = O\left(m_\tau \left(\frac{n}{r}\right)^{1/2} + \frac{n}{r}\right).$$

Summing this over all $O(r^2)$ cells τ, we obtain a total of

$$\sum_\tau I(P_\tau, L_\tau) = O\left(m \left(\frac{n}{r}\right)^{1/2} + nr\right)$$

incidences. This does not quite complete the count, because we also need to consider points that lie on the boundary of the cells of the cutting. A point p that lies in the relative interior of an edge e of the cutting lies on the boundary of at most two cells, and any line that passes through p, with the possible exception of the single line that contains e, crosses both cells. Hence, we may simply assign p to one of these cells, and its incidences (except for at most one) will be counted within the subproblem associated with that cell. Consider then a point p which is a vertex of the cutting, and let ℓ be a line incident to p. Then ℓ either crosses or bounds some adjacent cell τ. Since a line can cross the boundary of a cell in at most two points, we can charge the incidence (p, ℓ) to the pair (ℓ, τ), use the fact that no cell is crossed by more than n/r lines and that each cell is bounded by only a constant number of lines, and conclude that the number of incidences involving vertices of the cutting is at most $O(nr)$.

We have thus shown that

$$I(P, L) = O\left(m\left(\frac{n}{r}\right)^{1/2} + nr\right).$$

Choose $r = m^{2/3}/n^{1/3}$. This choice makes sense provided that $1 \leq r \leq n$. If $r < 1$, then $m < n^{1/2}$ and (3.13) implies that $I(P, L) = O(n)$. Similarly, if $r > n$ then $m > n^2$ and the symmetric version of (3.13) implies that $I(P, L) = O(m)$. If r lies in the desired range, we get $I(P, L) = O(m^{2/3}n^{2/3})$. Putting all these bounds together, we obtain the bound

$$I(P, L) = O(m^{2/3}n^{2/3} + m + n),$$

as required.

REMARK 3.1. *An equivalent statement is that, for a set P of m points in the plane, and for any integer $k \leq m$, the number of lines that contain at least k points of P is at most*

$$O\left(\frac{m^2}{k^3} + \frac{m}{k}\right).$$

Discussion. The cutting-based method is quite powerful, and can be extended in various ways. The crux of the technique is to derive somehow a weaker (but easier) bound on the number of incidences, construct a $(1/r)$-cutting of the set of curves, obtain the corresponding decomposition of the problem into $O(r^2)$ subproblems, apply the weaker bound within each subproblem, and sum up the bounds to obtain the overall bound. The work by Clarkson et al. [**245**] contains many such extensions.

Let us demonstrate this method to obtain an upper bound for the number of incidences between a set P of m points and a set C of n arbitrary circles in the plane. Here the forbidden subgraph property is that the incidence graph $H \subseteq P \times C$ does not contain $K_{3,2}$ as a subgraph, and thus (see [**585**])

$$I(P, C) = O(mn^{2/3} + n).$$

We construct a $(1/r)$-cutting for C, apply this weak bound within each cell τ of the cutting, and handle incidences that occur on the cell boundaries exactly as above, to obtain

$$I(P, C) = \sum_{\tau} I(P_\tau, C_\tau) = O\left(m\left(\frac{n}{r}\right)^{2/3} + nr\right).$$

We choose $r = m^{3/5}/n^{1/5}$, if $n^{1/3} \leq m \leq n^2$, and use the bound $O(m+n)$ otherwise, to obtain

$$I(P, C) = O(m^{3/5}n^{4/5} + m + n).$$

However, as we shall see later, in Section 5, this bound can be considerably improved.

The case of a set C of n unit circles is handled similarly, observing that in this case the intersection graph H does not contain $K_{2,3}$. This yields the same upper bound $I(P,C) = O(mn^{1/2} + n)$, as in (3.13). The analysis then continues exactly as in the case of lines, and yields the bound

$$I(P,C) = O(m^{2/3}n^{2/3} + m + n).$$

We can apply this bound to the Repeated Distances problem, recalling that the number of pairs of points in an n-element set of points in the plane that lie at distance exactly 1 from each other, is half the number of incidences between the points and the unit circles centered at them. Substituting $m = n$ in the above bound, we thus obtain that the number of repeated distances is at most $O(n^{4/3})$. This bound is far from the best known lower bound, mentioned in Section 2, and no improvement (except for simpler proofs and drastically smaller constants of proportionality) has been obtained since its original derivation in [**699**] in 1984.

As a matter of fact, this approach can be extended to any collection C of curves that have "d degrees of freedom", in the sense that any d points in the plane determine at most $t = O(1)$ curves from the family that pass through all of them, and any pair of curves intersect in only $O(1)$ points. The incidence graph does not contain $K_{d,t+1}$ as a subgraph, which implies that

$$I(P,C) = O(mn^{1-1/d} + n).$$

Combining this bound with a cutting-based decomposition (which works if the curves have constant description complexity) yields the bound

$$I(P,C) = O(m^{d/(2d-1)}n^{(2d-2)/(2d-1)} + m + n).$$

Note that this bound extrapolates the previous bounds for the cases of lines ($d = 2$), unit circles ($d = 2$), and arbitrary circles ($d = 3$). See [**598**] for a slight generalization of this result, using Székely's method, outlined in the following section.

4. Incidences via Crossing Numbers—Székely's Method

A graph G is said to be *drawn* in the plane if its vertices are mapped to distinct points in the plane, and each of its edges is represented by a Jordan arc connecting the corresponding pair of points. It is assumed that no edge passes through any vertex other than its endpoints, and that when two edges meet at a common interior point, they properly *cross* each other there, i.e., each curve passes from one side of the other curve to the other side. Such a point is called a *crossing*. In the literature, a graph drawn in the plane with the above properties is often called a *topological graph*. If, in addition, the edges are represented by straight-line segments, then the drawing is said to be a *geometric graph*. See Chapter 5 for a detailed exposition of this topic.

As we have indicated before, Székely discovered that the analysis outlined in the previous section can be substantially simplified, applying the following so-called Crossing Lemma for graphs drawn in the plane.

LEMMA 4.1 (Crossing Lemma; Leighton [**509**], Ajtai *et al.* [**61**]). *Let G be a simple graph drawn in the plane with V vertices and E edges. If $E > 4V$ then there are $\Omega(E^3/V^2)$ crossing pairs of edges.*

To establish the lemma, denote by $\mathrm{cr}(G)$ the minimum number of crossing pairs of edges in any "legal" drawing of G. Since G contains too many edges, it is not planar, and therefore $\mathrm{cr}(G) \geq 1$. In fact, using Euler's formula, a simple counting argument shows that $\mathrm{cr}(G) \geq E - 3V + 6 > E - 3V$. We next apply this inequality

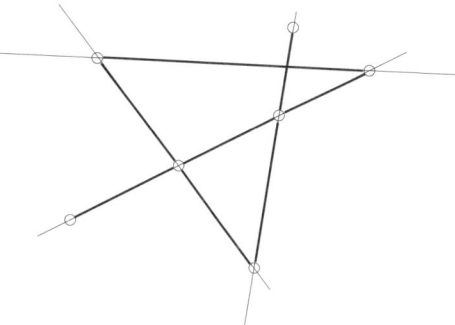

FIGURE 2. Székely's graph for points and lines in the plane.

to a random sample G' of G, which is an induced subgraph obtained by choosing each vertex of G independently with some probability p. By applying expectations, we obtain $\mathbf{E}[\mathrm{cr}(G')] \geq \mathbf{E}[E'] - 3\mathbf{E}[V']$, where E', V' are the numbers of edges and vertices in G', respectively. This can be rewritten as $\mathrm{cr}(G)p^4 \geq Ep^2 - 3Vp$, and choosing $p = 4V/E$ completes the proof of Lemma 4.1.

We remark that the constant of proportionality in the asserted bound, as yielded by the preceding proof, is $1/64$, but it has been improved by Pach and Tóth [**610**]. They proved that $\mathrm{cr}(G) \geq E^3/(33.75V^2)$ whenever $E \geq 7.5V$. In fact, the slightly weaker inequality $\mathrm{cr}(G) \geq E^3/(33.75V^2) - 0.9V$ holds without any extra assumption. We also note that it is crucial that the graph G be *simple* (i.e., any two vertices be connected by at most one edge), for otherwise no crossing can be guaranteed, regardless of how large E is.

Let P be a set of m points and L a set of n lines in the plane. We associate with P and L the following plane drawing of a graph G. The vertices of (this drawing of) G are the points of P. For each line $\ell \in L$, we connect each pair of points of $P \cap \ell$ that are consecutive along ℓ by an edge of G, drawn as the straight segment between these points (which is contained in ℓ). See Figure 2 for an illustration. Clearly, G is a simple graph, and, assuming that each line of L contains at least one point of P, we have $V = m$ and $E = I(P,L) - n$ (the number of edges along a line is smaller by 1 than the number of incidences with that line). Hence, either $E < 4V$, and then $I(P,L) < 4m + n$, or $\mathrm{cr}(G) \geq E^3/(64V^2) = (I(P,L) - n)^3/(64m^2)$. However, we have, trivially, $\mathrm{cr}(G) \leq \binom{n}{2}$, implying that $I(P,L) \leq 32^{1/3}m^{2/3}n^{2/3} + n \leq 3.18m^{2/3}n^{2/3} + n$.

Extensions: Many faces and unit circles. The simple idea behind Székely's proof is quite powerful, and can be applied to many variants of the problem, as long as the corresponding graph G is simple, or, alternatively, has a bounded edge multiplicity. For example, consider the case of incidences between a set P of m points and a set C of n unit circles. Draw the graph G exactly as in the case of lines, but only along circles that contain more than two points of P, to avoid loops and multiple edges along the same circle. We have $V = m$ and $E \geq I(P,C) - 2n$. In this case, G need not be simple, but the maximum edge multiplicity is at most two; see Figure 3. Hence, by deleting at most half of the edges of G we make it into a

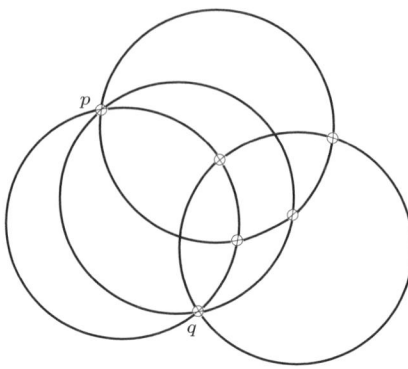

FIGURE 3. Székely's graph for points and unit circles in the plane: The maximum edge multiplicity is two—see the edges connecting p and q.

simple graph. Moreover, $\operatorname{cr}(G) \leq n(n-1)$, so we get $I(P,C) = O(m^{2/3}n^{2/3}+m+n)$, again with a rather small constant of proportionality.

We can also apply this technique to obtain an upper bound on the complexity of many faces in an arrangement of lines. Let P be a set of m points and L a set of n lines in the plane, so that no point lies on any line and each point lies in a distinct face of $\mathcal{A}(L)$. The graph G is now constructed in the following slightly different manner. Its vertices are the points of P. For each $\ell \in L$, we consider all faces of $\mathcal{A}(L)$ that are marked by points of P, are bounded by ℓ and lie on a fixed side of ℓ. For each pair f_1, f_2 of such faces that are consecutive along ℓ (the portion of ℓ between ∂f_1 and ∂f_2 does not meet any other marked face on the same side), we connect the corresponding marking points p_1, p_2 by an edge, and draw it as a polygonal path $p_1 q_1 q_2 p_2$, where $q_1 \in \ell \cap \partial f_1$ and $q_2 \in \ell \cap \partial f_2$. We actually shift the edge slightly away from ℓ so as to avoid its overlapping with edges drawn for faces on the other side of ℓ. The points q_1, q_2 can be chosen in such a way that a pair of edges meet each other only at intersection points of pairs of lines of L. See Figure 4. Here we have $V = m$, $E \geq K(P, L) - 2n$, and $\operatorname{cr}(G) \leq 2n(n-1)$ (each pair of lines can give rise to at most four pairs of crossing edges, near the same intersection point). Again, G is not simple, but the maximum edge multiplicity is at most two, because, if two faces f_1, f_2 are connected along a line ℓ, then ℓ is a common external tangent to both faces. Since f_1 and f_2 are disjoint convex sets, they can have at most two external common tangents. Hence, arguing as above, we obtain $K(P, L) = O(m^{2/3}n^{2/3} + m + n)$. We remark that the same upper bound can also be obtained via the partition technique, as shown by Clarkson et al. [245]. Moreover, in view of the discussion in Section 2, this bound is tight.

However, Székely's technique does not always apply. The simplest example where it fails is when we want to establish an upper bound on the number of incidences between points and circles of arbitrary radii. If we follow the same approach as for equal circles, and construct a graph analogously, we may now create edges with arbitrarily large multiplicities, as is illustrated in Figure 5. We will tackle this problem in the next section.

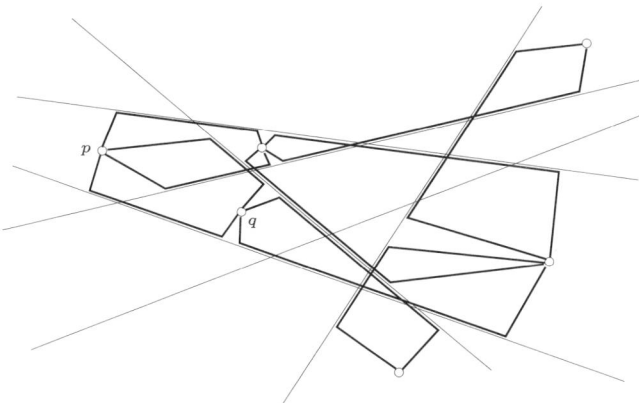

FIGURE 4. Székely's graph for face-marking points and lines in the plane. The maximum edge multiplicity is two—see, e.g., the edges connecting p and q.

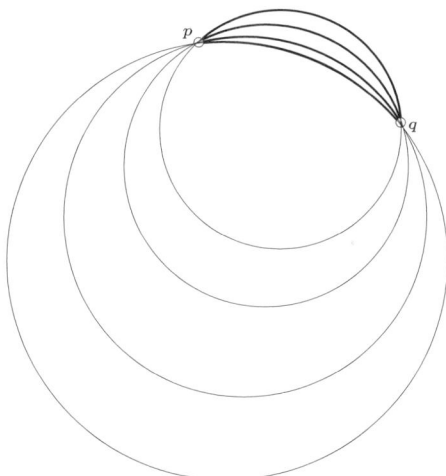

FIGURE 5. Székely's graph need not be simple for points and arbitrary circles in the plane.

Another case where the technique fails is when we wish to bound the total complexity of many faces in an arrangement of line *segments*. If we try to construct the graph in the same way as we did for full lines, the faces may not be convex any more, and we can create edges of high multiplicity; see Figure 6.

5. Improvements by Cutting into Pseudo-segments

Consider the case of incidences between points and circles of arbitrary radii. One way to overcome the technical problem in applying Székely's technique in this case is to cut the given circles into arcs so that any two of them intersect at most once. We refer to such a collection of arcs as a collection of *pseudo-segments*.

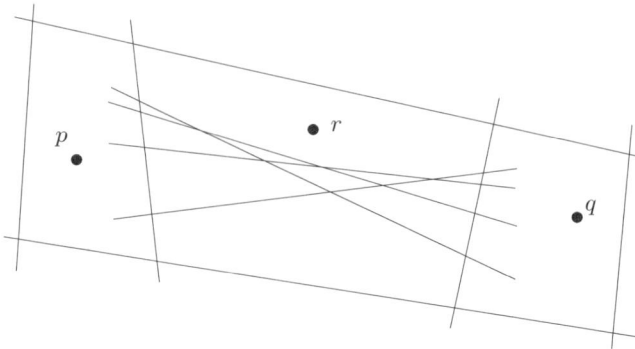

FIGURE 6. Székely's graph need not be simple for marked faces in an arrangement of segments in the plane: An arbitrarily large number of segments bound all three faces marked by the points p, q, r, so the edges (p, r) and (r, q) in Székely's graph have arbitrarily large multiplicity.

The first step in this direction has been taken by Tamaki and Tokuyama [**715**], who have shown that any collection C of n *pseudo-circles*, namely, closed Jordan curves, each pair of which intersect at most twice, can be cut into $O(n^{5/3})$ arcs that form a family of pseudo-segments. The union of two arcs that belong to distinct pseudo-circles and connect the same pair of points is called a *lens*. Let $\chi(C)$ denote the minimum number of points that can be removed from the curves of C, so that any two members of the resulting family of arcs have at most one point in common. Clearly, every lens must contain at least one of these cutting points, so Tamaki and Tokuyama's problem asks in fact for an upper bound on the number of points needed to "stab" all lenses. Equivalently, this problem can be reformulated, as follows.

Consider a hypergraph H whose vertex set consists of the edges of the arrangement $\mathcal{A}(C)$, i.e., the arcs between two consecutive crossings. Assign to each lens a *hyperedge* consisting of all arcs that belong to the lens. We are interested in finding the *transversal number* (or the size of the smallest "hitting set") of H, i.e., the smallest number of vertices of H that can be picked with the property that every hyperedge contains at least one of them. Based on Lovász' analysis [**519**] (see also [**585**]) of the greedy algorithm for bounding the transversal number from above (i.e., for constructing a hitting set), this quantity is not much bigger than the size of the largest *matching* in H, i.e., the maximum number of pairwise disjoint hyperedges. This is the same as the largest number of pairwise non-overlapping lenses, that is, the largest number of lenses, no two of which share a common edge of the arrangement $\mathcal{A}(C)$ (see Figure 7). Viewing such a family as a graph G, whose edges connect pairs of curves that form a lens in the family, Tamaki and Tokuyama proved that G does not contain $K_{3,3}$ as a subgraph, and this leads to the asserted bound on the number of cuts.

In order to establish an upper bound on the number of incidences between a set of m points P and a set of n circles (or pseudo-circles) C, let us construct a modified version G' of Székely's graph: its vertices are the points of P, and its edges

5. IMPROVEMENTS BY CUTTING INTO PSEUDO-SEGMENTS

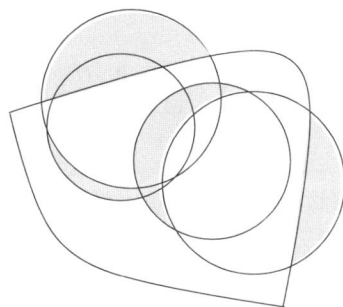

FIGURE 7. The boundaries of the shaded regions are nonoverlapping lenses in an arrangement of pseudo-circles. (Observe that the *regions* bounded by nonoverlapping lenses can overlap, as is illustrated here.)

connect adjacent pairs of points along the new pseudo-segment arcs. That is, we do not connect a pair of points that are adjacent along an original curve, if the arc that connects them has been cut by some point of the hitting set. Moreover, as in the original analysis of Székely, we do not connect points along pseudo-circles that are incident to only one or two points of P, to avoid loops and trivial multiplicities.

Clearly, the graph G' is simple, and the number E' of its edges is at least $I(P, C) - \chi(C) - 2n$. The crossing number of G' is, as before, at most the number of crossings between the original curves in C, which is at most $n(n-1)$. Using the Crossing Lemma (Lemma 4.1), we thus obtain

$$I(P, C) = O(m^{2/3}n^{2/3} + \chi(C) + m + n).$$

Hence, applying the Tamaki–Tokuyama bound on $\chi(C)$, we can conclude that

$$I(P, C) = O(m^{2/3}n^{2/3} + n^{5/3} + m).$$

An interesting property of this bound is that it is tight when $m \geq n^{3/2}$. In this case, the bound becomes $I(P, C) = O(m^{2/3}n^{2/3} + m)$, matching the lower bound for incidences between points and lines, which also serves as a lower bound for the number of incidences between points and circles or parabolas. However, for smaller values of m, the term $O(n^{5/3})$ dominates, and the dependence on m disappears. This can be rectified by combining this bound with a cutting-based problem decomposition, similar to the one used in the preceding section, and we shall do so shortly.

Before proceeding, though, we note that Tamaki and Tokuyama's bound is not tight. The best known lower bound is $\Omega(n^{4/3})$, which follows from the lower bound construction for incidences between points and lines. (That is, we have already seen that this construction can be modified so as to yield a collection C of n circles with $\Theta(n^{4/3})$ empty lenses. Clearly, each such lens requires a separate cut, so $\chi(C) = \Omega(n^{4/3})$.) Recent work by Alon *et al.* [**75**], Aronov and Sharir [**107**], Agarwal *et al.* [**41**], and Marcus and Tardos [**524**] has led to improved bounds, where the currently best upper bound, due to Marcus and Tardos, is $\chi(C) = O(n^{3/2} \log n)$, for families C of *pseudo-parabolas* (graphs of continuous totally defined functions, each pair of which intersect at most twice), or of *pseudo-circles* (closed Jordan curves with the same property).

With the aid of this improved bound on $\chi(C)$, the modification of Székely's method reviewed above yields, for a set C of n circles and a set P of m points,

$$I(P,C) = O(m^{2/3}n^{2/3} + n^{3/2} \log n + m).$$

As already noted, this bound is tight when it is dominated by the first or last terms, which happens when $m = \Omega(n^{5/4} \log^{3/2} n)$. For smaller values of m, we decompose the problem into subproblems, using the following so-called "dual" partitioning technique (see also Section 3). We map each circle $(x-a)^2 + (y-b)^2 = \rho^2$ in C to the dual point $(a, b, \rho^2 - a^2 - b^2)$ in 3-space, and map each point (ξ, η) of P to the dual plane $z = -2\xi x - 2\eta y + (\xi^2 + \eta^2)$. As is easily verified, each incidence between a point of P and a circle of C is mapped to an incidence between the dual plane and point. We now fix a parameter r, and construct a $(1/r)$-cutting of the arrangement of the dual planes, which partitions $\mathbb{R}^3$ into $O(r^3)$ cells (which is a tight bound in the case of planes), each crossed by at most m/r dual planes and containing at most n/r^3 dual points (the latter property, which is not an intrinsic property of the cutting, can be enforced by further partitioning cells that contain more than n/r^3 points). We apply, for each cell τ of the cutting, the preceding bound for the set P_τ of points of P whose dual planes cross τ, and for the set C_τ of circles whose dual points lie in τ. (Some special handling of circles whose dual points lie on boundaries of cells of the cutting is needed, as in Section 3, but we omit the treatment of this special case.) This yields the bound

$$I(P,C) = O(r^3) \cdot O\left(\left(\frac{m}{r}\right)^{2/3}\left(\frac{n}{r^3}\right)^{2/3} + \left(\frac{n}{r^3}\right)^{3/2} \log\left(\frac{n}{r^3}\right) + \frac{m}{r}\right)$$

$$= O\left(m^{2/3}n^{2/3}r^{1/3} + \frac{n^{3/2}}{r^{3/2}} \log\left(\frac{n}{r^3}\right) + mr^2\right).$$

Choose $r = n^{5/11}/m^{4/11} \log^{6/11} n$ in the last bound, assuming that m lies between $n^{1/3}$ and $n^{5/4} \log^{3/2} n$, to obtain

$$I(P,C) = O(m^{2/3}n^{2/3} + m^{6/11}n^{9/11} \log^{2/11}(m^3/n) + m + n).$$

It is not hard to see that this bound also holds for the complementary ranges of m.

6. Incidences in Higher Dimensions

It is natural to extend the study of incidences to instances involving points and curves or surfaces in higher dimensions. The case of incidences between points and (hyper)surfaces (mainly hyperplanes) has been studied earlier. Edelsbrunner *et al.* [**298**] considered incidences between points and planes in three dimensions. It is important to note that, without imposing some restrictions either on the set P of points or on the set H of planes, one can easily obtain $|P| \cdot |H|$ incidences, simply by placing all the points of P on a line, and making all the planes of H pass through that line. Some natural restrictions are to require that no three points be collinear, or that no three planes be collinear, or that the points be vertices of the arrangement $\mathcal{A}(H)$, and so on. Different assumptions lead to different bounds. For example, Agarwal and Aronov [**17**] proved an asymptotically tight bound $\Theta(m^{2/3}n^{d/3} + n^{d-1})$ for the number of incidences between n hyperplanes in d dimensions and $m > n^{d-2}$ vertices of their arrangement (see also [**298**]), as well as for the number of facets bounding m distinct cells in such an arrangement. Edelsbrunner and Sharir [**305**] considered the problem of incidences between points and hyperplanes in four dimensions, under the assumption that all points lie on the upper envelope of the hyperplanes. They obtained the bound $O(m^{2/3}n^{2/3}+m+n)$ for the number of such incidences, and applied the result to obtain the same upper bound on the number of bichromatic minimal distance pairs between a set of m blue points and a set of n red points in three dimensions. Another set of bounds and related results are obtained by Brass and Knauer [**173**], for incidences between m points and n planes in 3-space, and also for incidences in higher dimensions. See also the recent related work of Apfelbaum and Sharir [**85**].

The case of incidences between points and *curves* in higher dimensions has been studied only recently. There are two papers that address this problem. One of them, by Sharir and Welzl [**687**], studies incidences between points and lines in 3-space. The other, by Aronov *et al.* [**97**], is concerned with incidences between points and circles in higher dimensions. Both works were motivated by problems asked by Elekes. We briefly review these results in the following two subsections.

6.1. Points and lines in three dimensions. Let P be a set of m points and L a set of n lines in 3-space. Without making some assumptions on P and L, the problem is trivial, for the following reason. Project P and L onto some generic plane. Incidences between points of P and lines of L are bijectively mapped to incidences between the projected points and lines, so we have $I(P,L) = O(m^{2/3}n^{2/3} + m + n)$. Moreover, this bound is tight, as is shown by the planar lower bound construction. (As a matter of fact, this reduction holds in any dimension $d \geq 3$.)

There are several ways in which the problem can be made interesting. First, suppose that the points of P are *joints* in the arrangement $\mathcal{A}(L)$, namely, each point is incident to at least three *non-coplanar* lines of L. In this case, one has $I(P,L) = O(n^{5/3})$ [**687**]. Note that this bound is independent of m. In fact, it is known that the number of joints is at most $O(n^{112/69} \log^{6/23} n)$, which is $O(n^{1.6232})$ [**357**] (see also [**679**]); the best lower bound, based on lines forming a cube grid, is only $\Omega(n^{3/2})$. More details on joints can be found in Chapter 7.

For general point sets P, one can use a new measure of incidences, which aims to ignore incidences between a point and many incident coplanar lines. Specifically, we define the *plane cover* $\pi_L(p)$ of a point p to be the minimum number of planes that pass through p so that their union contains all lines of L incident to p, and define $I_c(P,L) = \sum_{p \in P} \pi_L(p)$. It is shown in [**687**] that

$$I_c(P,L) = O(m^{4/7}n^{5/7} + m + n),$$

which is smaller than the planar bound of Szemerédi and Trotter.

Another way in which we can make the problem "truly 3-dimensional" is to require that all lines in L be *equally inclined*, meaning that each of them forms a fixed angle (say, 45°) with the z-direction. In this case, every point of P that is incident to at least three lines of L is a joint, but this special case admits better upper bounds. Specifically, we have

$$I(P,L) = O(\min\left\{m^{3/4}n^{1/2}\log m, m^{4/7}n^{5/7}\right\} + m + n).$$

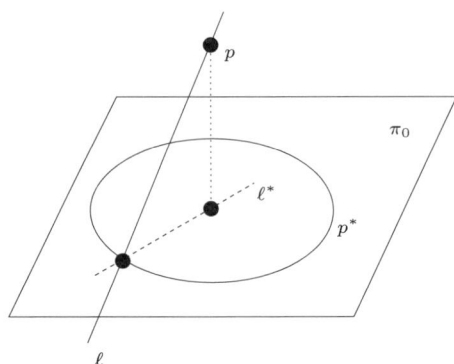

FIGURE 8. Transforming incidences between points and equally inclined lines to tangencies between circles in the plane.

The best known lower bound is
$$I(P, L) = \Omega(m^{2/3} n^{1/2}).$$
Let us briefly sketch the proof of the upper bound $O(m^{3/4} n^{1/2} \log m)$. For each $p \in P$ let C_p denote the (double) cone whose apex is p, whose symmetry axis is the vertical line through p, and whose opening angle is $45°$. Fix some generic horizontal plane π_0, and map each $p \in P$ to the circle $C_p \cap \pi_0$. Each line $\ell \in L$ is mapped to the point $\ell \cap \pi_0$, coupled with the projection ℓ^* of ℓ onto π_0. Note that an incidence between a point $p \in P$ and a line $\ell \in L$ is mapped to the configuration in which the circle dual to p is incident to the point dual to ℓ and the projection of ℓ passes through the center of the circle; see Figure 8. Hence, if a line ℓ is incident to several points $p_1, \ldots, p_k \in P$, then the dual circles $p_1^*, \ldots, p_k^*$ are all tangent to each other at the common point $\ell \cap \pi_0$. Viewing these tangencies as a collection of degenerate lenses, we can bound the overall number of these tangencies, which is equal to $I(P, L)$, by $O(n^{3/2} \log n)$. By a slightly more careful analysis, again based on cutting, one can obtain the bound $O(m^{3/4} n^{1/2} \log m)$.

6.2. Points and circles in three and higher dimensions. Let C be a set of n circles and P a set of m points in 3-space. Unlike in the case of lines, there is no obvious reduction of the problem to a planar one, because the projection of C onto some generic plane yields a collection of ellipses, rather than circles, which can cross each other at four points per pair. However, using a more refined analysis, Aronov et al. [**97**] have obtained the same asymptotic bound of $I(P, C) = O(m^{2/3} n^{2/3} + m^{6/11} n^{9/11} \log^{2/11}(m^3/n) + m + n)$ as in the plane (perhaps with a different constant of proportionality). The same bound applies in any dimension $d \geq 3$.

7. Applications

The problem of bounding the number of incidences between various geometric objects is elegant and fascinating, and it has been mostly studied for its own sake. However, it is closely related to a variety of questions in combinatorial and computational geometry. In this section, we briefly review some of these connections and applications. We elaborate more on some of these applications in Chapter 6.

7. APPLICATIONS

7.1. Algorithmic issues. There are two types of algorithmic problems related to incidences. The first group includes problems where we wish to actually determine the number of incidences between certain objects, e.g., between given sets of points and curves, or we wish to compute a collection of marked faces in an arrangement of curves or surfaces. The second group contains completely different questions whose solution requires tools and techniques developed for the analysis of incidence problems.

In the simplest problem of the first kind, known as Hopcroft's problem, we are given a set P of m points and a set L of n lines in the plane, and we ask whether there exists at least one incidence between P and L. As already noted in Chapter 2, the best running time known for this problem is $O(m^{2/3}n^{2/3} \cdot 2^{O(\log^*(m+n))} + (m+n)\log(m+n))$ [**534**] (see [**338**] for a matching lower bound). Similar running time bounds hold for the problems of counting or reporting all the incidences between P and L. The solutions are based on constructing cuttings of an appropriate size and thereby obtaining a decomposition of the problem into subproblems, each of which can be solved by a more brute-force approach. In other words, the solution can be viewed as an implementation of the cutting-based analysis of the combinatorial bound for $I(P, L)$, as presented in Section 3.

The case of incidences between a set P of m points and a set C of n circles in the plane is more interesting, because the analysis that leads to the current best upper bound on $I(P,C)$ is not easy to implement. In particular, suppose that we have already cut the circles of C into roughly $O(n^{3/2})$ pseudo-segments (an interesting and non-trivial algorithmic task in itself), and we now wish to compute the incidences between these pseudo-segments and the points of P. Székely's technique is non-algorithmic, so instead we would like to apply the cutting-based approach to these pseudo-segments and points. However, this approach, for the case of lines, after decomposing the problem into subproblems, proceeds by duality. Specifically, it maps the points in a subproblem to dual lines, constructs the arrangement of these dual lines, and locates in the arrangement the points dual to the lines in the subproblem. When dealing with the case of pseudo-segments, there is no obvious incidence-preserving duality that maps them to points and maps the points to pseudo-lines. Nevertheless, such a duality has been recently defined by Agarwal and Sharir [**55**] (refining an older and less efficient duality given by Goodman [**380**]), which can be implemented efficiently and thus yields an efficient algorithm for computing $I(P, C)$, whose running time is comparable with the bound on $I(P, C)$ given above. See also Agarwal et al. [**32**] for related results involving range searching with disks. A similar approach can be used to compute many faces in arrangements of pseudo-circles; see [**23**] and [**55**]. Algorithmic aspects of incidence problems have also been studied in higher dimensions; see, e.g., Brass and Knauer [**173**].

The cutting-based approach has by now become a standard tool in the design of efficient geometric algorithms in a variety of applications in range searching, geometric optimization, ray shooting, and many others. It is beyond the scope of this chapter to discuss these applications, and the reader is referred, e.g., to the survey of Agarwal and Erickson [**31**] and to the references therein. Some discussion of these applications can also be found in Chapter 2.

7.2. Distinct distances. The above techniques can be applied to obtain some nontrivial results concerning the Distinct Distances problem of Erdős [**323**] formulated in the Introduction of this chapter: what is the minimum number of distinct distances determined by n points in the plane? As we have indicated after presenting the proof of the Crossing Lemma (Lemma 4.1), Székely's idea can also be applied in several situations where the underlying graph is not *simple*, i.e., two vertices can be connected by more than one edge. However, for the method to work it is important to have an upper bound for the multiplicity of the edges. Székely [**709**] formulated the following natural generalization of Lemma 4.1.

LEMMA 7.1. *Let G be a multigraph drawn in the plane with V vertices, E edges, and with maximal edge-multiplicity M. Then there are $\Omega\left(\frac{E^3}{MV^2}\right) - O(M^2 V)$ crossing pairs of edges.*

Székely applied this statement to the Distinct Distances problem, and improved by a polylogarithmic factor the best previously known lower bound of Chung *et al.* [**240**] on the minimum number of distinct distances determined by n points in the plane. His new bound was $\Omega(n^{4/5})$. However, Solymosi and Tóth [**695**] have realized that, combining Székely's analysis of distinct distances with the Szemerédi-Trotter theorem for the number of incidences between m points and n lines in the plane, this lower bound can be substantially improved. They managed to raise the bound to $\Omega(n^{6/7})$. Later, Katz and Tardos have further improved this result, using the same general approach, but improving upon a key algebraic step of the analysis. In their latest paper [**468**], they combined their methods to prove that the minimum number of distinct distances determined by n points in the plane is $\Omega(n^{(48-14e)/(55-16e)-\varepsilon})$, for any $\varepsilon > 0$, which is $\Omega(n^{0.8641})$. This is the best known result so far. A close inspection of the general Solymosi–Tóth approach shows that, without any additional geometric idea, it can never lead to a lower bound better than $\Omega(n^{8/9})$. (We remind the reader that the upper bound, due to Erdős, is $O(n/\sqrt{\log n})$.)

7.3. Equal-area, equal-perimeter, and isosceles triangles. Let P be a set of n points in the plane. We wish to bound the number of triangles spanned by the points of P that have a *given area*, say 1. To do so, we note that if we fix two points $a, b \in P$, any third point $p \in P$ for which $\text{Area}(\Delta abp) = 1$ lies on the union of two fixed lines parallel to ab. Pairs (a,b) for which such a line ℓ_{ab} contains fewer than $n^{1/3}$ points of P generate at most $O(n^{7/3})$ unit area triangles. For the other pairs, we observe that the number of lines containing more than $n^{1/3}$ points of P is at most $O(n^2/(n^{1/3})^3) = O(n)$, which, as already mentioned, is an immediate consequence of the Szemerédi–Trotter theorem. The number of incidences between these lines and the points of P is at most $O(n^{4/3})$. We next observe that any line ℓ can be equal to one of the two lines ℓ_{ab} for at most n pairs a,b, because, given ℓ and a, there can be at most two points b for which $\ell = \ell_{ab}$. It follows that the lines containing more than $n^{1/3}$ points of P can be associated with at most $O(n \cdot n^{4/3}) = O(n^{7/3})$ unit area triangles. Hence, overall, P determines at most $O(n^{7/3})$ unit area triangles. This simple upper bound has recently been improved by Dumitrescu *et al.* [**281**] to $O(n^{44/19})$. Their proof also relies heavily on bounds for incidences between points and lines, and between points and certain kind of hyperbolas. The best known lower bound is $\Omega(n^2 \log \log n)$ (see [**174**]).

Next, consider the problem of estimating the number of *unit perimeter* triangles determined by P. Here we note that if we fix $a, b \in P$, with $|ab| < 1$, any third point $p \in P$ for which $\text{Perimeter}(\Delta abp) = 1$ lies on an ellipse whose foci are a and b and whose major axis is $1 - |ab|$. Clearly, any two distinct pairs of points of P give rise to distinct ellipses, and the number of unit perimeter triangles determined by P is equal to one third of the number of incidences between these $O(n^2)$ ellipses and the points of P. The set of these ellipses has four degrees of freedom, in the sense of Pach and Sharir [**598**] (see also Section 3), and hence the number of incidences between them and the points of P, and consequently the number of unit perimeter triangles determined by P, is at most

$$O(n^{4/7}(n^2)^{6/7}) = O(n^{16/7}).$$

Here the best known lower bound is very weak—only $\Omega(ne^{c\frac{\log n}{\log\log n}})$ [**174**].

Finally, consider the problem of estimating the number of *isosceles* triangles determined by P. Pach and Tardos [**604**] proved that the number of isosceles triangles induced by triples of an n-element point set in the plane is $O(n^{(11-3\alpha)/(5-\alpha)})$ (where the constant of proportionality depends on α), provided that $0 < \alpha < \frac{10-3e}{24-7e}$. In particular, the number of isosceles triangles is $O(n^{2.136})$. The best known lower bound is $\Omega(n^2 \log n)$ [**174**]. The proof proceeds through two steps, interesting in their own right.

(i) Let P be a set of n distinct points and let C be a set of ℓ distinct circles in the plane, with $m \leq \ell$ distinct centers. Then, for any $0 < \alpha < 1/e$, the number I of incidences between the points in P and the circles of C is

$$O\left(n + \ell + n^{\frac{2}{3}}\ell^{\frac{2}{3}} + n^{\frac{4}{7}}m^{\frac{1+\alpha}{7}}\ell^{\frac{5-\alpha}{7}} + n^{\frac{12+4\alpha}{21+3\alpha}}m^{\frac{3+5\alpha}{21+3\alpha}}\ell^{\frac{15-3\alpha}{21+3\alpha}} + n^{\frac{8+2\alpha}{14+\alpha}}m^{\frac{2+2\alpha}{14+\alpha}}\ell^{\frac{10-2\alpha}{14+\alpha}}\right),$$

where the constant of proportionality depends on α.

(ii) As a corollary, we obtain the following statement. Let P be a set of n distinct points and let C be a set of ℓ distinct circles in the plane such that they have at most n distinct centers. Then, for any $0 < \alpha < 1/e$, the number of incidences between the points in P and the circles in C is

$$O\left(n^{\frac{5+3\alpha}{7+\alpha}}\ell^{\frac{5-\alpha}{7+\alpha}} + n\right).$$

In view of the recent result of Katz and Tardos [**468**], mentioned above, both statements extend to all $0 < \alpha < \frac{10-3e}{24-7e}$, which easily implies the above bound on the number of isosceles triangles.

7.4. Congruent and similar simplices. Bounding the number of incidences between points and circles in higher dimensions can be applied to the following interesting question asked by Erdős and Purdy [**332, 333**] and discussed by Agarwal and Sharir [**54**]. Determine the largest number of simplices congruent to a fixed simplex σ, which can be spanned by an n-element point set $P \subset \mathbb{R}^d$.

Here we consider only the case when $P \subset \mathbb{R}^4$ and $\sigma = abcd$ is a 3-simplex. Fix three points $p, q, r \in P$ such that the triangle pqr is congruent to the face abc of σ. Then any fourth point $v \in P$ for which $pqrv$ is congruent to σ must lie on a circle whose plane is orthogonal to the triangle pqr, whose radius is equal to the height of σ from d, and whose center is at the foot of that height (relative to the triangle pqr). Hence, bounding the number of congruent simplices can be reduced to the problem of bounding the number of incidences between circles and points in 4-space. (The actual reduction is slightly more involved, because the same circle can arise for more than one triangle pqr; see [**54**] for details.) Using the bound of [**97**], mentioned in Section 6, one can deduce that the number of congruent 3-simplices determined by n points in 4-space is $O(n^{20/9+\varepsilon})$, for any $\varepsilon > 0$.

This is just one instance of a collection of bounds obtained in [**54**] for the number of congruent copies of a k-simplex in an n-element point set in $\mathbb{R}^d$, whose review is beyond the scope of this chapter. Recently, Agarwal *et al.* [**15**] have studied the related problem of determining the largest number of simplices similar to a fixed simplex σ, which can be spanned by an n-element point set $P \subset \mathbb{R}^d$, and have obtained initial nontrivial upper bounds.

See also Chapter 6 for a more detailed account of results on repeated patterns in point configurations.

CHAPTER 5

Crossing Numbers of Graphs: Graph Drawing and its Applications

Our ancestors drew their pictures (pictographs or, simply, "*graphs*") on walls of caves, nowadays we use mostly computer screens for this purpose. From the mathematical point of view, there is not much difference: both surfaces are "flat," they are topologically equivalent.

1. Crossings—the Brick Factory Problem

Let G be a finite graph with vertex set $V(G)$ and edge set $E(G)$. By a *drawing* of G, we mean a representation of G in the plane such that each vertex is represented by a distinct point and each edge by a simple (non-selfintersecting) continuous arc connecting the corresponding two points. If it is clear whether we talk about an "abstract" graph G or its planar representation, these points and arcs will also be called vertices and edges, respectively. For simplicity, we assume that in a drawing (a) no edge passes through any vertex other than its endpoints, (b) no two edges are tangent to each other (i.e., if two edges have a common interior point, then at this point they properly cross each other), and (c) no three edges cross at the same point.

Every graph has many different drawings. If G can be drawn in such a way that no two edges cross each other, then G is *planar*. According to an observation of K. Wagner [**737**] and I. Fáry [**350**] that also follows from a theorem of Steinitz [**703**], if G is planar then it has a drawing, in which every edge is represented by a straight-line segment.

It is well known that K_5, the *complete graph* with 5 vertices, and $K_{3,3}$, the *complete bipartite graph* with 3 vertices in each of its classes, are not planar. According to Kuratowski's theorem, a graph is planar if and only if it has no subgraph that can be obtained from K_5 or from $K_{3,3}$ by subdividing some (or all) of its edges with distinct new vertices. In the next section, we give a completely different representation of planar graphs (see Theorem 2.4).

If G is not planar then it cannot be drawn in the plane without crossing. Paul Turán [**726**] raised the following problem: find a drawing of G, for which the number of crossings is minimum. This number is called the *crossing number* of G and is denoted by $\text{CR}(G)$. More precisely, Turán's (still unsolved) original problem was to determine $\text{CR}(K_{n,m})$, for every $n, m \geq 3$. According to an assertion of Zarankiewicz, which was down-graded from theorem to conjecture [**406**], we have

$$\text{CR}(K_{n,m}) = \left\lfloor \frac{m}{2} \right\rfloor \cdot \left\lfloor \frac{m-1}{2} \right\rfloor \cdot \left\lfloor \frac{n}{2} \right\rfloor \cdot \left\lfloor \frac{n-1}{2} \right\rfloor,$$

but we do not even know the limits
$$\lim_{n\to\infty} \frac{\mathrm{CR}(K_{n,n})}{n^4}, \quad \lim_{n\to\infty} \frac{\mathrm{CR}(K_n)}{n^4}$$
(cf. [**490**, **649**]). It is not hard to show, however, that these limits exist and are positive.

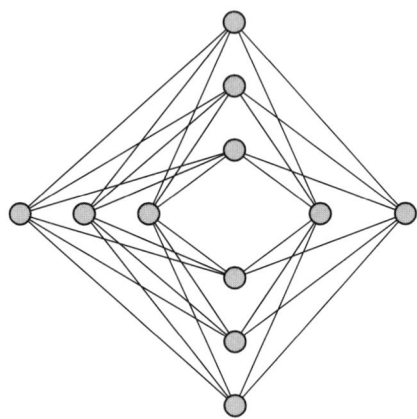

FIGURE 1. $K_{5,6}$ drawn with the minimum number of crossings.

Turán used to refer to the above question as the "brick factory problem", because it occurred to him at a factory yard, where, as forced labor during World War II, he moved wagons filled with bricks from kilns to storage places. According to his recollections, it was not a very tough job, except that they had to push much harder at the crossings. Had this been the only "practical application" of crossing numbers, much fewer people would have tried to estimate $\mathrm{CR}(G)$ during the past quarter of a century. In the early eighties, it turned out that the chip area required for the realization (VLSI layout) of an electrical circuit is closely related to the crossing number of the underlying graph [**509**]. This discovery gave an impetus to research in the subject.

2. Thrackles—Conway's Conjecture

A drawing of a graph is called a *thrackle*, if any two edges that do not share an endpoint cross precisely once, and any two edges that share an endpoint have no other point in common.

It is easy to verify that C_4, a cycle of length 4, cannot be drawn as a thrackle, but any other cycle can [**751**]. If a graph cannot be drawn as a thrackle, then the same is true for all graphs that contain it as a subgraph. Thus, a thrackle does not contain a cycle of length 4, and, according to an old theorem of Erdős in extremal graph theory, the number of its edges cannot exceed $n^{3/2}$, where n denotes the number of its vertices (see [**267**] and Chapter 4).

The following old conjecture states much more.

CONJECTURE 2.1 (J. Conway). *Every thrackle has at most as many edges as vertices.*

2. THRACKLES—CONWAY'S CONJECTURE

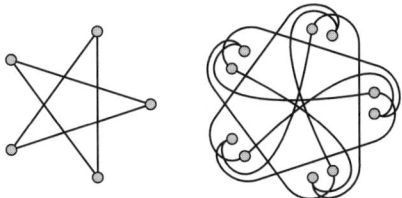

FIGURE 2. Cycles C_5 and C_{10} drawn as thrackles.

The first upper bound on the number of edges of a thrackle, which is linear in n, was found by Lovász et al. [**520**]: Every thrackle has at most twice as many edges as vertices. The constant two has been improved to one and a half.

THEOREM 2.2 (Cairns and Nikolayevsky [**189**]). *Every thrackle has at most one and a half times as many edges as vertices.*

Thrackle and planar graph are, in a certain sense, opposite notions: in the former any two edges intersect, in the latter there is no crossing pair of edges. Yet the next theorem shows how similar these concepts are.

A drawing of a graph is said to be a *generalized thrackle* if every pair of its edges intersect an odd number of times. Here the common endpoint of two edges also counts as a point of intersection. Clearly, every thrackle is a generalized thrackle, but not the other way around. For example, a cycle of length 4 can be drawn as a generalized thrackle, but not as a thrackle.

THEOREM 2.3 (Lovász et al. [**520**]). *A bipartite graph can be drawn as a thrackle if and only if it is planar.*

According to an old observation of Erdős, every graph has a bipartite subgraph which contains at least half of its edges. Clearly, every bipartite planar graph of $n \geq 3$ vertices has at most $2n - 4$ edges. Hence, Theorem 2.3 immediately implies that every thrackle with $n \geq 3$ vertices has at most $2(2n - 4) = 4n - 8$ edges. This bound is weaker than Theorem 2.2, but it is already linear in n.

In a drawing of a graph, a triple of internally vertex-disjoint paths $(P_1(u, v), P_2(u, v), P_3(u, v))$ between the same pair of vertices (u, v) is called a *trifurcation*. (The three paths cannot have any vertices in common, other than u and v, but they can cross at points different from their vertices.) A trifurcation $(P_1(u, v), P_2(u, v), P_3(u, v))$ is said to be a *converter* if the cyclic order of the initial pieces of P_1, P_2, and P_3 around u is opposite to the cyclic order of their final pieces around v.

THEOREM 2.4 (Lovász et al. [**520**]). *A graph is planar if and only if it has a drawing, in which every trifurcation is a converter.*

PROOF. The second half of the theorem is trivial: if a graph is planar, then it can be drawn without crossing, and, clearly, every trifurcation in this drawing is a converter. To establish the first half, by Kuratowski's theorem, it is sufficient to show that if every trifurcation in a graph G is a converter, then G does not contain a subdivision of $K_{3,3}$ or of K_5.

Suppose that G contains a subdivision of $K_{3,3}$ with vertex classes $\{u_1, u_2, u_3\}$ and $\{v_1, v_2, v_3\}$. Denote this subdivision by K. Deleting from K the point u_3 together with the three paths connecting it to the v_j's, we obtain a converter between u_1 and u_2. Similarly, deleting u_2 (u_1) we obtain a converter between u_1 and u_3 (u_2 and u_3, respectively). We say that the type of u_i is *clockwise* or *counterclockwise* according to the circular order of the initial segments of the paths $u_i v_1, u_i v_2, u_i v_3$ around u_i. It follows from the definition of the converter that any two u_i's must have opposite types, which is impossible.

The case when G contains a subdivision of K_5 is left to the reader. $\square$

3. Different Crossing Numbers?

As is illustrated by Theorem 2.4, the investigation of crossings in graphs often requires parity arguments. This phenomenon can be partially explained by the "banal" fact that if we start out from the interior of a simple (non-selfintersecting) closed curve in the plane, then we find ourselves inside or outside of the curve depending on whether we crossed it an even or an odd number of times.

Next we define three variants of the notion of crossing number.

(1) The *rectilinear crossing number*, LIN-CR(G), of a graph G is the minimum number of crossings in a drawing of G, in which every edge is represented by a straight-line segment.

(2) The *pairwise crossing number* of G, PAIR-CR(G), is the minimum number of crossing pairs of edges over all drawings of G. (Here the edges can be represented by arbitrary continuous curves, so that two edges may cross more than once, but every pair of edges can contribute to PAIR-CR(G) at most one.)

(3) The *odd-crossing number* of G, ODD-CR(G), is the minimum number of those pairs of edges which cross an odd number of times, over all drawings of G.

It readily follows from the definitions that

$$\text{ODD-CR}(G) \leq \text{PAIR-CR}(G) \leq \text{CR}(G) \leq \text{LIN-CR}(G).$$

Bienstock and Dean [**151**] exhibited a series of graphs with crossing number 4, whose rectilinear crossing numbers are arbitrary large. Pelsmajer, Schaefer, and Štefankovič [**625**] have shown that for any $\varepsilon > 0$ there exists a graph G with

$$\text{ODD-CR}(G) \leq \left(\frac{\sqrt{3}}{2} + \varepsilon\right) \text{PAIR-CR}(G).$$

A better construction was found by Tóth [**723**], with the constant $\frac{\sqrt{3}}{2}$ replaced by $\frac{3\sqrt{5}-5}{2}$. However, we cannot rule out the possibility that

CONJECTURE 3.1 ([**625**]). *There is a constant $\gamma > 0$ such that* ODD-CR$(G) \geq \gamma \cdot$ CR(G)*, for every graph G.*

CONJECTURE 3.2. *For every graph G, we have* PAIR-CR$(G) =$ CR(G).

The determination of the odd-crossing number can be rephrased as a purely combinatorial problem, thus the above two conjectures may offer a spark of hope that there exists an efficient approximation algorithm for estimating these crossing numbers.

According to a remarkable theorem of H. Hanani (alias Chojnacki) [**238**] and W. Tutte [**727**], if a graph G can be drawn in the plane so that any pair of its edges cross an even number of times, then it can also be drawn without any crossing. In other words, ODD-CR$(G) = 0$ implies that CR$(G) = 0$. Note that in this case, by the observation of Fáry mentioned in Section 1, we also have that LIN-CR$(G) = 0$.

The main difficulty in this problem is that a graph has so many essentially different drawings that the computation of any of the above crossing numbers, for a graph of only 15 vertices, appears to be a hopelessly difficult task even for very fast computers [**328**].

THEOREM 3.3 ([**377, 611, 658**]). *The computation of the crossing number, the pairwise crossing number, and the odd-crossing number are NP-complete problems.*

The growth rates of the three parameters in Theorem 3.3, CR(G), PAIR-CR(G), and ODD-CR(G), are not completely unrelated. (See also [**733**] and [**723**].) In addition to the trivial inequalities noted above, we have:

THEOREM 3.4 (Pach and Tóth [**611**]). *For any graph G, we have*
$$\mathrm{CR}(G) \le 2(\mathrm{ODD\text{-}CR}(G))^2.$$

The proof of the last statement is based on the following sharpening of the Hanani–Tutte theorem.

THEOREM 3.5 ([**611**]). *Any drawing of any graph in the plane can be redrawn in such a way that no edge, which originally crossed every other edge an even number of times, would participate in any crossing.*

PROOF OF THEOREM 3.4 USING THEOREM 3.5. Let $G = (V, E)$ be a simple graph drawn in the plane with $\lambda = $ ODD-CR(G) pairs of edges that cross an odd number of times. Let $E_0 \subset E$ denote the set of edges in this drawing which cross every other edge an even number of times. Since every edge not in E_0 crosses at least one other edge an odd number of times, we obtain that
$$|E \setminus E_0| \le 2\lambda.$$
By Theorem 3.5, there exists a drawing of G, in which no edge of E_0 is involved in any crossing. Pick a drawing with this property such that the total number of crossing points between all pairs of edges not in E_0 is minimal. Notice that in this drawing, any two edges cross at most once. Therefore, the number of crossings is at most
$$\binom{|E \setminus E_0|}{2} \le \binom{2\lambda}{2} \le 2\lambda^2,$$
and Theorem 3.4 follows. $\square$

In [**599**], the original form of the Hanani–Tutte theorem was applied to answer a question about the "complexity" of the boundary of the union of geometric figures [**471**]. A very elegant proof of a slight generalization of Theorem 3.5 was found by Pelsmajer *et al.* [**624**]. It is conjectured that Theorem 3.5 can be strengthened so that the conclusion remains true for every edge that in the original drawing meets all other edges *not incident to its endpoints* an even number of times.

In the original definition of the crossing number we assume that no three edges pass through the same point. Of course, this can be always achieved by slightly perturbing the drawing. Equivalently, we can say that k-fold crossings are permitted, but they are counted $\binom{k}{2}$ times.

G. Rote, M. Sharir, and others asked what happens if multiple crossings are counted only *once*? To what extent does this modification effect the notion of crossing number? It is important to assume here that *no tangencies are allowed* between the edges. Indeed, otherwise, given a complete graph with n vertices, one can easily draw it with only *one* crossing point p so that every pair of vertices is connected by an edge passing through p.

Let $\text{CR}'(G)$ denote the *degenerate crossing number* of G, that is, the minimum number of crossing *points* over all drawings of G satisfying this condition, where k-fold crossings are also allowed, for any k. Of course, we have

$$\text{CR}'(G) \leq \text{CR}(G),$$

and the two crossing numbers are not necessarily equal. For example, Kleitman [**490**] proved that the crossing number of the complete bipartite graph $K_{5,5}$ with five vertices in its classes is 16. On the other hand, the degenerate crossing number of $K_{5,5}$ in the plane is at most 15. Another example is depicted in Figure 3.

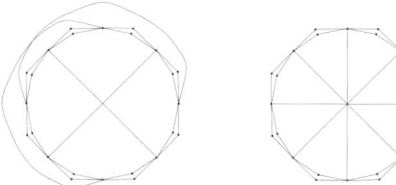

FIGURE 3. $\text{CR}(G) = 2$, $\text{CR}'(G) = 1$.

Let $n = n(G)$ and $e = e(G)$ denote the number of vertices and the number of edges of a graph G. As shown in Lemma 4.1 of Chapter 4 (the Crossing Lemma; see also Lemma 6.4 of Chapter 2), Ajtai, Chvátal, Newborn, Szemerédi [**61**] and, independently, Leighton [**509**] proved that if $e \geq 4n$ then

$$\text{CR}(G) \geq \frac{1}{64} \frac{e^3(G)}{n^2(G)}.$$

As we have seen in Chapter 4, this statement has many interesting applications in combinatorial geometry. Does it remain asymptotically true for the *degenerate* crossing number? It is not hard to show [**613**] that the answer is "no" if we permit drawings in which two edges may cross an arbitrary number of times. More precisely, any graph G with $n(G)$ vertices and $e(G)$ edges has a proper drawing with fewer than $e(G)$ crossings, where each crossing point that belongs to the interior of several edges is counted only once. That is, we have $\text{CR}'(G) < e(G)$. The order of magnitude of this bound cannot be improved if $e \geq 4n$.

Therefore, we restrict our attention to so-called *simple* drawings, i.e., to proper drawings in which two edges are allowed to cross at most once. Let $\text{CR}^*(G)$ denote the minimum number of crossing points over all simple drawings, where several edges may cross at the same point. One can prove that in this sense the degenerate crossing number of very "dense" graphs is at least $\Omega(e^3/n^2)$. More precisely, we have

THEOREM 3.6 (Pach and Tóth [**613**]). *There exists a constant $c^* > 0$ such that*
$$\mathrm{CR}^*(G) \geq c^* \frac{e^4(G)}{n^4(G)}$$
holds for any graph G with $e(G) \geq 4n(G)$.

It is a challenging question to decide whether here the term $\frac{e^4(G)}{n^4(G)}$ can be replaced by $\frac{e^3(G)}{n^2(G)}$, just like in the Crossing Lemma.

4. Straight-Line Drawings

For "straight-line thrackles," Conway's conjecture discussed in Section 2 had been settled by H. Hopf and E. Pannwitz [**445**] and (independently) by Paul Erdős much before the problem was raised!

If every edge of a graph is drawn by a straight-line segment, then we call the drawing a *geometric graph* [**581, 582, 585**]. We assume for simplicity that no three vertices of a geometric graph are collinear and that no segment representing an edge passes through any vertex other than its endpoints.

THEOREM 4.1 (Hopf–Pannwitz–Erdős theorem). *If every two edges of a geometric graph intersect (in an endpoint or an internal point), then it can have at most as many edges as vertices.*

PROOF (Perles). We say that an edge uv of a geometric graph is a *leftmost* edge at its endpoint u if turning the ray uv around u through 180 degrees in the counterclockwise direction, it never contains any other edge uw. For each vertex u, delete the leftmost edge at u, if such an edge exists. We claim that at the end of the procedure, no edge is left. Indeed, if at least one edge uv remains, it follows that we did not delete it when we visited u and we did not delete it when we visited v. Thus, there exist two edges uw and vz such that the ray uw can be obtained from uv by a counterclockwise rotation about u through less than 180 degrees, and the ray vz can be obtained from vu by a counterclockwise rotation about v through less than 180 degrees. This implies that uw and vz cannot intersect, which contradicts our assumption. □

The systematic study of extremal problems for geometric graphs was initiated by Avital and Hanani [**119**], Erdős, Perles, and Kupitz [**503**]. In particular, they asked the following question: what is the maximum number of edges of a geometric graph of n vertices, which does not have k pairwise disjoint edges? (Here, by "disjoint" we mean that they cannot cross and cannot even share an endpoint.) Denote this maximum by $e_k(n)$.

Using this notation, the above theorem says that $e_2(n) = n$, for every $n > 2$. Alon and Erdős [**72**] proved that $e_3(n) \leq 6n$. This bound was first reduced by a factor of two [**379**], and not long ago Černý [**197**] showed that $e_3(n) = 2.5n + O(1)$. It had been an open problem for a long time to decide whether $e_k(n)$ is linear in n for every fixed $k > 3$. Pach and Törőcsik [**609**] were the first to show that this is indeed the case. More precisely, they used Dilworth's theorem for partial orders to prove that $e_k(n) \leq (k-1)^4 n$. This bound was improved successively by G. Tóth and P. Valtr [**724**], and by Tóth [**721**].

THEOREM 4.2 (Tóth [**721**]). *For every k and every n, we have*
$$e_k(n) \leq 2^{10}(k-1)^2 n.$$

It is very likely that the dependence of $e_k(n)$ on k is also (roughly) linear.

Analogously, one can try to determine the maximum number of edges of a geometric graph with n vertices, which does not have k pairwise crossing edges. Denote this maximum by $f_k(n)$. It follows from Euler's polyhedral formula that, for $n > 2$, every planar graph with n vertices has at most $3n-6$ edges. Equivalently, we have $f_2(n) = 3n - 6$.

THEOREM 4.3 (Agarwal *et al.* [**19**]). $f_3(n) = O(n)$.

Better proofs and generalizations were found in [**5, 362, 590, 592**].

Recently, Ackerman [**4**] proved that $f_4(n) = O(n)$. Plugging this bound into the result of [**594**], we obtain $f_k(n) = O(n \log^{2k-8} n)$, for any $k > 4$. The exponent of the logarithmic factor was reduced to *one* by Valtr.

THEOREM 4.4 (Valtr [**731**]). *For any $k > 4$, we have $f_k(n) = O(n \log n)$.*

It can be conjectured that $f_k(n) = O(n)$. Moreover, it cannot be ruled out that there exists a constant c such that $f_k(n) \leq ckn$, for every k and n. However, we cannot even decide whether every complete geometric graph with n vertices contains at least (a positive) constant times n pairwise crossing edges. We are ashamed to admit that the strongest result in this direction is the following

THEOREM 4.5 (Aronov *et al.* [**95**]). *Every complete geometric graph with n vertices contains at least $\lfloor \sqrt{n/12} \rfloor$ pairwise crossing edges.*

Several *Ramsey-type* results for geometric graphs, closely related to the subject of this section, were established in [**459, 460, 461**]. In [**266**], some of these results have been generalized to *geometric hypergraphs* (systems of simplices).

5. Angular Resolution and Slopes

It is one of the major goals of graphic visualization to improve the readability of diagrams. If the angle between two adjacent edges is too small, it causes "blob effects" and it is hard to tell those edges apart. The minimum angle between two edges in a straight-line drawing of a graph G is called the *angular resolution* of the drawing. Of course, if the maximum degree of a vertex of G is d, then the angular resolution of any drawing of G is at most $\frac{2\pi}{d}$. It was shown by Formann *et al.* [**360**] that every graph G of maximum degree d can be drawn by straight-line edges with angular resolution at least constant times $\frac{1}{d^2}$ and that this bound is best possible up to a logarithmic factor. For planar graphs, one can achieve the asymptotically optimal resolution $\Omega\left(\frac{1}{d}\right)$, but then the optimal drawing is not necessarily crossing-free. In the case we insist on crossing-free (planar) straight-line drawings, Malitz and Papakostas [**522**] proved that there exists a constant $\alpha > 0$ such that any planar graph of maximum degree d permits a good drawing of angular resolution at least α^d.

Wade and Chu [**736**] defined the *slope number* $\text{sl}(G)$ of G as the smallest number of distinct edge slopes used in a straight-line drawing of G. Dujmović et al. [**276**] asked whether the slope number of a graph of maximum degree d can be arbitrarily large. The following short argument of Pach and Pálvölgyi shows that the answer is yes for $d \geq 5$.

THEOREM 5.1 (Pach and Pálvölgyi [**586**], Barát et al. [**129**]). *For every $d \geq 5$, there exists a sequence of graphs of maximum degree d such that their slope numbers tend to infinity.*

PROOF. Define a "frame" graph F on the vertex set $\{1,\ldots,n\}$ by connecting vertex 1 to 2 by an edge and connecting every $i > 2$ to $i-1$ and $i-2$. Adding a perfect matching M between these n points, we obtain a graph $G_M := F \cup M$. The number of different matchings is at least $(n/3)^{n/2}$. Let G denote the huge graph obtained by taking the union of disjoint copies of all G_M. Clearly, the maximum degree of the vertices of G is five. Suppose that G can be drawn using at most S slopes, and fix such a drawing.

For every edge $ij \in M$, label the points in G_M corresponding to i and j by the slope of ij in the drawing. Furthermore, label each frame edge ij ($|i-j| \leq 2$) by its slope. Notice that no two components of G receive the same labeling. Indeed, up to translation and scaling, the labeling of the edges uniquely determines the positions of the points representing the vertices of G_M. Then the labeling of the vertices uniquely determines the edges belonging to M. Therefore, the number of different possible labelings, which is $S^{|F|+n} < S^{3n}$, is an upper bound for the number of components of G. On the other hand, we have seen that the number of components (matchings) is at least $(n/3)^{n/2}$. Thus, for any S we obtain a contradiction, provided that n is sufficiently large. $\square$

A more complicated proof has been found independently by Barát, Matoušek, and Wood [**129**].

With some extra care one can obtain

THEOREM 5.2 ([**277, 586**]). *For any $d \geq 5$ and $\varepsilon > 0$, there exist graphs G with n vertices of maximum degree d, whose slope numbers satisfy*

$$\text{sl}(G) \geq \max\{n^{\frac{1}{2}-\frac{1}{d-2}-\varepsilon}, n^{1-\frac{8+\varepsilon}{d+4}}\}.$$

On the other hand, for *cubic* graphs we have

THEOREM 5.3 ([**476, 559**]). *Any connected graph of maximum degree three can be drawn with edges of at most four different slopes.*

This leaves open the annoying question whether graphs of maximum degree *four* can have arbitrarily large slope numbers.

6. An Application in Computer Graphics

It is a pleasure for the mathematician to see her research generate some interest outside her narrow field of studies. During the past twenty-five years, combinatorial geometers have been fortunate enough to experience this feeling quite often. Automated production lines revolutionized *robotics*, and started an avalanche of questions whose solution required new combinatorial geometric tools [**680**]. *Computer* *graphics*, whose group of users encompasses virtually everybody from engineers to film-makers, has had a similar effect on our subject [**146**].

Most graphics packages available on the market contain some (so-called *warping* or *morphing*) program suitable for deforming figures or pictures. Originally, these programs were written for making commercials and animated movies, but today they are widely used.

An important step in programs of this type is to fix a few basic points of the original picture (say, the vertices of the straight-line drawing of a planar graph), and then to choose new locations for these points. We would like to redraw the graph without creating any crossing. In general, now we cannot insist that the edges be represented by segments, because such a drawing may not exist. Our goal is to produce a drawing with polygonal edges, in which the total number of segments is small. The complexity and the running time of the program is proportional to this number.

THEOREM 6.1 (Pach and Wenger [**614**]). *Every planar graph with n vertices can be re-drawn in such a way that the new positions of the vertices are arbitrarily prescribed, and each edge is represented by a polygonal path consisting of at most $24n$ segments. There is an $O(n^5)$–time algorithm for constructing such a drawing.*

Badent *et al.* [**120**] have strengthened this theorem by constructing a drawing in which every edge consists of at most $3n + 3$ segments, and the running time of their algorithm is only $O(n^2 \log n)$. The next result shows that Theorem 6.1 cannot be substantially improved.

THEOREM 6.2 (Pach and Wenger [**614**]). *For every n, there exist a planar graph G_n with n vertices and an assignment of new locations for the vertices such that in any polygonal drawing of G_n there are at least $n/100$ edges, each composed of at least $n/100$ segments.*

The proof of this theorem is based on a generalization of a result of Leighton [**509**], found independently by Pach *et al.* [**594**] and by Sýkora *et al.* [**707**]. It turned out to play a crucial role in the solution of several extremal and algorithmic problems related to graph embeddings.

The *bisection width* of a graph is the minimum number of edges whose removal splits the graph into two pieces such that there are no edges running between them and the larger piece has at most twice as many vertices as the smaller. The following result can be proved using a weighted version of the Lipton–Tarjan separator theorem for planar graphs [**516**].

THEOREM 6.3 ([**594**, **707**]). *Let G be a graph of n vertices whose degrees are $d_1, d_2, \ldots, d_n$. Then the bisection width of G is at most*

$$1.58 \left(16\mathrm{CR}(G) + \sum_{i=1}^{n} d_i^2 \right)^{1/2}.$$

In the next section, following Pach *et al.* [**602**], we use the last result to give an unusual proof of the Crossing Lemma of Leighton and Ajtai *et al.* (Lemma 4.1 of Chapter 4). For similar applications of Theorem 6.3, see [**363**, **364**, **594**]. Kolman and Matoušek [**491**] have established a similar relationship between the bisection width and the pairwise crossing number of a graph.

7. An Unorthodox Proof of the Crossing Lemma

Let $b(G)$ denote the bisection width of G. By repeated application of Theorem 6.3, we obtain

COROLLARY 7.1. *Let G be a graph of n vertices with degrees $d_1, d_2, \ldots, d_n$. Then, for any edge disjoint subgraphs $G_1, G_2, \ldots, G_j \subseteq G$, we have*

$$\sum_{i=1}^{j} b(G_i) \leq 1.58 j^{1/2} \left(16 \operatorname{CR}(G) + \sum_{k=1}^{n} d_k^2 \right)^{1/2}.$$

PROOF. Let d_{ik} denote the degree of the k-th vertex in G_i. Applying Theorem 6.3 to each G_i separately and adding up the resulting inequalities, we obtain

$$\sum_{i=1}^{j} b^2(G_i) \leq (1.58)^2 \left(16 \sum_{i=1}^{j} \operatorname{CR}(G_i) + \sum_{i=1}^{j} \sum_{k=1}^{n} d_{ik}^2 \right)$$

$$\leq (1.58)^2 \left(16 \operatorname{CR}(G) + \sum_{k=1}^{n} d_k^2 \right).$$

Therefore, we have

$$\left(\sum_{i=1}^{j} b(G_i) \right)^2 \leq j \sum_{i=1}^{j} b^2(G_i) \leq (1.58)^2 j \left(16 \operatorname{CR}(G) + \sum_{k=1}^{n} d_k^2 \right),$$

as required. □

COROLLARY 7.2 (Pach and Tardos [**606**]). *Let G be a graph of n vertices with degrees $d_1, d_2, \ldots, d_n$. Then, for any $1 < s \leq n$, one can remove at most*

$$8.6 \left(\frac{n}{s} \right)^{1/2} \left(16 \operatorname{CR}(G) + \sum_{i=1}^{n} d_i^2 \right)^{1/2}$$

edges from G so that every connected component of the resulting graph has fewer than s vertices.

PROOF. Partition G by subsequently subdividing each of its large components into two roughly equal halves as follows. Start the procedure by deleting $b(G)$ edges of G so that it falls into two parts, each having at most $\frac{2}{3}|V(G)| = \frac{2}{3}n$ vertices. As long as there exists a component $H \subset G$ whose size is at least s, by the removal of $b(H)$ edges, *cut* it into two smaller components, each of size at most $(2/3)|V(H)|$. When no such components are left, stop.

Let $\mathcal{H}$ denote the family of all components arising at *any* level of the above procedure (e.g., we have $G \in \mathcal{H}$ if G is connected). Define the *order* of any element $H \in \mathcal{H}$ as the largest integer k, for which there is a chain

(7.14) $$H_0 \subsetneq H_1 \subsetneq \ldots \subsetneq H_k$$

in $\mathcal{H}$ such that $H_k = H$. For any k, let $\mathcal{H}_k$ denote the set of all elements of $\mathcal{H}$ of order k. Thus, $\mathcal{H}_0$ is the set of the components in the final decomposition.

For any fixed k, the elements of $\mathcal{H}_k$ are pairwise (vertex) disjoint. Recall that in a chain (7.14) we have $|V(H_1)| \geq s$ and the ratio of the sizes of any two consecutive members is at least $3/2$. Therefore, the number of vertices in any element of $\mathcal{H}_k$ is at least $(3/2)^{k-1}s$, which in turn implies that for $k \geq 1$

$$j_k := |\mathcal{H}_k| \leq \frac{n}{(3/2)^{k-1}s} = \frac{(2/3)^{k-1}n}{s}.$$

Applying Corollary 7.1 to the subgraphs in $\mathcal{H}_k$, we obtain that the total number of edges removed from them, when they are first subdivided during our procedure, is at most

$$1.58 \cdot (2/3)^{(k-1)/2} \left(\frac{n}{s}\right)^{1/2} \left(16\mathrm{CR}(G) + \sum_{i=1}^{n} d_i^2\right)^{1/2},$$

Summing up over all $k \geq 1$, we conclude that the total number of edges deleted during the whole procedure does not exceed the number claimed. □

Let G be a graph with $n(G) = n$ vertices and $e(G) = e > 4n$ edges. We want to prove that its crossing number $\mathrm{CR}(G)$ satisfies

$$\mathrm{CR}(G) \geq \gamma \frac{e^3}{n^2},$$

for an absolute constant $\gamma > 0$.

Let d denote the average degree of the vertices of G, that is, let $d = 2e/n$. Consider a drawing of G in which the number of crossings is minimum. We modify this drawing into a drawing of another graph, G', with maximum degree d, as follows. One by one we visit each vertex v of G, and if its degree $d(v)$ is larger than d, then we split v it into $\lceil d(v)/d \rceil$ vertices, each lying very close to the original location of v and each of them incident to at most d consecutive edges originally terminating at v. In the new drawing, no two edges cross in a small neighborhood of the old vertex v, including the new vertices it gave rise to, and outside of this neighborhood the new drawing is identical with the old one. Obviously, the resulting graph G' has the same number of edges as the original one and its number of vertices satisfies

$$n(G') = \sum_{v \in V(G)} \left\lceil \frac{d(v)}{d} \right\rceil < \sum_{v \in V(G)} \left(\frac{d(v)}{d} + 1\right) = 2n.$$

The number of crossings in the new drawing is precisely the same as in the original one, that is, $\mathrm{CR}(G)$.

Applying Corollary 7.2 to G', we obtain that for every $s > 1$ one can remove

$$e^* = 8.6 \left(\frac{2n}{s}\right)^{1/2} \left(16\mathrm{CR}(G') + \sum_{v' \in V(G')} d^2(v')\right)^{1/2}$$

$$< 8.6 \left(\frac{2n}{s}\right)^{1/2} \left(16\mathrm{CR}(G) + 2n \left(\frac{2e}{n}\right)^2\right)^{1/2}$$

edges from G' so that every connected component of the resulting graph has fewer than s vertices.

Set $s := e/n$. After the removal of at most e^* edges, the remaining graph has at most $\frac{n}{s}\binom{s}{2} < \frac{e}{2}$ edges, so that we have $e^* > e/2$. This yields

$$(8.6)^2 \frac{2n^2}{e} \left(16\mathrm{CR}(G) + \frac{8e^2}{n}\right) > \frac{e^2}{4}, \quad \text{or}$$

$$2\mathrm{CR}(G) + \frac{e^2}{n} > \frac{1}{5000}\frac{e^3}{n^2}.$$

Consequently, either $2\mathrm{CR}(G) > 10^{-4}(e^3/n^2)$, in which case we are done, or $e^2/n > 10^{-4}(e^3/n^2)$. In the latter case, we have $e < 10^4 n$, and the relation $\mathrm{CR}(G) = \Omega(e^3/n^2)$ follows from the easy observation that

$$\mathrm{CR}(G) \geq e - 3n + 6,$$

for every graph G with $n > 2$ vertices and e edges (see, e.g., [**610**] and the proof of Lemma 4.1 of Chapter 4).

The above argument can be easily modified to obtain better bounds for graphs without some forbidden subgraphs. Assume, for example, that G has no cycle of length *four*. According to an old theorem of Erdős, then G has at most $2n^{3/2}$ edges. Repeating essentially the same argument as above, we can argue that after the removal of e^* edges, each component C of the remaining graph has at most $2|C|^{3/2}$ edges. Setting $s = e^2/(4n)^2$, we conclude that $e^* > e/2$, and the proof can be completed analogously. We obtain

THEOREM 7.3 (Pach *et al.* [**602**]). *Let G be a graph with n vertices and $e \geq 4n$ edges, which does not contain a cycle of length four. Then the crossing number of G is at least $\gamma e^4/n^3$, where $\gamma > 0$ is a suitable constant. This result is tight, apart from the value of γ.*

The proof generalizes to other bipartite forbidden subgraphs in the place of the cycle of length *four*.

CHAPTER 6

Extremal Combinatorics: Repeated Patterns and Pattern Recognition

1. Models and Problems

Today massive amounts of pictures and other geometric data is collected by digital cameras, laser scanners, electron microscopes, telescopes, etc. A digital survey underway (Sloan Digital Sky Survey) maps in detail one-quarter of the entire sky, determining the positions of more than 100 million celestial objects. To find regularities in this huge digital archive, one needs new mathematical and computational methods. This challenge has not found the mathematics community completely unprepared. Several extremal problems raised by Erdős, Hadwiger, and others, that half-a-century ago may have appeared to be curiosities in recreational mathematics, have turned out to be instrumental in this area of research. They have motivated the discovery of important combinatorial tools, including the probabilistic method, Szemerédi's regularity lemma, and Ramsey Theory.

We discuss some extremal problems on repeated geometric patterns in finite point sets in Euclidean space. Throughout this chapter, a *geometric pattern* is an equivalence class of point sets in d-dimensional space under some fixed geometrically defined equivalence relation. Given such an equivalence relation and the corresponding concept of patterns, one can ask several natural questions:

(1) *What is the maximum number of occurrences of a given k-element pattern among all subsets of an n-point set?*

(2) *How does the answer to the previous question depend on the particular pattern?*

(3) *What is the minimum number of distinct k-element patterns determined by a set of n points?*

These questions make sense for many specific choices of the underlying set and the equivalence relation. Hence it is not surprising that several basic problems of combinatorial geometry can be studied in this framework [**585**].

In the simplest and historically first examples, due to Erdős [**323**], the pattern is a pair of points in the plane and the defining equivalence relation is the isometry (congruence). That is, two point pairs, $\{p_1, p_2\}$ and $\{q_1, q_2\}$, determine the same pattern if and only if $|p_1 - p_2| = |q_1 - q_2|$. In this case, (1) becomes the well known *Repeated Distance problem*: What is the maximum number pairs at a fixed distance apart determined by n points in the plane? It follows by scaling that the answer does not depend on the particular distance (pattern). For most other equivalence relations, this is not the case: different patterns may have different maximal multiplicities. For $k = 2$, question (3) becomes the *problem of Distinct*

Distances: What is the minimum number of distinct distances that must occur among n points in the plane? In spite of many efforts, we have no satisfactory answers to these questions. The best known results are the following (see also Chapter 4).

THEOREM 1.1 (Spencer, Szemerédi, and Trotter [**699**]). *Let $f(n)$ denote the maximum number of times the same distance can be repeated between n points in the plane. We have*
$$ne^{\Omega\left(\frac{\log n}{\log \log n}\right)} \leq f(n) \leq O\left(n^{4/3}\right).$$

THEOREM 1.2 (Katz and Tardos [**468**]). *Let $g(n)$ denote the minimum number of distinct distances determined by n points in the plane. We have*
$$\Omega\left(n^{0.8641}\right) \leq g(n) \leq O\left(\frac{n}{\sqrt{\log n}}\right).$$

In Theorems 1.1 and 1.2, the lower and upper bounds, respectively, are due to Erdős [**323**] and are conjectured to be asymptotically sharp. See more about these questions in Section 3 and Chapter 4.

Erdős and Purdy [**331, 334**] initiated the investigation of the analogous problems with the difference that, instead of pairs, we consider *triples* of points, and call two of them *equivalent* if the corresponding triangles have the same angle, or area, or perimeter. This leads to questions about the maximum number of equal angles, or unit-area or unit-perimeter triangles, that can occur among n points in the plane, and to questions about the minimum number of distinct angles, triangle areas, and triangle perimeters, respectively. Erdős's Repeated Distance problem and his problem of Distinct Distances have motivated a lot of research in extremal graph theory. The above mentioned questions of Erdős and Purdy, and, in general, problems (1), (2), and (3) for larger than 2-element patterns, require the extension of graph theoretic methods to hypergraphs. This appears to be one of the most important trends in modern combinatorics.

It is most natural to define two sets to be *equivalent* if they are congruent or similar to, or translates, homothets or affine images of each other. This justifies the choice of the word "pattern" for the arising equivalence classes. Indeed, the algorithmic aspects of these problems have also been studied in the context of geometric pattern matching [**54, 63, 172**]. A typical algorithmic question is the following.

(4) *Design an efficient algorithm for finding all occurrences of a given pattern in a set of n points.*

It is interesting to compare the equivalence classes that correspond to the same relation applied to sets of different sizes. If A and A' are equivalent under congruence (or under some other group of transformations mentioned above), and a is a point in A, then there exists a point $a' \in A'$ such that $A \setminus \{a\}$ is equivalent to $A' \setminus \{a'\}$. On the other hand, if A is equivalent (congruent) to A' and A is large enough, then usually its possible extensions are also determined: for each a, there exist only a small number of distinct elements a' such that $A \cup \{a\}$ is equivalent to $A' \cup \{a'\}$. Therefore, in order to bound the number of occurrences of a large pattern, it is usually sufficient to study small pattern fragments.

We have indicated above that one can rephrase many extremal problems in combinatorial geometry as questions of type (1) (so-called *Turán-type* questions). Similarly, many *Ramsey-type* geometric coloring problems can also be formulated in this general setting.

(5) *Is it possible to color space with k colors such that there is no monochromatic occurrence of a given pattern?*

For point pairs in the plane under congruence, we obtain the famous Hadwiger–Nelson problem [**407, 409**]: What is the smallest number of colors $\chi(\mathbb{R}^2)$ needed to color all points of the plane so that no two points at unit distance apart get the same color?

FIGURE 1. Seven-coloring of the plane showing that $\chi(\mathbb{R}^2) \leq 7$.

THEOREM 1.3 (Moser and Moser [**556**], Hadwiger [**409**]). $4 \leq \chi(\mathbb{R}^2) \leq 7$.

Another instance of question (5) is the following open problem from [**326**]: Is it possible to color all points of the three-dimensional Euclidean space with three colors so that no color class contains two points at distance one and the midpoint of the segment determined by them? It is known that four colors suffice, but there is no such coloring with two colors. In fact, Erdős *et al.* [**326**] proved that, for every d, the Euclidean d-space can be colored with four colors without creating a monochromatic triple of this kind.

2. A Simple Sample Problem: Equivalence under Translation

We illustrate our framework by analyzing the situation when two point sets are considered equivalent if and only if they are translates of each other. In this case, we know the (almost) complete solution to problems (1)–(5) in Section 1.

THEOREM 2.1. *Any set B of n points in d-dimensional space has at most $n+1-k$ subsets that are translates of a fixed set A of k points. This bound is attained if and only if $A = \{p, p+v, \ldots, p+(k-1)v\}$ and $B = \{q, q+v, \ldots, q+(n-1)v\}$ for some $p, q, v \in \mathbb{R}^d$.*

PROOF. Notice first that any linear mapping φ that keeps all points of B distinct also preserves the number of translates: if $A + t \subset B$, then $\varphi(A) + \varphi(t) \subset \varphi(B)$. Thus, we can use any generic projection into $\mathbb{R}$, and the question reduces to a one-dimensional problem: Given real numbers $a_1 < \cdots < a_k$, $b_1 < \cdots < b_n$, what is the maximum number of values t such that $t + \{a_1, \ldots, a_k\} \subset \{b_1, \ldots b_n\}$. Clearly, $a_1 + t$ must be one of $b_1, \ldots, b_{n-k+1}$, so there are at most $n + 1 - k$ translates. If there are $n + 1 - k$ translates $t_i + \{a_1, \ldots, a_k\}$ that occur in $\{b_1, \ldots b_n\}$, for translation vectors $t_1 < \cdots < t_{n-k+1}$, then $t_i = b_i - a_1 = b_{i+1} - a_2 = b_{i+j} - a_{1+j}$, for $i = 1, \ldots, n - k + 1$ and $j = 0, \ldots, k - 1$. But then $a_2 - a_1 = b_{i+1} - b_i = a_{j+1} - a_j = b_{i+j} - b_{i+j-1}$, so all differences between consecutive a_j and b_i are the same. For higher dimensional sets, this holds for every one-dimensional projection, which guarantees the claimed structure. In other words, the maximum is attained only for sets of a very special type, which answers question (1). □

An asymptotically tight answer to (2), describing the dependence on the particular pattern, was obtained in [**172**].

THEOREM 2.2 (Brass [**172**]). *Let A be a set of points in d-dimensional space, such that the rational affine space spanned by A has dimension k. Then the maximum number of translates of A that can occur among n points in d-dimensional space is $n - \Theta(n^{\frac{k-1}{k}})$.*

Any set of the form $\{p, p+v, \ldots, p+(k-1)v\}$ spans a 1-dimensional rational affine space. An example of a set spanning a two-dimensional rational affine space is $\{0, 1, \sqrt{2}\}$, so for this set there are at most $n - \Theta(n^{1/2})$ possible translates. This bound is attained, e.g., for the set $\{i + j\sqrt{2} \mid 1 \leq i, j \leq \sqrt{n}\}$.

In this case, it is also easy to answer question (3), i.e., to determine the minimum number of distinct patterns (translation-inequivalent subsets) determined by an n-element set.

THEOREM 2.3. *Any set of n points in d-dimensional space has at least $\binom{n-1}{k-1}$ distinct k-element subsets, no two of which are translates of each other. This bound is attained only for sets of the form $\{p, p+v, \ldots, p+(n-1)v\}$ for some $p, v \in \mathbb{R}^d$.*

By projection, it is again sufficient to prove the result on the line. Let $f(n, k)$ denote the minimum number of translation inequivalent k-element subsets of a set of n real numbers. Considering the set $\{1, \ldots, n\}$, we obtain that $f(n, k) \leq \binom{n-1}{k-1}$, since every equivalence class has a unique member that contains 1. To establish the lower bound, observe that, for any set of n real numbers, there are $\binom{n-2}{k-2}$ distinct subsets that contain both the smallest and the largest numbers, and none of them is translation equivalent to any other. On the other hand, there are at least $f(n-1, k)$ translation inequivalent subsets that do not contain the last element. So we have $f(n, k) \geq f(n-1, k) + \binom{n-2}{k-2}$, which, together with $f(n, 1) = 1$, proves the claimed formula. To verify the structure of the extremal set, observe that, in the one-dimensional case, an extremal set minus its first element, as well as the same set minus its last element, must again be extremal sets, and for $n = k + 1$ it follows from Theorem 1.1 that all extremal sets must form arithmetic progressions. Thus, the whole set must be an arithmetic progression, which holds, in higher dimensional cases, for each one-dimensional projection. The corresponding algorithmic question (4) has a natural solution: Given two sets, $A = \{a_1, \ldots, a_k\}$ and $B = \{b_1, \ldots, b_n\}$, we can fix any element of A, say, a_1, and try all possible image points b_i. Each of them specifies a unique translation $t = b_i - a_1$, so we simply have to test for each set $A + (b_i - a_1)$ whether it is a subset of B. This takes $\Theta(kn \log n)$ time. The running time of this algorithm is not known to be optimal.

PROBLEM 2.4. *Does there exist a $o(kn)$-time algorithm for finding all real numbers t such that $t + A \subset B$, for every pair of input sets, A and B, consisting of k and n reals, respectively?*

The Ramsey-type question (5) is trivial for translates: Given any set A of at least two points $a_1, a_2 \in A$, we can two-color $\mathbb{R}^d$ without generating any monochromatic translate of A. Indeed, the space can be partitioned into arithmetic progressions with difference $a_2 - a_1$, and each of them can be colored separately with alternating colors.

3. Equivalence under Congruence in the Plane

Problems (1)–(5) are much more interesting and difficult under congruence. In the plane, considering two-element subsets, the congruence class of a pair of points is determined by their distance. Questions (1) and (3) become the Erdős's famous problems, mentioned in Section 1.

PROBLEM 3.1. *What is the maximum number of times the same distance can occur among n points in the plane?*

PROBLEM 3.2. *What is the minimum number of distinct distances determined by n points in the plane?*

The best known results concerning these questions are summarized in Theorems 1.1 and 1.2, respectively. There are several different proofs known for the currently best upper bound in Theorem 1.1 (see [**245, 585, 607, 699, 709**]), which, obviously, does not depend on the particular distance (congruence class). This answers question (2). As for the lower bound of Katz and Tardos [**468**] in Theorem 1.2, it represents the latest improvement over a series of previous results [**136, 239, 240, 555, 695, 709**]. Again, see Chapter 4 for more details.

The algorithmic question (4) can now be stated as follows.

PROBLEM 3.3. *How fast can we find all unit distance pairs among n points in the plane?*

Some of the methods developed to establish the $O(n^{4/3})$ bound for the number of unit distances can also be used to design an algorithm for finding all unit distance pairs in time $O(n^{4/3} \log n)$ (similarly to the algorithms for detecting point-line incidences [**534**]).

The corresponding Ramsey-type question (5) for patterns of size two is the famous Hadwiger–Nelson problem, see Theorem 1.3 above.

PROBLEM 3.4. *What is the minimum number of colors necessary to color all points of the plane so that no pair of points at unit distance receive the same color?*

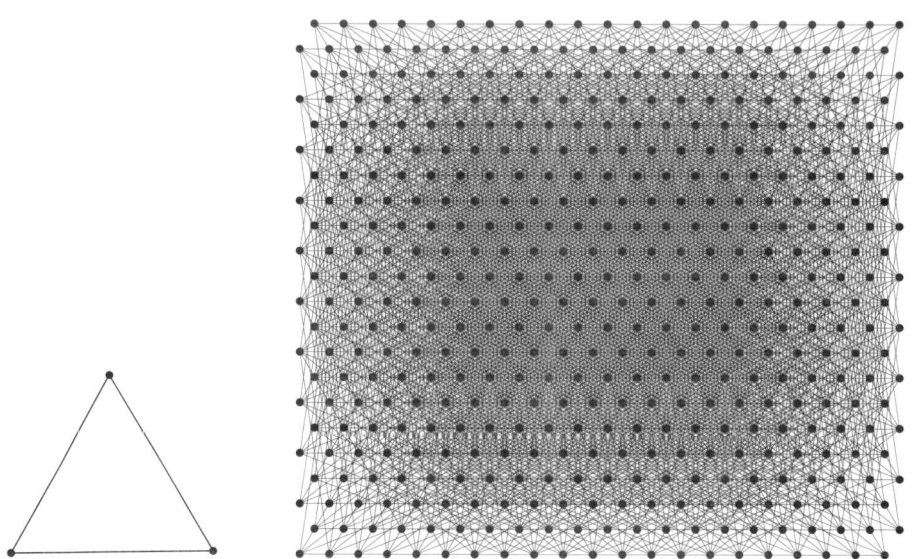

FIGURE 2. A unit equilateral triangle and a lattice section containing many congruent copies in parallel positions.

If we ask the same questions for patterns of size k rather than point pairs, but still in the plane, concerning question (1) it remains true that given a pattern $A = \{a_1, \ldots, a_k\}$, any congruent image of A is already determined, up to reflection, by the images of a_1 and a_2. Thus, the maximum number of congruent copies of a set is at most twice the maximum number of (ordered) unit distance pairs. Depending on the given set, this maximum number may be smaller, but no results of this kind are known. As n tends to infinity, the square and triangular lattice constructions that realize $ne^{\frac{c \log n}{\log \log n}}$ unit distances among n points also contain asymptotically the same number of congruent copies of *any* fixed set that is a subset of a square or triangular lattice. However, it is likely that this asymptotics cannot be attained for most other patterns.

PROBLEM 3.5. *Does there exist, for every finite set A, a positive constant $c(A)$ with the following property? For every n, there is a set of n points in the plane containing at least $ne^{\frac{c(A) \log n}{\log \log n}}$ congruent copies of A.*

Question (3) on the minimum number of distinct congruence classes of k-element subsets of a point set, is strongly related to the problem of Distinct Distances, just like the maximum number of pairwise congruent subsets was related to the Repeated Distance problem. For if we consider ordered k-tuples instead of

k-subsets (counting each subset $k!$ times), then two such k-tuples are certainly incongruent if their first two points determine distinct distances. For each distance s, fix a point pair that determines s. Clearly, any two different ways one can extend it by filling the remaining $k-2$ positions results in incongruent k-tuples. This leads to a lower bound of $\Omega(n^{k-2+0.8641})$ for the minimum number of distinct congruence classes of k-element subsets. Since a regular n-gon has $O(n^{k-1})$ pairwise incongruent k-element sets, this problem becomes less interesting for large k.

The algorithmic question (4) can also be reduced to the corresponding problem on unit distances. Given the sets A and B, we first fix $a_1, a_2 \in A$ and use our algorithm developed for detecting unit distance pairs to find all possible image pairs $b_1, b_2 \in B$ whose distance is the same as that of a_1 and a_2. Then we check for each of these pairs whether the rigid motion that takes a_i to b_i ($i=1,2$) maps the whole set A into a subset of B. This takes $O^*(n^{4/3}k)$ time, and we cannot expect any substantial improvement in the dependence on n, unless we apply a faster algorithm for finding unit distance pairs. (In what follows, we write O^* to indicate suppression of lower order factors, i.e., $O^*(n^\alpha) = o(n^{\alpha+\varepsilon})$ for every $\varepsilon > 0$.)

Many problems of Euclidean Ramsey theory can be interpreted as special cases of question (5) in our model. We particularly like the following problem raised in [**327**].

PROBLEM 3.6 (Erdős et al. [**326**]). *Is it true that, for any triple $A = \{a_1, a_2, a_3\} \subset \mathbb{R}^2$ that does not span an equilateral triangle, and for any coloring of the plane with two colors, one can always find a monochromatic congruent copy of A?*

It has been conjectured [**327**] that the answer to this question is in the affirmative. It is easy to see that the statement is not true for equilateral triangles A. Indeed, decompose the plane into half-open parallel strips whose width is equal to the height of A, and color them red and blue, alternately. On the other hand, the seven-coloring of the plane, with no two points at unit distance whose colors are the same, shows that any given pattern can be avoided with seven colors (see Figure 1). Nothing is known about coloring with three colors.

PROBLEM 3.7. *Does there exist a triple $A = \{a_1, a_2, a_3\} \subset \mathbb{R}^2$ such that any three-coloring of the plane contains a monochromatic congruent copy of A?*

4. Equivalence under Congruence in Higher Dimensions

All questions discussed in the previous section can also be asked in higher dimensions. There are two notable differences. In the plane, the image of a fixed pair of points was sufficient to specify a congruence. Therefore, the number of congruent copies of any larger set was bounded from above by the number of congruent pairs. In d-space, however, one has to specify d image points to determine a congruence, up to reflection. Hence, estimating the maximum number of congruent copies of a k-point set is a different problem for each $k = 2, \ldots, d$.

The second difference from the planar case is that starting from *four* dimensions, there exists another type of construction, discovered by Lenz, that provides asymptotically best answers to some of the above questions. For $k = \lfloor \frac{d}{2} \rfloor$, choose k concentric circles of radius $\frac{1}{\sqrt{2}}$ in pairwise orthogonal planes in $\mathbb{R}^d$ through the origin and distribute a total of n points on them as evenly as possible. Then any two points from distinct circles are at distance one, so the number of unit distance pairs is $(\frac{1}{2} - \frac{1}{2k} + o(1))n^2$, which is a positive fraction of all point pairs. It is known [**324**] that here the constant of proportionality cannot be improved. Similarly, in this construction, any choice of three points from distinct circles span a unit equilateral triangle, so if $d \geq 6$, a positive fraction of all triples can be mutually congruent. In general, for each $k \leq \lfloor \frac{1}{2}d \rfloor$, Lenz' construction shows that a positive fraction of all k-element subsets can be mutually congruent. Obviously, this gives the correct order of magnitude for question (1). With some extra work, even the exact maxima can be determined, as shown for $k = 2, d = 4$ in [**170, 739**].

Even for $k > \frac{d}{2}$, we do not know any construction better than Lenz', but for these parameters the problem is not trivial: now one is forced to select several points from the same circle, and only one of them can be selected freely. So, for $d = 3$, in the interesting versions of (1), we have $k = 2$ or 3 (now there is no Lenz construction). For $d \geq 4$, the cases $\lfloor \frac{d}{2} \rfloor < k \leq d$ are nontrivial.

PROBLEM 4.1. *What is the maximum number of unit distances among n points in three-dimensional space?*

Here, the currently best bounds are $\Omega(n^{4/3} \log \log n)$ [**324**] and $O^*(n^{3/2})$ [**245**].

PROBLEM 4.2. *What is the maximum number of pairwise congruent triangles spanned by a set of n points in three-dimensional space?*

Here the lower bound is $\Omega(n^{4/3})$ [**3, 329**], and the upper bound is $O^*(n^{5/3})$ [**54**], improving previous bounds in [**63, 171**]. For higher dimensions, Lenz' construction or, in the odd-dimensional cases, a combination of Lenz' construction with the best known three-dimensional point set [**3, 329**], is most likely optimal. The only results in this direction, given in [**54**], are for $d \leq 7$ and do not quite attain this bound.

PROBLEM 4.3. *Is it true that, for any $\lfloor \frac{d}{2} \rfloor \leq k \leq d$, the maximum number of congruent k-dimensional simplices among n points in d-dimensional space is $O(n^{\frac{d}{2}})$ if d is even, and $O(n^{\frac{d}{2} - \frac{1}{6}})$ if d is odd?*

Very little is known about question (2) in this setting. For point pairs, scaling again shows that all 2-element patterns can occur the same number of times. For 3-element patterns (triangles), the above mentioned $\Omega(n^{4/3})$ lower bound in [**329**] was originally established only for right-angle isosceles triangles. It was later extended in [**3**] to any fixed triangle. However, the problem is already open for full-dimensional simplices in 3-space. An especially interesting special case is the following.

PROBLEM 4.4. *What is the maximum number of orthonormal bases that can be selected from n distinct unit vectors?*

The upper bound $O(n^{4/3})$ is simple, but the construction of [**329**] that gives $O(n^{4/3})$ orthogonal pairs does not extend to orthogonal triples.

Question (3) on the minimum number of distinct patterns is largely open. For 2-element patterns, we obtain higher-dimensional versions of the problem of Distinct Distances. Here the upper bound $O(n^{2/d})$ is realized, e.g., by a cubic section of the d-dimensional integer lattice. The general lower bound of $\Omega(n^{1/d})$ was observed already in [**323**]. For $d = 3$, this was subsequently improved to $\Omega^*(n^{\frac{77}{141}})$ [**99**] and to $\Omega(n^{0.564})$ [**697**]. For large values of d, Solymosi and Vu [**697**] got very close to finding the best exponent by establishing the lower bound $\Omega\left(n^{\frac{2}{d}-\frac{2}{d(d+2)}}\right)$. This extends, in the same way as in the planar case, to a bound of $\Omega\left(n^{k-2+\frac{2}{d}-\frac{2}{d(d+2)}}\right)$ for the minimum number of distinct k-point patterns of an n-element set, but even for triangles, nothing better is known. Lenz-type constructions are not useful in this context, because they span $\Omega(n^{k-1})$ distinct k-point patterns, as do regular n-gons.

As for the algorithmic question (4), it is easy to find all congruent copies of a given k-point pattern A in an n-point set. For any $k \geq d$, this can be achieved in $O(n^d k \log n)$ time: fix a d-tuple C in A, and test all d-tuples of the n-point set B, whether they could be an image of C. If yes, test whether the congruence specified by them maps all the remaining $k-d$ points to elements of B. It is very likely that there are much faster algorithms, but, for general d, the only published improvement is by a factor of $\log n$ [**647**].

The Ramsey-type question (5) includes a number of problems of Euclidean Ramsey theory, as special cases.

PROBLEM 4.5. *Is it true that, for every two-coloring of the three-dimensional space, there are four vertices of the same color that span a unit square?*

It is easy to see that if we divide the plane into half-open strips of width one and color them alternately by two colors, then no four vertices that span a unit square will receive the same color. On the other hand, it is known that any two-coloring of four-dimensional space will contain a monochromatic unit square [**327**]. Actually, the (vertex set of a) square is one of the simplest examples of a *Ramsey-set*, i.e., a set B with the property that, for every positive integer c, there is a constant $d = d(c)$ such that under any c-coloring of the points of $\mathbb{R}^d$ there exists a monochromatic congruent copy of B. All boxes and all triangles [**365**] are known to be Ramsey. It is a longstanding open problem to decide whether all finite subsets of finite dimensional spheres are Ramsey. If the answer is in the affirmative, this would provide a perfect characterization of Ramsey sets, since all Ramsey sets are known to be subsets of a sphere [**326**].

The simplest non-spherical example, consisting of an equidistant sequence of three points along the same line, was mentioned at the end of the Section 1.

5. Equivalence under Similarity

If we consider problems (1)–(5) with similarity (congruence and scaling) as the equivalence relation, again we find that many of the arising questions have been extensively studied. Since any two point pairs are similar to each other, we can restrict our attention to patterns of size at least three. The first interesting instance of question (1) is to determine or to estimate the maximum number of pairwise similar triangles spanned by n points in the plane. This problem was almost completely solved in [**319**]. For any given triangle, the maximum number of similar triples in a set of n points in the plane is $\Theta(n^2)$. If the triangle is equilateral, we even have fairly good bounds on the multiplicative constants hidden in the Θ-notation [**2**]. In this case, most likely, suitable sections of the triangular lattice are close to being extremal. In general, the following construction from [**319**] always gives a quadratic number of similar copies of a given triangle $\{a, b, c\}$.

Interpreting a, b, c as complex numbers $0, 1, z$, consider the points $\frac{i_1}{n}z$, $\frac{i_2}{n} + (1 - \frac{i_2}{n})z$, and $\frac{i_3}{n}z + (1 - \frac{i_3}{n})z^2$. Then any triangle $(\beta - \alpha)z$, $\alpha + (1 - \alpha)z$, $\beta z + (1 - \beta)z^2$ is similar to $0, 1, z$, which can be checked by computing the quotient of the sides. (See [1], for a related "stability" result.)

The answer to question (1) in the plane, for k-point patterns, $k > 3$, is more or less the same as for $k = 3$. Certain patterns, including all k-element subsets of a regular triangular lattice, permit $\Theta(n^2)$ similar copies, and in this case a suitable section of the triangular lattice is probably close to being extremal. For some other patterns, the order $\Theta(n^2)$ cannot be attained. All patterns of the former type were completely characterized in [504]: for any pattern A of $k \geq 4$ points, one can find n points containing $\Theta(n^2)$ similar copies of A if and only if the cross ratio of every quadruple of points in A, interpreted as complex numbers, is algebraic. Otherwise, the maximum is slightly subquadratic. This result also answers question (2).

In higher dimensions, the situation is entirely different: we do not have good bounds for question (1) in any nontrivial case. The first open question is to determine the maximum number of triples in a set of n points in 3-space, which induce pairwise similar triangles. The trivial upper bound, $O(n^3)$, was first reduced to $O(n^{2.2})$ in [63], and then to $O(n^{13/6})$ in [15]. On the other hand, we do not have any better lower bound than $\Omega(n^2)$, which is already valid in the plane. These estimates extend to similar copies of k-point patterns, $k > 3$, provided that they are planar.

PROBLEM 5.1. *What is the maximum number of pairwise similar triangles induced by n points in three-dimensional space?*

For full-dimensional patterns, no useful constructions are known. The only lower bound we are aware of follows from subsets of lattices: in 3 dimensions, they span $\Omega(n^{4/3})$ similar copies of the full-dimensional simplex formed by its basis vectors or, in fact, of any k-element subset of lattice points. However, to attain this bound, we do not need to allow rotations: the lattice spans $\Omega(n^{4/3})$ *homothetic* copies.

PROBLEM 5.2. *In three-dimensional space, what is the maximum number of quadruples in an n-point set that span pairwise similar tetrahedra?*

For higher dimensions and for larger pattern sizes, the best known lower bound follows from Lenz' construction for congruent copies, which again does not use the additional freedom of scaling. Some nontrivial upper bounds were recently obtained by Agarwal et al. [15]: The maximum number of mutually similar $(d-2)$-simplices (resp., $(d-1)$-simplices) in an n-element point set in $\mathbb{R}^d$ is $O(n^{d-8/5})$ (resp., $O^*(n^{d-72/55})$).

Since, for $d \geq 3$, we do not know the answer to question (1) on the maximum number occurrences, there is no hope that we would be able to answer question (2) on the dependence of this maximum number on the pattern.

Question (3) on the minimum number of pairwise inequivalent patterns under similarity is an interesting problem even in the plane.

PROBLEM 5.3. *What is the minimum number of similarity classes of triangles spanned by a set of n points in the plane?*

There is a trivial lower bound of $\Omega(n)$: if we choose two arbitrary points, and consider all of their $n-2$ possible extensions to a triangle, then among these triangles each (oriented) similarity class will be represented only at most three times. This has been improved by Solymosi and Tardos [**694**] to $\Omega(n^2/\log n)$. On the other hand, as shown by the example of a regular n-gon, the number of similarity classes of triangles can be $O(n^2)$.

For higher dimensions and for larger sets, our knowledge is more limited. In 3-dimensional space, for instance, we do not even have an $\Omega(n)$ lower bound for the number of similarity classes of triangles, while the best known upper bound, $O(n^2)$, remains the same. For four-element patterns, we have a linear lower bound (fix any triangle, and consider its extensions), but we have no upper bound better than $O(n^3)$ (consider again a regular n-gon). Here we have to be careful in the statement of the problem: are we counting similarity classes of *full-dimensional* simplices only, or all similarity classes of possibly degenerate four-tuples? A regular $n-1$-gon with an additional point on its axis has only $\Theta(n^2)$ similarity classes of full-dimensional simplices, but $\Theta(n^3)$ similarity classes of four-tuples. In dimensions larger than three, nothing nontrivial is known.

In the plane, the algorithmic question (4) of finding all similar copies of a fixed k-point pattern is not hard: it can be achieved in time $O(n^2 k \log n)$, which is tight up to the $\log n$-factor, because the output complexity is this large in the worst case. For dimensions three and higher, we have no nontrivial algorithmic results. Obviously, the problem can always be solved in $O(n^d k \log n)$ time, by testing all possible d-tuples of the underlying set, but this is probably far from optimal.

The Ramsey-type question (5) has a negative answer, for any finite number of colors, even for homothetic copies. Indeed, for any finite set A and for any coloring of space with a finite number of colors, one can always find a monochromatic set similar (even homothetic) to A. This follows from the Hales–Jewett theorem [**411**], which implies that every coloring of the integer lattice $\mathbf{Z}^d$ with a finite number of colors contains a monochromatic homothetic copy of the lattice cube $\{1,\ldots,m\}^d$ (Gallai–Witt theorem [**639, 750**]).

6. Equivalence under Homothety or Affine Transformations

For homothety-equivalence, questions (1) and (2) have been completely answered in all dimensions [**172, 319, 501**]. The maximum number of homothetic copies of a set that can occur among n points is $\Theta(n^2)$; the upper bound $O(n^2)$ is always trivial, since the image of a set under a homothety is specified by the images of two points; and a lower bound of $\Omega(n^2)$ is attained by the homothetic copies of $\{1,\ldots,k\}$ in $\{1,\ldots,n\}$. The maximum order is attained only for this 1-dimensional example. If the dimension of the rational affine space induced by a given pattern A is k, then the maximum number of homothetic copies of A that can occur among n points is $\Theta(n^{1+\frac{1}{k}})$, which answers question (2).

Question (3) on the minimum number of distinct homothety classes of k-point subsets among n points, seems to be still open. As in the case of translations, by projection, we can restrict our attention to the one-dimensional case, where a sequence of equidistant points $\{0,\ldots,n-1\}$ should be extremal. This gives $\Theta(n^{k-1})$ distinct homothety classes. To see this, notice that as the size of the sequence increases from $n-1$ to n, the number of additional homothety classes, which were not already present in $\{0,\ldots n-2\}$, is $\Theta(n^{k-2})$. (The increment certainly includes

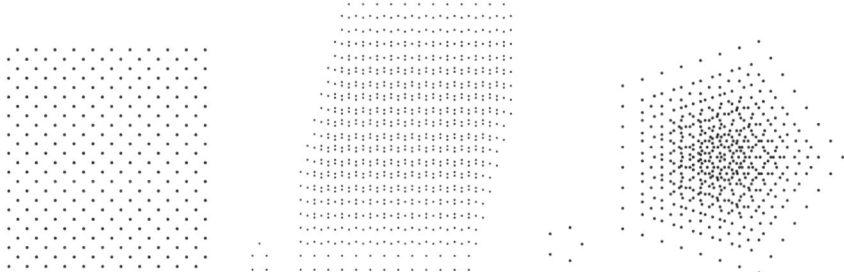

FIGURE 3. Three five-point patterns of different rational dimensions and three sets containing many of their translates.

the classes of all k-tuples that containing 0 and $n-1$, and a third number coprime to $n-1$.) Unfortunately, the pigeonhole principle gives only an $\Omega(n^{k-2})$ lower bound for the number of pairwise dissimilar k-point patterns spanned by a set of n numbers.

PROBLEM 6.1. *What is the minimum number of distinct homothety classes among all k-element subsets of a set of n numbers?*

The algorithmic question (4) was settled in [**172, 501**]. All homothetic copies of a given full-dimensional k-point pattern in an n-element set in d-space can be found in $O(n^{1+\frac{1}{d}} k \log n)$ time. This is asymptotically tight up to the $\log n$-factor. As mentioned in the previous section, the answer to the corresponding Ramsey-type question (5), is in the negative: one cannot avoid monochromatic homothetic copies of any finite pattern with any finite number of colors.

The situation is very similar for affine images. The maximum number of affine copies of a set among n points in d-dimensional space is $\Theta(n^{d+1})$. The upper bound is trivial, since an affine image is specified by the images of $d+1$ points. On the other hand, the d-dimensional "lattice cube," $\{1, \ldots, n^{\frac{1}{d}}\}^d$, contains $\Omega(n^{d+1})$ affine images of $\{0,1\}^d$ or of any other small lattice-cube of fixed size.

However, the answers to questions (2) and (3) are not so clear.

PROBLEM 6.2. *Do there exist, for every full-dimensional pattern A in d-space, n-element sets containing $\Omega(n^{d+1})$ affine copies of A?*

PROBLEM 6.3. *What is the minimum number of affine equivalence classes among all k-element subsets of a set of n points in d-dimensional space?*

For the algorithmic question (4), the brute force method of trying all possible $(d+1)$-tuples of image points is already worst-case optimal. The Ramsey-type question (5) has again a negative answer, since every homothetic copy is also an affine copy.

7. Other Equivalence Relations for Triangles in the Plane

For triples in the plane, several other equivalence relations have been studied. An especially interesting example is the following. Two ordered triples are considered equivalent if they determine the same angle. It was proved in [**597**] that the maximum number of triples in a set of n points in the plane that determine the same angle $\alpha \neq 0$ is $\Theta(n^2 \log n)$. This order of magnitude is attained for a dense set of angles α. For every other angle α, distribute as evenly as possible $n-1$ points on two rays that emanate from the origin and enclose angle α, and place the last point at the origin. Clearly, the number of triples determining angle α is $\Omega(n^2)$, which "almost" answers question (2). As for the minimum number of distinct angles determined by n points in the plane, Erdős conjectured that the answer to the following question is in the affirmative.

PROBLEM 7.1. *Is it true that every set of n points in the plane, not all on a line, determine at least $n-2$ distinct angles?*

This number is attained by a regular n-gon and by several other configurations. The corresponding algorithmic question (4) is easy: list, for each point p of the set, all lines ℓ through p, together with the points on ℓ. Then we can find all occurrences of a given nonzero angle in time $O(n^2 \log n + a)$, where a is the number of occurrences of that angle. Thus, by the above bound from [**597**], the problem can be solved in $O(n^2 \log n)$ time, which is optimal. The negative answer to the Ramsey-type question (5) again follows from the analogous result for homothetic copies: no coloring with a finite number of colors can avoid a given angle. In three and four dimensions, Apfelbaum and Sharir [**84**] have shown that the maximum number of occurrences of an angle in a set of n points in $\mathbb{R}^3$ is $O(n^{7/3})$, and that this bound is tight for a right angle, and that the maximum number of occurrences of an angle, other than $\pi/2$, in a set of n points in $\mathbb{R}^4$ is $O(n^{3/2} \lambda_6(n))$.

Another natural equivalence relation classifies triangles according to their areas.

PROBLEM 7.2. *What is the maximum number of unit-area triangles that can be determined by n points in the plane?*

The old upper bound $O(n^{7/3})$, established in [**597**], has been improved to $O(n^{44/19})$ by Dumitrescu *et al.* [**281**] (see also Chapter 4). On the other hand, it was pointed out in [**331**] that a section of the integer lattice gives the lower bound $\Omega(n^2 \log \log n)$. By scaling, we see that all areas allow the same multiplicities, which answers (2). Following earlier work by Erdős and Purdy [**333**], Burton and Purdy [**187**], and Dumitrescu and Tóth [**284**], Pinchasi [**631**] solved question (3) in this case.

THEOREM 7.3 (Pinchasi [**631**]). *Every set of n points in the plane, not all on a line, spans at least $\lfloor \frac{n-1}{2} \rfloor$ triangles of pairwise different areas.*

This bound is attained by placing on two parallel lines two equidistant point sets whose sizes differ by at most one. This construction was conjectured to be extremal for a long time [**334, 705**]. For analogous results in *three* dimensions, see [**285**].

The corresponding algorithmic question (4) is to find all unit-area triangles. Again, this can be done in $O(n^2 \log n + a)$ time, where a denotes the number of unit area triangles. First, dualize the points to lines, and construct their arrangement, together with a point location structure. Next, for each pair (p,q) of original points, consider the two parallel lines that contain all points r such that pqr is a triangle of unit area. These lines correspond to points in the dual arrangement, for which we can perform a point location query to determine all dual lines containing them.

They correspond to points in the original set that together with p and q span a triangle of area one. Each such query takes $O(\log n)$ time.

Concerning the Ramsey-type question (5), it is easy to see that, for any 2-coloring of the plane, there is a monochromatic triple that spans a triangle of unit area. The same statement may hold for any coloring with a finite number of colors.

PROBLEM 7.4. *Is it true that, for any coloring of the plane with a finite number of colors, there is a monochromatic triple that spans a triangle of unit area?*

Dumitrescu *et al.* [**281**] also establish the upper bound $O^*(n^{17/7})$ on the number of unit-area triangles determined by n points in $\mathbb{R}^3$, and study several variants of the problem, such as the maximum number of triangles of maximal or minimal area, in two and three dimensions.

The *perimeter* of triangles was also discussed in the paper [**597**], and later in [**600**], where an upper bound of $O(n^{16/7})$ was established, but there is no nontrivial lower bound. The lattice section has $\Omega(ne^{\frac{c \log n}{\log \log n}})$ pairwise *congruent* triangles, which, of course, also have the same perimeter, but this bound is probably far from being sharp.

PROBLEM 7.5. *What is the maximum number of unit-perimeter triangles spanned by n points in the plane?*

By scaling, all perimeters are equivalent, answering (2). By the pigeonhole principle, we obtain an $\Omega(n^{5/7})$ lower bound for the number of distinct perimeters, which has recently been slightly improved to $\Omega(n^{97/129})$ by Bien [**150**], but again this is probably far from the truth.

PROBLEM 7.6. *What is the minimum number of distinct perimeters assumed by all $\binom{n}{3}$ triangles spanned by a set of n points in the plane?*

Here neither the algorithmic question (4) nor the Ramsey-type problem (5) has an obvious solution. For the latter question, it is clear that with a sufficiently large number of colors, one can avoid unit perimeter triangles: color the plane "cellwise," where each cell is too small to contain a unit perimeter triangle, and two cells of the same color are far apart. The problem of determining the minimum number of colors required seems to be similar to the question addressed by Theorem 1.3.

CHAPTER 7

Lines in Space:
From Ray Shooting to Geometric Transversals

1. Introduction

In this chapter we address certain combinatorial and algorithmic problems about lines in three-dimensional Euclidean space. Algorithmic problems of this kind arise in numerous applications, including the hidden surface removal and ray tracing problems in computer graphics, motion planning, placement and assembly problems in robotics, object recognition using three-dimensional range data in computer vision, interaction of solids and of surfaces in solid modeling and CAD, and terrain analysis and reconstruction in geography. In addition, lines in 3-space present many challenging problems in their own right, whose solution requires the use of many sophisticated tools from the theory of higher-dimensional arrangements that has been reviewed in the preceding chapters.

Not surprisingly, straight lines, one of the simplest types of geometric objects, already present many conceptual difficulties when studied in a spatial context. In fact, as we will see below, lines in space are modeled best by non-linear objects. For a classical treatment of the geometry of lines in 3-space see the book by Sommerville [**693**], the book by Pottmann and Wallner [**636**], or various kinematics texts [**167, 448**].

We begin this chapter by studying several combinatorial problems involving *arrangements of lines in space*. By the arrangement of a given set of lines we mean the partitioning of the space of all lines according to their interaction with the given lines; this is a special case of a more general situation—see below. We first provide a combinatorial and algorithmic analysis of what we call an *orientation class* of a collection of lines in space, i.e., the subspace of oriented lines having a specified orientation with each of the given lines. (Informally, the orientation of a pair of lines is the manner in which they "turn around" each other, as a right-handed or a left-handed screw). We show how to express the "above/below" relationship of lines in space by means of the orientation relationship, and use this reduction to analyze various problems concerning the vertical relationship of lines in space. Even though (as discussed below) the "natural complexity" of an arrangement of n lines in space is $\Theta(n^4)$, there are many natural problems involving lines in space, which can be solved much faster, typically in nearly-quadratic time, and sometimes even faster.

In order to introduce and summarize in more detail results involving lines in space, we review some basic geometric properties of these lines. A line in space requires four real parameters to specify it, so it is natural to study arrangements of lines in space within an appropriate 4-dimensional parametric space. Unfortunately, any reasonable such representation introduces non-linear surfaces. For example, in many standard parametrizations, the space of all lines that meet a given line is a quadric in 4-space. To obtain a combinatorial representation of an arrangement of n lines in space, in a sense that captures the way in which other lines interact with these lines, one therefore has to construct an arrangement of n quadrics in 4-space. This arrangement has complexity $O(n^4)$, which is usually unacceptable for practical applications. The same reasoning applies when we want to partition the set of lines in space according to their interaction with more complex objects, say, balls. In this case, each ball induces a surface in $\mathbb{R}^4$, which is the locus of all points representing lines tangent to that ball. The arrangement of these surfaces decompose $\mathbb{R}^4$ into cells, so that for all points within a cell, all the lines in 3-space that they represent meet the same set of balls (and if the balls are pairwise disjoint, in the same order too). These observations indicate why "naive" solutions to visibility and related problems involving arbitrary collections of lines (or other simply shaped objects) in space produce bounds like $O(n^4)$ or worse. Moreover, for some of these problems, no better solutions are known to date—see below.

Fortunately, there are two lucky breaks that we are able to exploit, which lead to improved solutions in many applications. The first is that there is an alternative way to represent lines, using *Plücker coordinates* (see, e.g., [**167, 448, 704**]; the original reference is [**633**]). Using these coordinates, one can transform (oriented) lines into either points or hyperplanes in homogeneous 6-space (more precisely, in oriented projective 5-space), in such a way that the property of one line ℓ_1 intersecting another line ℓ_2 is transformed into the property of the Plücker point of ℓ_1 lying on the Plücker hyperplane of ℓ_2 (or vice versa). Thus, at the cost of passing to five dimensions, we can linearize the incidence relationship between lines. This Plücker machinery is developed in Section 2.

In studying arrangements of lines in space, it is more important to analyze the *relative orientation* of two lines rather than only the incidence between them, as the latter is a degenerate case of the former. Intuitively, it distinguishes between right-handed pairs and left-handed pairs of oriented lines. We develop this concept of relative orientation in Section 3 and show how to efficiently determine whether a query line is of a particular orientation class with respect to n given lines. We give a method that takes preprocessing and storage of $O(n^{2+\varepsilon})$ and allows query time of $O(\log n)$. In the process, we show that the total combinatorial description complexity of any particular orientation class is in the worst case $\Theta(n^2)$. We get this bound by mapping our lines to hyperplanes in oriented projective 5-space using Plücker coordinates. Our orientation class then corresponds to a convex polyhedron defined by the intersection of n halfspaces supported by these hyperplanes. Our second lucky break now comes from the Upper Bound Theorem (see, e.g., [**543, 759**]), stating that the complexity of such a polyhedron is only $O(n^{\lfloor 5/2 \rfloor}) = O(n^2)$.

For many applications, however, we need to analyze the relationship of one line lying above or below another. In Section 3.1 we show how, by adding certain auxiliary lines, we can express "above/below" relationships by means of orientation relationships. This leads, using standard machinery, to an efficient algorithm for testing whether a query line lies above the n given lines, using nearly-quadratic preprocessing time and storage, and $O(\log n)$ query time. The algorithm can also output the line of $\mathcal{L}$ that lies "immediately below" ℓ, i.e., the first line of $\mathcal{L}$ that is hit as ℓ is translated downwards. We also observe in Section 3.2 that the worst case combinatorial complexity of the "upper envelope" of n lines is $\Theta(n^3)$—the main observation being that such an envelope can be expressed as the union of n orientation classes of the kind discussed in the previous paragraph. With some modifications, we can also apply this technique to the batched version of the problem: Given m blue lines and n red lines, determine whether all blue lines lie above all red lines (we call this the "towering property"), and, if so, find for each blue line the red line lying immediately below it (in the above sense). The resulting algorithm runs in $O((m+n)^{4/3+\varepsilon})$ time, for any $\varepsilon > 0$.

Having introduced this basic machinery involving Plücker coordinates, we then go on to review other topics involving lines in $\mathbb{R}^3$. We first address, in Section 4, the problem of detecting and eliminating all cycles in the "depth order" of n lines in space, namely, the relation consisting of those pairs (ℓ_1, ℓ_2) where ℓ_1 passes below ℓ_2. The goal is to eliminate all cycles by cutting the lines into a small number of pieces, for which the depth order is indeed a (partial) order. The goal is to achieve this using a subquadratic number of cuts (a quadratic number is trivial). This seems to be a hard problem, and so far there are only very few results that provide some partial solutions. A related (and simpler) problem is to obtain bounds on the number of *joints* in a line arrangement in 3-space, namely, points incident to at least three non-coplanar lines. In spite of some recent progress, there is still a gap between the best known upper and lower bounds. See also Chapter 4.

We then study, in Section 5, the *ray shooting* problem, a classic in computer graphics and modeling. The basic problem is to preprocess a set T of n triangles in space, so that, given a query ray ρ, the first triangle of T hit by ρ can be found efficiently. Using a clever trick of Agarwal and Matoušek [**37**], the problem is reduced to testing whether a query segment intersects any triangle in T. This latter problem can be solved by standard range searching techniques. We review the basic results, and discuss some related visibility problems.

The next topic that we study, in Section 6, is the theory of *line transversals* in 3-space, which is a special topic in the general theory of geometric transversals, as reviewed in the surveys [**386, 622, 745**]. A line transversal to a set $\mathcal{F}$ of compact convex sets in $\mathbb{R}^3$ is a line that intersects every member of $\mathcal{F}$. Analyzing the structure of the space $T^3(\mathcal{F})$ of all line transversals to $\mathcal{F}$ has been the subject of extensive study, which has generated many interesting (and hard) problems. We review some of those, such as bounding the number of *geometric permutations* of $\mathcal{F}$ (the number of orders in which a transversal line can meet the members of $\mathcal{F}$), bounding the number of connected components (isotopy classes) of $T^3(\mathcal{F})$, and several other problems. Here we have not attempted a fully comprehensive survey, and many topics, such as Helly-type and Hadwiger-type theorems for line transversals, have not been treated; see the surveys [**386, 622, 745**] for more details.

We conclude in Section 7 with a list of the most obvious open problems involving lines in space.

2. Geometric Preliminaries

The main geometric object studied in this chapter is a line in 3-space. Such a line ℓ can be specified by four real parameters in many ways. For example, we can take two fixed parallel planes (e.g., $z = 0$ and $z = 1$) and specify ℓ by its two intersection points with these planes. We can therefore represent all lines in

3-space, except those parallel to the two given planes, as points in four dimensions. An important observation is in order: The space of lines in 3-space is projective (a special case of the so-called *Grassmannian manifold* [**442**]), so any attempt to represent *all* lines in 3-space as points in *real* 4-space is doomed to failure—there will always be a sub-manifold (of lower dimension) of lines that will have to be excluded (like the horizontal lines in the above parametrization). We will tend to ignore this issue, since it will not affect the combinatorial or algorithmic analysis of most of the problems that we will study.

However, as already noted, even simple relationships between lines, such as incidence between a pair of lines, become non-linear in 4-space. More specifically, a collection $\mathcal{L} = \{\ell_1, \ldots, \ell_n\}$ of n lines induces a corresponding collection of hypersurfaces $\mathcal{S} = \{s^1, \ldots, s^n\}$ in 4-space, where s^i represents the locus of all lines that intersect, or are parallel to, ℓ_i (it is easily checked that each s^i is a quadratic hypersurface, if we use the above parametrization). The *arrangement* $\mathcal{A} = \mathcal{A}(\mathcal{S})$ induced by these hypersurfaces represents the arrangement of the lines in $\mathcal{L}$, in the sense that each (4-dimensional) cell of $\mathcal{A}$ represents an isotopy class of lines in 3-space (i.e., any such line in the class can be moved continuously to any other line in the same class without crossing, or becoming parallel to, any line in $\mathcal{L}$).

This arrangement can be understood in three dimensions as follows. Given three lines in general position[1] in 3-space, they define a quadratic ruled surface, called a *regulus*, which is the union of all lines incident with the given three lines. (More precisely, the regulus is the 1-parameter family of lines incident to the three given lines; it becomes a 1-dimensional curve in the 4-dimensional parametric space of lines. With a slight abuse of notation, we also refer to the ruled surface spanned by these lines as a regulus.) A fourth line will in general cut this surface in at most two points. Thus four lines in general position will have at most two lines incident with all four of them. These quadruplets of lines with a common stabber correspond to vertices of the arrangement $\mathcal{A}$. (In other words, a line moving within an isotopy class comes to rest when it is in contact with four of the given arrangement lines—each of them removing one of the four degrees of freedom that the moving line has. Note that some of these contacts can be at infinity, corresponding to the moving line becoming parallel to one of the given lines.) Similarly, the edges of $\mathcal{A}$ correspond to motion of a line while incident with three given lines, in the regulus fashion described earlier. In the general case each vertex of $\mathcal{A}$ has eight incident edges. Higher dimensional faces of $\mathcal{A}$ can be obtained similarly, by letting the common stabber move away from two, three, or all four of the lines defining a vertex. This shows that the number of these higher dimensional faces of $\mathcal{A}$ is, in each case, related by at most a constant factor to the number of vertices of $\mathcal{A}$. This last statement remains valid even if the given lines are not in general position, as follows from a standard perturbation argument.

By the discussion in the preceding paragraph, we can conclude that the combinatorial complexity of the arrangement of n quadratic surfaces in 4-space that represent n lines in 3-space is $O(n^4)$, and this bound is attainable. In particular, $\mathcal{A}$ has $O(n^4)$ vertices, where each such vertex represents a line that meets four of the lines in $\mathcal{L}$.

---

[1] We take this to mean that the lines are pairwise non-intersecting and non-parallel. For more lines, we add the condition that no five of our lines can be simultaneously incident with another line (not necessarily of our collection).

One approach to many problems involving lines in space is to construct $\mathcal{A}$ explicitly, in time close to $O(n^4)$, examine each of its cells, and look for a (point representing a) line in that cell that satisfies some desired property. Here is a simple example: Let T be a set of n triangles in 3-space. We want to find a line that *stabs* the maximum number of triangles in T. To solve this problem, let $\mathcal{L}$ denote the set of the $3n$ lines supporting the edges of the triangles in T, and let $\mathcal{A} = \mathcal{A}(\mathcal{L})$ denote the corresponding arrangement in 4-space. For each 4-dimensional cell c of $\mathcal{A}$, all points in c represent lines that stab the same subset of triangles: When a line ℓ starts/stops crossing a triangle, it touches (the line containing) one of its edges, in which case the point that represents ℓ will have to cross one of the surfaces of $\mathcal{A}$, and then leave c. We thus obtain $O(n^4)$ cells c, and can easily compute the number of triangles crossed by lines in each c, but we can observe that when we cross from one cell c to a neighboring cell c', the number of stabbed triangles changes by at most ± 1, so by traversing the cells of $\mathcal{A}$ in an appropriate fashion, we can find the optimal line in a total of close to $O(n^4)$ time. No faster algorithm for this problem is known. The same approach can be used to solve Open Problem (f) mentioned in Section 7, in time close to $O(n^4)$.

However, in many problems faster solutions do exist. To obtain them, we exploit another representation of lines, using *Plücker coordinates and coefficients* (see [**167**, **448**, **704**] for a review of these concepts). Let ℓ be an oriented line, and a, b two points on ℓ such that the line is oriented from a to b. Let $[a_0, a_1, a_2, a_3]$ and $[b_0, b_1, b_2, b_3]$ be the homogeneous coordinates of a and b, with $a_0, b_0 > 0$ being the homogenizing weights. (By this we mean that the Cartesian coordinates of a are $(a_1/a_0, a_2/a_0, a_3/a_0)$ when $a_0 \neq 0$; the case $a_0 = 0$ represents points at infinity.) The Plücker coordinates of ℓ are the six real numbers

$$\pi(\ell) = [\pi_{01}, \pi_{02}, \pi_{12}, \pi_{03}, \pi_{13}, \pi_{23}],$$

where $\pi_{ij} = a_i b_j - a_j b_i$ for $0 \leq i < j \leq 3$. Similarly, the Plücker coefficients of ℓ are

$$\varpi(\ell) = \langle \pi_{23}, -\pi_{13}, \pi_{03}, \pi_{12}, -\pi_{02}, \pi_{01} \rangle,$$

i.e., the Plücker coordinates listed in reverse order with two signs flipped. The most important property of Plücker coordinates and coefficients is that incidence between lines is a bilinear predicate. Specifically, ℓ^1 is incident to ℓ^2 if and only if their Plücker coordinates π^1, π^2 satisfy the relationship

(2.15) $$\pi^1_{01}\pi^2_{23} - \pi^1_{02}\pi^2_{13} + \pi^1_{12}\pi^2_{03} + \pi^1_{03}\pi^2_{12} - \pi^1_{13}\pi^2_{02} + \pi^1_{23}\pi^2_{01} = 0.$$

This formula follows from expanding the four-by-four determinant whose rows are the coordinates of four distinct points a, b, c, d, with a, b on ℓ^1 and c, d on ℓ^2. This determinant is equal to 0 if and only if the two lines are incident (or parallel). In general, the absolute value of the quantity in equation (2.15) is[2] six times the volume of the tetrahedron $abcd$, and its sign gives the orientation of the tetrahedron $abcd$. As long as ℓ^1 is oriented from a to b and ℓ^2 from c to d, this sign is independent of the choice of the four points (and of the choice of the scalar factors multiplying their homogeneous coordinates, as long as these factors remain positive), and defines the *relative orientation* of the pair ℓ^1, ℓ^2, which we denote by $\ell^1 \diamond \ell^2$ [**704**].

It is easily checked that any positive scalar multiple of $\pi(\ell)$ is also a valid set of Plücker coordinates for the same oriented line ℓ, corresponding to a different choice

---

[2]It also equals the product $|ab| \cdot |cd| \cdot D \sin \alpha$, where D is the distance between ℓ^1 and ℓ^2, and α is the angle between the two lines.

of the defining points a and b, and/or to a positive scaling of their homogeneous coordinates. Also, any negative multiple of $\pi(\ell)$ is a representation of ℓ with the opposite orientation. Therefore, we can regard the Plücker coordinates $\pi(\ell)$ as the homogeneous coordinates of a point in projective oriented 5-space $\mathcal{P}^5$, which is a double covering of ordinary projective 5-space.[3] Dually, we can regard the Plücker coefficients $\varpi(\ell)$ as the homogeneous coefficients of an oriented hyperplane of $\mathcal{P}^5$. Equation (2.15) merely states that line ℓ^1 is incident to line ℓ^2 if and only if the Plücker point $\pi(\ell^1)$ lies on the Plücker hyperplane $\varpi(\ell^2)$ (or the other way around). In fact, the relative orientation $\ell^1 \diamond \ell^2$ of the two lines is $+1$ if $\pi(\ell^1)$ lies on the positive side of the hyperplane $\varpi(\ell^2)$, and -1 if it lies on the negative side.

We observe that not every point of $\mathcal{P}^5$ is the Plücker image of some line. Specifically, the real six-tuple (π_{ij}) is such an image if and only if it satisfies the quadratic equation

$$(2.16) \qquad \pi_{01}\pi_{23} - \pi_{02}\pi_{13} + \pi_{12}\pi_{03} = 0,$$

which states (using (2.15)) that every line is incident to itself. Thus among the six Plücker coordinates *two* are redundant. Equation (2.16) defines a four-dimensional subset of $\mathcal{P}^5$, called the *Plücker hypersurface* (or the *Grassmannian* or the *Klein hypersurface*) Π. Notice that the relative orientation of a line relative to a sixtuplet of numbers that does not correspond to a point on the Plücker hypersurface still makes perfect sense—simply plug the appropriate numbers into the left-hand side of Equation (2.15). It turns out that such "imaginary" lines do have a natural geometric interpretation in 3-space. They are known as *linear complexes*, and their properties are studied in the literature [**349, 454**].

We close this review by emphasizing that the projective representation of lines in space via Plücker coordinates is indeed a representation of all lines, including those at infinity.

3. The Orientation of a Line Relative to n Given Lines

We wish to analyze the set $\overline{C}(\mathcal{L}, \sigma)$, consisting of all lines ℓ in 3-space that have specified orientations $\sigma = (\sigma^1, \sigma^2, \ldots, \sigma^n)$ relative to n given lines $\mathcal{L} = (\ell^1, \ell^2, \ldots, \ell^n)$. (We call this set the *orientation class* σ relative to $\mathcal{L}$.) Translated to Plücker space, the definition says that point $\pi(\ell)$ has to lie on side σ^i of hyperplane $\varpi(\ell^i)$, for every $i = 1, \ldots, n$, and therefore inside the convex polyhedron $C(\mathcal{L}, \sigma)$ in $\mathcal{P}^5$ that is the intersection of those n halfspaces. The orientation class $\overline{C}(\mathcal{L}, \sigma)$ is thus the intersection of the polyhedron $C(\mathcal{L}, \sigma)$ and the Plücker hypersurface Π. Note that since Π is of degree 2, it can interact in at most "a constant fashion" with each feature of the polyhedron $C(\mathcal{L}, \sigma)$.

For the purpose of this section, we consider the polyhedron $C(\mathcal{L}, \sigma)$ to be an adequate description of the orientation class $\overline{C}(\mathcal{L}, \sigma)$. Computing the class then means computing all the features of this polyhedron, i.e., all its faces (of any dimension). The number of such features is the *combinatorial complexity* of the class—intersecting with Π can only increase this number by a constant factor. By the Upper Bound Theorem (see, e.g., [**543, 759**]), this complexity is only

---

[3]The points of $\mathcal{P}^5$ can be viewed as the oriented lines through the origin of $\mathbb{R}^6$, with the geometric structure induced by the linear subspaces of $\mathbb{R}^6$; or, equivalently, as the points of the 5-dimensional sphere $\mathbb{S}^5$, with the geometric structure induced by its great spheres. See reference [**704**] for more details on the theory of oriented projective spaces.

$O(n^{\lfloor 5/2 \rfloor}) = O(n^2)$. It is not difficult to find configurations of lines $\mathcal{L}$ that attain this bound. Consider the regulus (actually hyperbolic paraboloid) $z = xy$ and two families of $n/2$ generating lines lying on the regulus. One family consists of lines from one of the two rulings of the regulus, and the other of lines from the other ruling. By perturbing the lines of one family to be slightly off the regulus, we can make this a non-degenerate arrangement. It is simple to check that in every elementary square defined by two successive lines from one ruling and two successive lines from the other ruling there corresponds a line incident to all four of the lines defining the square and passing above all the rest. A more detailed construction of this kind will be given in Section 3.2.

A possible data structure for representing the polyhedron $C(\mathcal{L}, \sigma)$ is its face-incidence lattice, as described in [288]. Seidel's output-sensitive convex hull algorithm [666] constructs this representation in $O(\log n)$ amortized time per face. Alternatively, one may use the optimal but complex convex hull algorithm of Chazelle [209] or the simpler randomized incremental algorithm of Clarkson and Shor [246], to obtain the polyhedron C in $O(n^2)$ (deterministic or expected) time.

THEOREM 3.1. *The set of all lines in 3-space that have specified orientations to n given lines has combinatorial complexity $\Theta(n^2)$ in the worst case, and can be calculated in time $O(n^2)$.*

It was shown by Neil White (see [542]) that the intersection of the convex polyhedron $C(\mathcal{L}, \sigma)$ and the Plücker hypersurface Π may consist of many connected components. In other words, an orientation class relative to the fixed lines in $\mathcal{L}$ may contain multiple distinct isotopy classes; see also [386]. We note that the vertices of those isotopy classes are intersections of the Plücker hypersurface Π with the edges of the polyhedron $C(\mathcal{L}, \sigma)$. Since Π is a quadratic hypersurface, there are at most two such intersections per edge, and therefore the total number of vertices in all those isotopy classes is only $O(n^2)$. In other words, there are at most $O(n^2)$ lines that touch four of the lines of $\mathcal{L}$ and have specified orientations with all the others. A slightly more complicated argument shows that there are at most $O(n^2)$ isotopy classes in one orientation class. We do not know if this bound can be attained.

The above discussion easily leads to an efficient algorithm for deciding whether a given query line ℓ in 3-space lies in a particular orientation class σ relative to a set $\mathcal{L}$ of n fixed lines. This simply amounts to checking whether the Plücker point of ℓ lies in the polyhedron $C(\mathcal{L}, \sigma)$. Using, e.g., the algorithm of Matoušek and Schwarzkopf [538], this can be done using $O(n^2/\log^{2-\delta} n)$ storage and preprocessing time, for any $\delta > 0$, so that a query can be answered in $O(\log n)$ time.

THEOREM 3.2. *Given n lines in space and an orientation class σ, we can preprocess these lines by a procedure whose running time and storage is $O(n^2/\log^{2-\delta} n)$, for any $\delta > 0$, so that, given any query line ℓ, we can determine, in $O(\log n)$ time, whether ℓ lies in the orientation class σ with respect to the given lines.*

We note that a simple modification of this data structure allows us to actually *compute* in $O(\log n)$ time the orientation class of a line ℓ relative to n fixed lines, rather then merely test whether ℓ is in a predetermined class. The modification consists of locating the Plücker point of ℓ in the zone of the Plücker hypersurface Π in the entire arrangement of the Plücker hyperplanes of $\mathcal{L}$ (see Section 5). This leads to a data structure of size $O(n^{4+\varepsilon})$, which can be constructed in $O(n^{4+\varepsilon})$

time, for any $\varepsilon > 0$, that can be used to compute the orientation class of a given query line in logarithmic time.

3.1. Testing whether a line lies above n given lines. We will now consider a particular case of the general problem discussed in the previous subsection, which turns out to have significant applications on its own. We will be concerned with the property of one line lying above or below another. Formally, ℓ^1 lies above ℓ^2 if there exists a vertical line that meets both lines, and its intersection with ℓ^1 is higher than its intersection with ℓ^2. We are assuming that neither ℓ^1 nor ℓ^2 is vertical, and that the two lines are not parallel. Our previous non-degeneracy assumptions already exclude concurrent or parallel lines; whenever we are discussing the "above/below" relation, we also exclude vertical lines from consideration.

We can express this notion in terms of the relative orientation of these lines, as follows. Assume the lines ℓ^1 and ℓ^2 have been oriented in an arbitrary way, and consider their (oriented) orthogonal projections $\bar{\ell}^1$, $\bar{\ell}^2$ onto the xy-plane, seen from above. Observe that ℓ^1 is above ℓ^2 if and only if either

   the direction of $\bar{\ell}^1$ is clockwise to that of $\bar{\ell}^2$ and $\ell^1 \diamond \ell^2 = +1$, or
   the direction of $\bar{\ell}^1$ is counterclockwise to that of $\bar{\ell}^2$ and $\ell^1 \diamond \ell^2 = -1$.

Now let us introduce the line at infinity λ^2 that is parallel to ℓ^2 and passes through zenith point $z_\infty = (0, 0, 0, 1)$, the point at positive infinity on the z-axis. We orient the line λ^2 so that its projection on the xy-plane has the same direction as the projection of ℓ^2. It is easy to check that the direction of $\bar{\ell}^1$ is clockwise of $\bar{\ell}^2$ if and only if $\ell^1 \diamond \lambda^2 = -1$. Therefore, we conclude that ℓ^1 is above ℓ^2 if and only if

$$(3.17) \qquad \ell^1 \diamond \ell^2 = -\ell^1 \diamond \lambda^2.$$

Intuitively, ℓ^1 passes above ℓ^2 if and only if ℓ^1 passes "between" the lines ℓ^2 and λ^2. Thus, to express the fact that one line lies above another we need to check consistency between the signs of two linear inequalities. This fact complicates the analysis of the above/below relationship, in particular when many lines are involved.

Now let $\mathcal{L}$ be a collection of n lines in 3-space, and consider the set $\mathcal{U}(\mathcal{L})$, *the upper envelope of $\mathcal{L}$*, consisting of all lines ℓ that pass above every line of $\mathcal{L}$. We introduce the corresponding set of auxiliary lines at infinity $\Lambda = \{\lambda^1, \lambda^2, \ldots, \lambda^n\}$, with each λ^i parallel to the corresponding ℓ^i and passing through the point z_∞. Then, according to (3.17), a line ℓ is above all lines in $\mathcal{L}$ if and only if $\ell \diamond \ell^i = -\ell \diamond \lambda^i$; that is, if the orientation class of ℓ relative to the set $\mathcal{L}$ is exactly opposite to its orientation with respect to the set Λ.

Therefore, the set $\mathcal{U}(\mathcal{L})$ is the union of all orientation classes $\overline{C}(\mathcal{L} \cup \Lambda, \sigma \cdot \bar{\sigma})$, where $\sigma \cdot \bar{\sigma}$ is a sign sequence of the form $(\sigma^1, \sigma^2, \ldots, \sigma^n, -\sigma^1, -\sigma^2, \ldots, -\sigma^n)$. Luckily for us, only n of these classes are non-empty. To see why, let us assume that the x- and y-coordinate axes have been rotated and the lines oriented so that the projection of ℓ^1 coincides with the negative y-axis, and all other lines (including the query line ℓ) point towards increasing x. Let us assume also that the lines $\ell^2, \ldots, \ell^n$ are sorted in order of increasing xy-slope. It is easy to see that if the xy-slope of ℓ lies between those of ℓ^k and ℓ^{k+1}, then its orientation class relative to the set Λ is $(-^k +^{n-k})$. Therefore, we conclude that there are only n orientation classes relative to Λ.

This observation leads to a fast algorithm for deciding whether a query line ℓ passes above n fixed lines $\mathcal{L}$. For each of the n valid orientation classes $\sigma_k = (-^k +^{n-k})$, we build a data structure $\overline{C}_k(\mathcal{L}) = \overline{C}(\mathcal{L} \cup \Lambda, \sigma_k \cdot \bar{\sigma}_k)$, as described in the first part of Section 3. Then, to test a given query line ℓ, we first use binary search to locate its xy-slope among the slopes of the n given lines. This information determines the orientation class σ_k of ℓ relative to the lines in Λ. Once this has been found, we use the data structure $\overline{C}_k(\mathcal{L})$ to test whether ℓ has the opposite orientation class $\bar{\sigma}_k$ relative to the lines in $\mathcal{L}$.

This straightforward algorithm uses space approximately cubic in n. To reduce the amount of space, we will merge all the n data structures $\overline{C}_k(\mathcal{L})$ into a single data structure $\overline{C}^*(\mathcal{L})$, as follows. Assume all lines in $\mathcal{L}$ have been sorted by their xy-slope and oriented as described above. Let m be a parameter, to be chosen later. Partition $\mathcal{L}$ into m subsets $\mathcal{L}_1, \ldots, \mathcal{L}_m$, each subset consisting of approximately n/m consecutive lines in slope order. Prepare the data structures $\overline{C}_j^p(\mathcal{L}) = \overline{C}(\mathcal{L}_j^p, (++\cdots+))$ and $\overline{C}_j^s(\mathcal{L}) = \overline{C}(\mathcal{L}_j^s, (---\cdots-))$ for each prefix set $\mathcal{L}_j^p = \bigcup_{1 \leq k < j} \mathcal{L}_k$ ($1 \leq j \leq m$) and each suffix set $\mathcal{L}_j^s = \bigcup_{j < k \leq m} \mathcal{L}_k$ ($2 \leq j \leq m$). The storage and preprocessing time for these steps amount to $O(mn^{2+\varepsilon})$, for any $\varepsilon > 0$. Then, recursively build the data structure $\overline{C}^*(\mathcal{L}_j)$ for each subset $\mathcal{L}_j$ (possibly with a different value of m). Therefore $\overline{C}^*(\mathcal{L})$ is a data structure tree whose degree is m and whose depth will be $O(\log n / \log m)$. Testing a query line ℓ now proceeds as follows. As before, we use binary search to locate the xy-slope of ℓ between the slopes of two lines ℓ^k and ℓ^{k+1} of $\mathcal{L}$. (This step has to be performed only once). Let $\mathcal{L}_j$ be the subset containing the line ℓ^k. By construction, the xy-slope of ℓ is greater than the slopes of all lines in $\mathcal{L}_j^p$ and less than the slopes of all lines in $\mathcal{L}_j^s$. Then we can test, in $O(\log n)$ time, whether ℓ lies above all lines in these two subsets, using the data structures $\overline{C}_j^p$ and $\overline{C}_j^s$. If ℓ does not lie above all these lines we stop immediately; otherwise we recursively test ℓ against $\mathcal{L}_j$ using the data structure $\overline{C}^*(\mathcal{L}_j)$. If we set $m = \lceil n^\nu \rceil$, for some fixed and very small $\nu > 0$, the entire procedure takes time $O((\log n)^2 / \log m)) = O(\log n)$. The storage and preprocessing time amount to $O(m(n^2/\log^{2-\delta} n)(\log n)/\log m)$, which is $O(n^{2+\varepsilon})$, for any $\varepsilon > 0$.

We can also provide a modified version of this procedure, having the same complexity bounds, that can determine, for each query line ℓ lying above all lines of $\mathcal{L}$, which is the first line of $\mathcal{L}$ that ℓ will hit when translated vertically downwards. The key observation is that translation of ℓ downwards corresponds to motion of $\pi(\ell)$ along a straight line, say $\rho(\ell)$, on the Plücker hypersurface Π: the coordinates $\pi(\ell)$ change linearly with the altitude of ℓ, and are given by

$$[\pi_{01}, \pi_{02}, \pi_{12}, \pi_{03}, \pi_{13} - t\pi_{01}, \pi_{23} - t\pi_{02}],$$

where t is a parameter denoting the altitude. As ℓ moves vertically, it will become incident with another line ℓ' exactly when $\rho(\ell)$ crosses the plane $\varpi(\ell')$, and the crossing point can be computed in constant time. Moreover, the crossing point determines the line $\rho(\ell)$ uniquely, since it corresponds to a unique line in 3-space and the inverse of the downwards translation is a unique upwards translation. Note that as t tends to infinity (which corresponds to lifting the line up by an infinite amount), its Plücker image tends to a point of the form $[0, 0, 0, 0, -\pi_{01}, -\pi_{02}]$. These limit points constitute a line τ in $\mathcal{P}^5$, and correspond to lines at infinity of 3-space passing through the zenith point. In other words, finding the first line that ℓ hits amounts to performing a ray shooting query within the corresponding polyhedron in Plücker space.

To perform such queries efficiently, we slightly modify the construction of the data structure $\overline{C}(\mathcal{L})$. We construct it in a hierarchical recursive manner. In each step we take a random sample $\mathcal{R}$ of r of the hyperplanes $\varpi(\ell^i)$, and construct the convex polyhedron $C(\mathcal{R})$. Instead of decomposing $C(\mathcal{R})$ into simplices, as is usually done, we partition its interior by a set of hypersurfaces with the property that no line $\rho(\ell)$ crosses one of these hypersurfaces, and the resulting cells still have constant complexity. Specifically, take a decomposition of the boundary of $C(\mathcal{R})$ into simplices, and back-project from each such simplex s along the lines ρ that terminate at points on s. The collection of these back-projections yield a decomposition of $C(\mathcal{R})$ into $O(r^2)$ cells. We argue that the combinatorial complexity of each cell is a constant independent of r. Indeed, the base of each cell is a 4-dimensional simplex, the walls of the cell are a lifting of the boundary of this simplex along the lines $\rho(\ell)$, and the roof of the cell is some interval on the line τ. Because of the way these cells are constructed, to each cell c there corresponds a unique line $\ell(c)$ of $\mathcal{R}$ that is first hit as we translate downwards any line whose Plücker point lies in c. Again, the ε-net theory tells us that we can find a subset of $O(\frac{n}{r}\log r)$ lines of $\mathcal{L}$ such that the downwards translation of any line ℓ, with $\pi(\ell)$ in c, will not meet any other line of $\mathcal{L}$ until it reaches λ.

Therefore, if we use this modified cell decomposition of $C(\mathcal{R})$ when constructing the data structure $\overline{C}^*(\mathcal{L})$, then when we test a line ℓ for being above the n given lines, we can at the same time locate the nearest line below ℓ.

THEOREM 3.3. *Given n lines in space, we can preprocess them by a procedure whose running time and storage is $O(n^{2+\varepsilon})$, for any $\varepsilon > 0$, so that, given any query line ℓ, we can determine, in $O(\log n)$ time, whether ℓ lies above all the given lines, and, if so, which is the first line of $\mathcal{L}$ that ℓ will hit when translated downwards.*

3.2. The complexity of the upper envelope of n lines. In the previous subsection we saw that the upper envelope $\mathcal{U}(\mathcal{L})$ of a set of n lines in 3-space is the union of n orientation classes relative to the set $\mathcal{L} \cup \Lambda$. Each of these classes can be described as a polyhedron of $\mathcal{P}^5$ with at most $O(n^2)$ features. Therefore, the combinatorial complexity of $\mathcal{U}(\mathcal{L})$ is at most $O(n^3)$.

Notice that each of these n selected orientation classes relative to $\mathcal{L} \cup \Lambda$ defines a single isotopy class. This is so because any two lines ρ^1, ρ^2 in this class point in the same sector defined by the lines of $\mathcal{L}$ down in the xy-plane. Thus we can always continuously move ρ^1 to ρ^2 by first lifting it up high enough, then rotating it to align it with ρ^2, and then dropping it down onto ρ^2. In particular, this implies that in each of these n orientation classes, there are at most $O(n^2)$ lines that touch four lines of $\mathcal{L}$, and lie above all the remaining ones. (Each such line is the intersection of the Plücker hypersurface Π with an edge of polyhedron $C(\mathcal{L}, \sigma)$; since Π is a quadric, there are at most two such intersections per edge.)

It is also possible to exhibit a set of n lines that attains this cubic bound. Details of such a construction are given in [**220**]. That is, we have:

THEOREM 3.4. *The maximum combinatorial complexity of the entire upper envelope of n lines in space, as defined above, is $\Theta(n^3)$.*

3.3. Testing the towering property. In this subsection we exhibit a reasonably efficient deterministic algorithm for testing whether n blue lines $b_1, \ldots, b_n$

in 3-space lie above m other red lines $r_1, \ldots, r_m$; this is what we call the "towering property". The algorithm runs in time $O\left((m+n)^{4/3+\varepsilon}\right)$, for any $\varepsilon > 0$, a substantial improvement over the obvious $O(mn)$ method.

We first consider the case where the xy-slope of every red line is at least as large as that of any blue line. In that case, if we map the blue lines (oriented so as to have xy-projections going from left to right) to points $\lambda_1, \ldots, \lambda_n$ in $\mathcal{P}^5$ via Plücker coordinates, and the red lines (similarly oriented) to hyperplanes $\rho_1, \ldots, \rho_m$ in $\mathcal{P}^5$ via Plücker coefficients, then the towering property is equivalent to asserting that all n blue points lie in the convex polyhedron $\overline{C}$ obtained by intersecting the appropriate halfspaces bounded by the m red hyperplanes, as given by equation (2.15).

This latter property is easy to test, using standard space decomposition techniques: We choose a sufficiently large constant r, sample a subset $\mathcal{R}$ of r random elements of the Plücker hyperplanes, construct the intersection of the corresponding halfspaces, get a convex polyhedron $\overline{C}_\mathcal{R}$ of $O(r^2)$ complexity, and decompose it into $O(r^2)$ simplices. If any Plücker point lies outside $\overline{C}_\mathcal{R}$, the property does not hold and we stop. Otherwise, we distribute the points in the cells of the decomposition—by refining some cells as needed, we may also assume that each cell contains at most n/r^2 points.

We then need to solve $O(r^2)$ subproblems, each involving the points in some cell and the $O\left(\frac{m}{r}\log r\right)$ hyperplanes crossing the cell. We do this by simply exchanging the roles of points and hyperplanes, mapping the red lines into their Plücker points and the blue lines into their Plücker hyperplanes. We apply, to each of these $O(r^2)$ dual problems, the same sampling-and-decomposition approach that was used above, obtaining a grand total of $O(r^4)$ subproblems, each involving $O\left(\frac{n}{r^3}\log r\right)$ blue lines, and $O\left(\frac{m}{r^3}\log r\right)$ red lines. Assuming r to be a constant, the overhead in preparing these subproblems is only $O(m+n)$.

Unwinding the recursion in this manner, we obtain a total running time of $O\left((m+n)^{4/3+\varepsilon}\right)$, for any $\varepsilon > 0$ (where the choice of r depends on the desired ε).

Let us now return to the general towering problem and relax all assumptions on the slopes of the projections. Compute the median slope among the projections onto the xy-plane of all the red and blue lines together. This partitions the red lines into two sets, R_1 and R_2, and the blue lines into two sets, B_1 and B_2 such that each line in $R_2 \cup B_2$ projects onto the xy-plane into a line of slope at least as large as that of any projected line of $R_1 \cup B_1$; furthermore, the sizes of $R_1 \cup B_1$ and $R_2 \cup B_2$ are roughly equal. Now, we solve the towering problem recursively with respect to R_1 vs. B_1 and then for R_2 vs. B_2. If no negative answer has been produced yet, then we may apply the previous algorithm to the pairs (R_1, B_2) and (R_2, B_1). The correctness of the procedure follows from the fact that all pairs of red and blue lines are (implicitly) checked. Moreover, the asymptotic bound on the running time is not affected by this extra layer of processing, as is easy to check. (See [**220**] for more details, and for a slightly different approach.)

An additional computation similar to that detailed at the end of the previous section allows us to determine, within the same bounds, the red line immediately below each blue line, and thus also, if the towering property holds, the smallest vertical distance between the two groups of lines. So we obtain the following theorem:

THEOREM 3.5. *Given n blue lines and m red lines in space, one can test whether all the blue lines pass above all the red lines (the towering property) in time and*

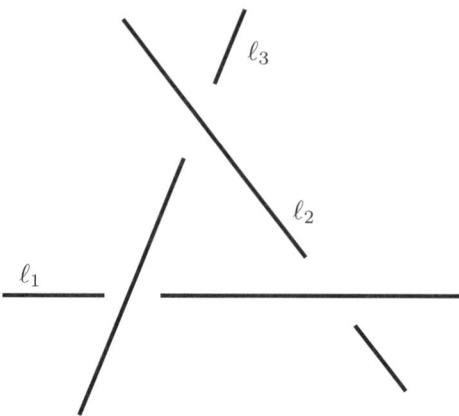

FIGURE 1. A triangular cycle, viewed from below.

space $O((m+n)^{4/3+\varepsilon})$, for any $\varepsilon > 0$. If so, within the same time bound, we can actually find the first red line below each blue line.

Separating lines by translation. We close this part of the chapter, by mentioning the following curious fact, whose proof is given in [**220**].

THEOREM 3.6. *There exists a set of nine mutually disjoint lines in 3-space that cannot be taken apart by continuously translating a proper subset off to infinity.*

4. Cycles and Depth Order

Let $\mathcal{L}$ be a collection of n lines in $\mathbb{R}^3$ in *general position*. In particular, assume that no line in $\mathcal{L}$ is vertical, that no two lines in $\mathcal{L}$ intersect or are parallel, and that the xy-projections of the lines are all distinct. For any pair ℓ, ℓ' of lines in $\mathcal{L}$, we say, as in Section 3.1, that ℓ passes *above* ℓ' (equivalently, ℓ' passes *below* ℓ) if the unique vertical line λ that meets both ℓ and ℓ' intersects ℓ at a point that lies higher than its intersection with ℓ'. We denote this relation as $\ell' \prec \ell$. The relation $\prec$ is total, but in general it does not have to be transitive, and it can contain *cycles* of the form $\ell_1 \prec \ell_2 \prec \cdots \prec \ell_k \prec \ell_1$. We refer to k as the *length* of the cycle. Cycles of length 3 are called *triangular*. See Figure 1.

If we cut the lines of $\mathcal{L}$ at a finite number of points, we obtain a collection of lines, segments, and rays. We can extend the definition of the relation $\prec$ to the new collection in an obvious manner, except that it is now only a partial relation. Our goal is to cut the lines in such a way that $\prec$ becomes a *partial order*, and to do so with a small number of cuts. In this case we call $\prec$ a *depth order*. We note that it is trivial to construct a depth order with $\Theta(n^2)$ cuts: Simply cut each line near every point whose xy-projection is a crossing point with another projected line. A long standing conjecture is that one can always construct a depth order with a *subquadratic* number of cuts.

Background. The main motivation for studying this problem comes from *hidden surface removal* (HSR) in computer graphics. We are given a collection of objects in $\mathbb{R}^3$, say pairwise disjoint triangles, and a viewing point, placed for convenience at $z = -\infty$, and we wish to compute and render all visible portions of the input

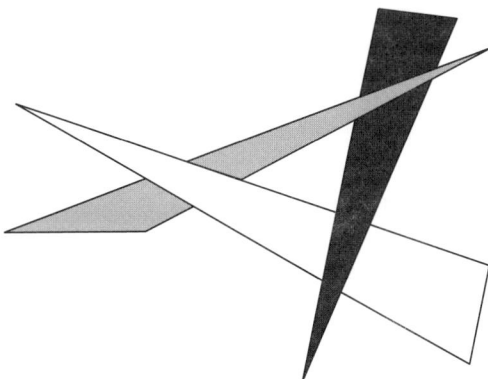

FIGURE 2. A depth cycle defined by three triangles.

objects; that is, for each object o we wish to compute the subset of all points p on o for which the downward-directed ray emanating from p meets no other object.

Until the 1970s, HSR was considered one of computer graphics' most important problems, and has received a substantial amount of attention; see [**706**] for a survey of the ten leading HSR algorithms circa 1974. Since then it has been solved in hardware, using the *z-buffer* technique [**196**], which produces a "discrete" solution to the problem, by computing the nearest object at each pixel of the image. Nevertheless, there is still considerable interest in obtaining an *object-space* representation of the visible scene, which is a combinatorial description of the visible portions, independent of the pixel locations and resolution.

These considerations motivated an extensive study of hidden-surface removal in computational geometry, culminating in the early 1990s with a number of algorithms that provide both conceptual simplicity and satisfactory running-time bounds. See de Berg [**139**] and Dorward [**275**] for overviews of these developments, and Overmars and Sharir [**682**] for a simple HSR algorithm with good theoretical running-time bounds.

A common feature of most HSR algorithms is that they rely on the existence of a consistent *depth order* for the input objects, which is defined as in the case of lines: A pair of objects A, B satisfy $A \prec B$ if there exists a point on B so that the downward-directed ray emanating from it meets A. The relation is well defined for convex and pairwise disjoint objects. These algorithms begin by sorting the objects either front-to-back (e.g., the Overmars–Sharir algorithm [**682**]) or back-to-front (e.g., the classical Painter's Algorithm [**706**]). They then process the objects in this order, either painting the new object on top of the previous ones, in the latter case, or finding the portions not covered by the previous objects, in the former case.

A large number of algorithms have been developed for *testing* whether the depth relationship $\prec$ in a collection of triangles contains a cycle with respect to a specific viewpoint; see de Berg [**139**] and the references therein. However, while these algorithms help detect cycles, they do not provide strategies for dealing with cycles.

One such common strategy is to eliminate all depth cycles, by cutting the objects into portions that do not form cycles, and running an HSR algorithm on the resulting collection of pieces. In 1980, Fuchs *et al.* [**368**] introduced *binary space*

FIGURE 3. A collection of line segments that forms a grid; the near-horizontal segments are "blue" and the near-vertical segments are "red".

partition (BSP) trees, which can be used to perform the desired cutting. However, a BSP tree may force up to a quadratic number of cuts [**619**], which brings us back to the original challenge: Devise an algorithm that, given a specific viewpoint and a collection of n triangles in $\mathbb{R}^3$, removes all depth cycles defined by this collection with respect to the viewpoint, using a subquadratic number of cuts. This has been an open problem since 1980.

We only consider the simpler problem mentioned above, by restricting the input to lines in space, rather than triangles. Note that since any cycle defined by a collection of *line segments* is also a cycle in the collection of lines spanned by these segments, the case of line segments is simpler than the case of lines, and we thus concentrate on the latter case. (The case of triangles, though, is more involved, since a depth cycle among triangles does not necessarily imply a depth cycle among their edges.)

The work of Solan [**693**] and of Har-Peled and Sharir [**427**] supply algorithms that achieve the above goal, provided a subquadratic number of cuts is always sufficient. In particular, these works present algorithms that, given a collection $\mathcal{L}$ of n lines (or segments) in 3-space, perform close to $O(n\sqrt{C})$ cuts (the precise bound is $O(n^{1+\varepsilon}\sqrt{C})$ for [**693**] and $O(n\sqrt{C}\alpha(n)\log n)$ for [**427**]) that eliminate all cycles defined by $\mathcal{L}$ as seen from $z = -\infty$, where C is the minimal required number of such cuts for $\mathcal{L}$. That is, if we can provide a subquadratic bound on the minimum number of cuts that suffice to eliminate all cycles defined by a collection of lines, then the aforementioned algorithms are guaranteed to find a collection of such cuts of (potentially larger but still) subquadratic size.

Such an upper bound however remains elusive. One major progress in this direction is due to Chazelle *et al.* [**214**], who in 1992 have analyzed the following special case of the problem. A collection of line segments in the plane is said to form a *grid* if it can be partitioned into two subcollections of "red" and "blue" segments, such that all red (resp., blue) segments are pairwise disjoint, and all red (resp., blue) segments intersect all blue (resp., red) segments in the same order; see Figure 3. Chazelle *et al.* [**214**] have shown that if the xy-projections of a collection

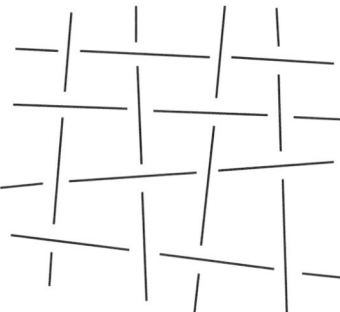

FIGURE 4. The unrealizable complete 4×4 weaving.

of n segments in 3-space form a grid, then all cycles defined by this collection (again, as seen from $z = -\infty$) can be eliminated with $O(n^{9/5})$ cuts.

A more recent contribution, due to Aronov *et al.* [**96**] makes the first step towards obtaining subquadratic general upper bounds on the number of cuts that are sufficient to eliminate all cycles defined by an arbitrary collection of lines in space. Specifically, it shows that all *triangular cycles*, which are cycles formed by triples of lines, can be eliminated with $O(n^{2-1/69} \log^{16/69} n)$ cuts. This allows adapting the technique of Har-Peled and Sharir [**427**] or of Solan [**693**], to yield an algorithm that eliminates all such cycles using close to $O(n^{2-1/138})$ cuts.

The best known lower bound for the number of required cuts is $\Omega(n^{3/2})$, due to Chazelle *et al.* [**214**]; it is obtained from an appropriately rotated and slightly perturbed grid of lines (see Figure 6; an appropriate slight perturbation of the lines depicted there will create a triangular cycle near each of the "joints").

Both upper bound proofs in [**96, 214**] use, as a main ingredient, the unrealizability of certain "weaving patterns" of lines in space. Specifically, a *weaving* is a finite collection of lines drawn in the plane, such that at each intersection of a pair of lines, it is specified which of the two is "above" the other. A weaving Ψ is said to be *realizable* if there is a collection $\mathcal{L}$ of lines in 3-space (called the *realization* of Ψ) whose xy-projection forms a collection of lines that is combinatorially equivalent to the one that defines Ψ, and the lines in $\mathcal{L}$ adhere to the above-below constraints specified by Ψ. Otherwise, Ψ is said to be unrealizable. A growing, albeit still relatively small, body of work deals with the analysis and classification of realizable and unrealizable weavings [**96, 374, 591, 646**]. While it can be shown that, for a sufficiently large number of lines, most weavings are unrealizable, showing that specific (small-size) weavings are unrealizable is a rather nontrivial problem. Figures 4 and 5 show the unrealizable weavings that are used in the proofs of [**96, 214**], respectively.

4.1. Joints. Let $\mathcal{L}$ be a set of n lines in space. A *joint* of $\mathcal{L}$ is a point in $\mathbb{R}^3$ where at least three non-coplanar lines ℓ, ℓ', ℓ'' of $\mathcal{L}$ meet. We denote the joint by the triple (ℓ, ℓ', ℓ'') (noting that a joint may have more than one such label).

Let $\mathcal{J}_\mathcal{L}$ denote the set of joints of $\mathcal{L}$, and put $J(n) = \max |\mathcal{J}_\mathcal{L}|$, taken over all sets $\mathcal{L}$ of n lines in space. A trivial upper bound on $J(n)$ is $O(n^2)$, as a joint is a point of intersection of (more than) two lines, but it was shown by Sharir [**679**], following a weaker subquadratic bound in [**214**], that $J(n)$ is only $O(n^{23/14} \text{polylog}(n)) =$

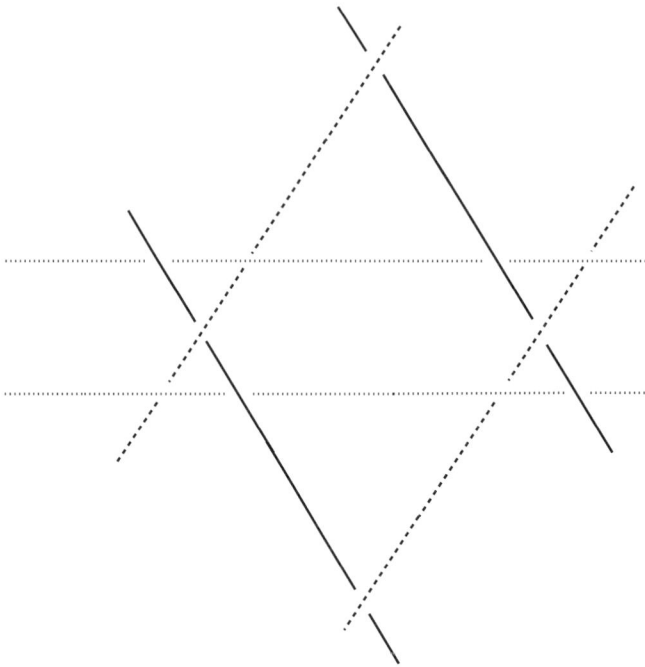

FIGURE 5. The unrealizable Star-of-David weaving.

$O(n^{1.643})$. An easy construction, based on lines forming an $n^{1/2} \times n^{1/2} \times n^{1/2}$ portion of the integer grid, shows that $|\mathcal{J}_\mathcal{L}|$ can be $\Omega(n^{3/2})$ (see Figure 6 and [**214**]).

Sharir's bound has been recently improved by Feldman and Sharir [**357**] to $O(n^{112/69} \log^{6/23} n) = O(n^{1.6232})$. The proof proceeds by mapping the lines of $\mathcal{L}$ into points and/or hyperplanes in projective 5-space, using Plücker coordinates. This is followed by a two-stage decomposition process, which partitions the problem into subproblems, using *cuttings* of arrangements of appropriate subsets of the Plücker hyperplanes. One then bounds the number of joints within each subproblem, and sums up the resulting bounds to obtain the above bound. The proof adapts and applies some of the tools used by Sharir and Welzl [**687**] and later enhanced by Aronov *et al.* [**96**], related mainly to the connection between joints and *reguli* spanned by the lines of $\mathcal{L}$.

One of the main motivations for studying joints of a set $\mathcal{L}$ of lines in space is their connection to cycles of $\mathcal{L}$. Joints can be regarded as a degenerate case of cycles. In fact, a slight random perturbation of the lines in $\mathcal{L}$ turns any joint incident to $O(1)$ lines into a cycle with some constant probability, implying that the number of joints is strongly related to the number of *elementary cycles*, whose xy-projections form single faces in the arrangement of the projected $\mathcal{L}$.

The problem of joints is considerably simpler than the corresponding problems involving cycles, as witnessed by the much sharper upper bounds cited above (or just by the ability to prove *any* subquadratic bound). Still, it is a rather challenging open problem to tighten the gap between the upper and lower bounds. One hopes that better insights into the joints problem would lead to tools that could also

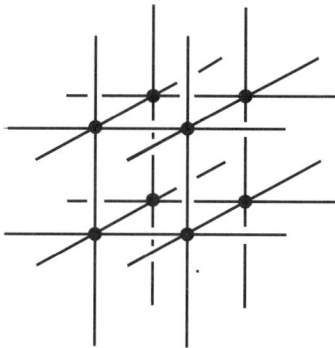

FIGURE 6. The lower bound construction for joints, illustrated with $n = 12$ lines.

be used to obtain subquadratic bounds for elementary cycles, and for many other problems that involve lines in space. Sharir and Welzl [**687**] have shown that the number of incidences between the points in $\mathcal{J}_\mathcal{L}$ and the lines in $\mathcal{L}$ is $O(n^{5/3})$; see also Chapter 4.

5. Ray Shooting and Other Visibility Problems

The ray shooting problem is to preprocess a set of objects such that the first object hit by a query ray can be determined efficiently. The ray shooting problem has received considerable attention in the past because of its applications in computer graphics and other geometric problems. The planar case has been studied thoroughly. Optimal solutions, which answer a ray shooting query in $O(\log n)$ time using $O(n)$ space, have been proposed for some special cases [**213, 223, 439**]. For an arbitrary collection of segments in the plane, the best known algorithm answers a ray shooting query in time $O(\frac{n}{\sqrt{s}} \log^{O(1)} n)$, using $O(s^{1+\varepsilon})$ space and preprocessing [**11, 46, 126**], for any $\varepsilon > 0$, where s is a parameter that can vary between n and n^2.

The three-dimensional ray shooting problem seems much harder, and it seems to be still far from being fully solved. It constitutes a nice and useful application of the theory of lines in space. Most studies of this problem consider the case where the given set is a collection of triangles. If these triangles are the faces of a convex polyhedron, then an optimal algorithm (with $O(n)$ storage and $O(\log n)$ query time) can be obtained using the hierarchical decomposition scheme of Dobkin and Kirkpatrick [**272**]. If the triangles form a polyhedral terrain (an xy-monotone piecewise-linear surface), then the technique of Chazelle *et al.* [**218, 220**] yields an algorithm that requires $O(n^{2+\varepsilon})$ space and answers ray shooting queries in $O(\log n)$ time; we will give more details concerning this technique in the next subsection on visibility problems. The best known algorithm for the general ray shooting problem (involving triangles) is due to Agarwal and Matoušek [**38**]; it answers a ray shooting query in time $O(\frac{n^{1+\varepsilon}}{s^{1/4}})$, with $O(s^{1+\varepsilon})$ space and preprocessing, where the parameter s can range between n and n^4. A variant of this technique was presented in [**49**] for the case of ray shooting amid a collection of convex polyhedra. See also [**49, 457**] for another variant of the problem for the case of convex polyhedra, with performance which depends both on the number of polyhedra and on the overall number of their facets. Thus, to achieve fast (logarithmic) query time, we need about n^4 storage, "in accordance" with the general observations made earlier in this chapter. Alternatively, to achieve near-linear storage, the best query time that we can achieve in general is close to $O(n^{3/4})$.

Here is a brief review of this technique. Let T be the set of n input triangles. The first tool is to reduce the problem, using *parametric search* [37], to the problem of testing whether a query segment (an arbitrary "prefix" of the query ray) intersects any triangle in T. Next, we take each triangle $t \in T$ and orient the lines $\lambda_1, \lambda_2, \lambda_3$ supporting its edges in a cyclic manner, so that a line ℓ intersects t if and only if it has the same (positive or negative) orientation with respect to each of the λ_i's. Hence a segment ab intersects t if and only if (i) a and b lie on different sides of the plane supporting t, and (ii) the line containing ab has the same orientation with respect to each of $\lambda_1, \lambda_2, \lambda_3$. Testing whether T contains a triangle that satisfies (i) and (ii) can be done using a multi-level halfspace range searching data structure, as described, e.g., in [38].

Selecting triangles that satisfy (i) can be done using two levels of a 3-dimensional halfspace range searching structure (such as the efficient partition trees of [529]). Selecting triangles that satisfy (ii) can be done by first transforming the problem into Plücker's 5-space. Since we only deal now with triangles that satisfy (i), we need to find all triangles that are crossed by the *line* ℓ supporting the query segment (or original ray), or, more precisely, determine whether any such triangle exists. In Plücker space, this amounts to finding all triangles t for which the Plücker point of ℓ lies, say, below the three Plücker hyperplanes of the lines supporting the edges of t. We perform this task using three more levels of a 5-dimensional halfspace range searching data structure. There is however a twist: The points that we query with lie on the 4-dimensional Plücker hypersurface Π, so, when we construct the partition tree of [529], using an arrangement of some subset of r hyperplanes, we do not need to consider all the cells of that arrangement, but only those that lie in the zone of Π. Since the overall complexity of that zone is $O(r^4 \log r)$ [100] (see Section 5), the partition tree has performance that is similar to that of partition trees in four dimensions. Putting everything together, and omitting several additional technical details (for which see [38, 46]), we obtain a solution that has the performance asserted above.

On the other hand, there are certain special cases of the 3-dimensional ray shooting problem which can be solved more efficiently. De Berg [139] presents some of these instances. For example, if the objects are planes or halfplanes, ray shooting amid them can be performed in time $O(\frac{n^{1+\varepsilon}}{s^{1/3}})$, with $O(s^{1+\varepsilon})$ space and preprocessing, for $n \leq s \leq n^3$; see [37] for details. If the objects are horizontal fat triangles or axis-parallel polyhedra, ray shooting can be performed in time $O(\log n)$ using $O(n^{2+\varepsilon})$ space; see [145] for details. If the objects are spheres, ray shooting can be performed in time $O(\log^4 n)$ with $O(n^{3+\varepsilon})$ space; see [554].

Sharir and Shaul [683] have recently considered several special cases of the ray shooting problem, including the case of arbitrary fat triangles and the case of triangles stabbed by a common line. They present an improved solution for the case where only near-linear storage is allowed, improving the query time to $O(n^{2/3+\varepsilon})$, using $O(n^{1+\varepsilon})$ space and preprocessing. Curiously, at the other end of the trade-off, no better solutions than the general solution that requires $O(n^{4+\varepsilon})$ storage to guarantee logarithmic query time, are known for these special cases.

5.1. Other visibility problems. Besides the ray shooting problem, computer graphics and modeling, as well as several other areas, are teeming with visibility problems in 3-space of all shapes and colors. We briefly mention some of them:

Visibility from a fixed point. In this kind of problems, we are typically given some polyhedral scene consisting of n non-crossing triangles, and a fixed viewing point v, and our goal is either to compute the portions of these triangles that are seen from v (the so-called *hidden surface removal* problem), or to preprocess the scene for efficient ray shooting queries, for rays emanating from v. In other variants, v is not fixed, but its location is constrained, say, to lie on a fixed line.

By projecting the given triangles centrally from v onto some sphere (or a cube, or a pair of planes), we obtain a 2-dimensional arrangement of the projected triangles, and each face of the arrangement determines uniquely the triangle seen from v when looking in directions that lead to that face. Hence, the *visibility map*, which is the decomposition of the sphere of projection into regions, each being a maximal connected visible portion of a single triangle, has combinatorial complexity $O(n^2)$, and this bound is easily attained. However, in practice the complexity of the map is much smaller, so output-sensitive algorithms for constructing visibility maps are of interest. See O'Rourke [**577**] for a survey of some algorithms of this kind. When the input triangles form a *polyhedral terrain* (the graph of a continuous piecewise linear bivariate function), the visibility map can still have quadratic complexity, but, as shown by Cole and Sharir [**248**], it can be preprocessed into a data structure that uses only $O(n\alpha(n)\log n)$ storage, in $O(n\alpha(n)\log n)$ time, so that a ray shooting query from v can be answered in $O(\log n)$ time.

Aspect graphs. When we vary continuously the point of view v, the visibility map changes continuously, but its combinatorial structure generally remains unchanged, except at certain critical viewing points, for which a vertex of the visibility map coincides with an edge, or three edges become concurrent. In these cases, the visibility map undergoes a discrete combinatorial change. The *aspect graph* of the original 3-dimensional scene is the collection of all combinatorially different visibility maps (or *views*) of the given scene, as we vary the point of view v. This definition depends on how are we allowed to move v. Two main special cases are *orthographic aspect graphs*, where v lies on the plane at infinity, and *perspective aspect graphs*, where v can lie anywhere in 3-space. In the former (resp., latter) case, we also decompose the sphere $\mathbb{S}^2$ of viewing directions (resp., the entire 3-space of viewing points) into maximal connected regions, so that, for each region K, the visibility maps for all directions in K (resp., points in K) are combinatorially the same. The number of such regions is the complexity of the respective aspect graph. For background and a survey of earlier research on aspect graphs, see [**168**]; see also [**577**] for a more recent survey.

For unrestricted scenes, the size of their orthographic (resp., perspective) aspect graph is $O(n^6)$ (resp., $O(n^9)$), and these bounds can be attained [**577, 632**]. In certain special cases better bounds can be obtained. For example, for polyhedral terrains, the bounds are $O(n^{5+\varepsilon})$ in the orthographic case, and $O(n^{8+\varepsilon})$ in the perspective case, and both bounds are nearly worst-case tight [**47**].

Here is a sketch of the argument. Let Σ be the given terrain. A pair of parallel rays (ρ_1, ρ_2) is called *critical* if for each $i = 1, 2$, the source point of ρ_i lies on an edge a_i of Σ, ρ_i passes through three edges of Σ (including a_i), and ρ_i does not intersect the (open) region lying below Σ. It can be shown that the number of topologically different orthographic views of Σ is $O(n^5)$ plus the number of critical pairs of parallel rays. Agarwal and Sharir [**47**] define, for each pair (a_1, a_2) of edges of Σ, a collection $\mathcal{F}_{a_1,a_2}$ of n trivariate functions, so that every pair (ρ_1, ρ_2) of critical rays, where ρ_i emanates from a point on a_i (for $i = 1, 2$), corresponds to a vertex of the minimization diagram $\mathcal{M}_{\mathcal{F}_{a_1,a_2}}$. They also show that the graphs of the functions in $\mathcal{F}_{a_1,a_2}$ satisfy assumptions (A1)–(A2) of Chapter 2. Using Theorem 3.1 of Chapter 2 and summing over all pairs of edges of Σ, we can conclude that the number of critical pairs of rays, and thus the number of topologically different orthographic views of Σ, is $O(n^{5+\varepsilon})$, for any $\varepsilon > 0$. Using a more careful analysis, Halperin and Sharir [**419**] proved that the number of different orthographic views is $n^5 2^{O(\sqrt{\log n})}$. de Berg *et al.* [**144**] have constructed a terrain for which there are $\Omega(n^5 \alpha(n))$ topologically different orthographic views.

For perspective views of Σ, it is argued in [**47**] that the number of such views is proportional to the number of triples of rays emanating from a common point, each of which passes through three edges of Σ before intersecting the open region lying below Σ. Following a similar approach to the one sketched above, the problem is reduced to the analysis of lower envelopes of $O(n^3)$ families of 5-variate functions, each family consisting of $O(n)$ functions that satisfy assumptions (A1)–(A2) of Chapter 2. This leads to an overall bound of $O(n^{8+\varepsilon})$ for the number of topologically different perspective views of Σ. This bound is also known to be almost tight in the worst case, as follows from another lower-bound construction given by de Berg *et al.* [**144**].

There are some other special cases where better bounds can be obtained. de Berg *et al.* [**144**] showed that if Σ is a set of k pairwise-disjoint convex polytopes with a total of n vertices, then the number of orthographic views is $O(n^4 k^2)$; the best known lower bound is $\Omega(n^2 k^4)$. For translates of a cube, Aronov *et al.* [**101**] established the bounds $O(n^{4+\varepsilon})$ for orthographic views and $O(n^{6+\varepsilon})$ for perspective views; both bounds are nearly tight in the worst case.

While all these bounds are huge, they do not reflect the real nature of the problem, because they are non-local: When the viewing point v crosses one critical surface σ, some local feature of the visibility map changes in a discrete manner, and when v crosses another critical surface σ', another visible feature, at a different location, undergoes a discrete change, and similarly for a third critical surface σ''. So near the intersection point $\sigma \cap \sigma' \cap \sigma''$, all eight combinations of old and new features are possible, and this is what makes the size of aspect graphs so large. The real issue is to encode in a more compact form all these different views, so that searching in each of them can still be made efficient.

Guarding a 3-dimensional scene. This is a natural extension of the extensively studied topic of "art gallery" problems, where some 2-dimensional scene has to be guarded by the minimal number of guards, so that each point in the scene is visible from at least one guard. Many instances of this problem are known to be NP-hard, already in the plane [**729**], and the situation does not get easier in 3-space.

One specific topic that has attracted some attention is that of guarding a polyhedral terrain by *watchtowers*. The goal is to erect a small set of watchtowers (vertical segments whose bottom endpoints lie on the terrain), so that collectively their top endpoints guard the entire terrain. The problem is interesting because a single watchtower can watch the entire terrain if it is sufficiently high. The goal is to optimize the solution, either by specifying the number of watchtowers and aiming at minimizing their (common) height, or by specifying their height and aiming at minimizing their number. The latter problem is NP-hard (already for height zero) [**248**], while the former can be solved in polynomial time for any fixed number k of watchtowers. For a single watchtower, the problem can be solved in $O(n \log n)$ time [**758**] (see also [**675**]). For two watchtowers, a recent work of Agarwal *et al.* [**25**] gives an algorithm with running time $O(n^{11/3+\varepsilon})$ for the case where the watchtowers can be erected only at vertices of the terrain.

6. Transversal Theory

Let $\mathcal{F}$ be a family of convex sets in $\mathbb{R}^3$. A line ℓ is a *transversal* of $\mathcal{F}$ if it intersects every member of $\mathcal{F}$. The set of all line transversals of $\mathcal{F}$ is called the *transversal space* of $\mathcal{F}$, and is denoted by $T^3(\mathcal{F})$. There has been extensive study of various topological, combinatorial, and other structural properties of spaces of line transversals. We will survey here only some of them, and ignore others, such as Helly-type and Hadwiger-type theorems for line transversals. Also, the study of line transversals is a special topic of the broader theory of geometric transversals, which will not be addressed either. See the survey papers [**386, 622, 745**].

Concerning the topological and combinatorial structure of $T^3(\mathcal{F})$, we note that there are two equivalence relations between line transversals of $\mathcal{F}$: *isotopy* (lying in the same connected component of $T^3(\mathcal{F})$), and inducing a common *geometric permutation* (meeting the objects of $\mathcal{F}$ in the same linear order). For the latter relation to be well defined, one assumes that the objects in $\mathcal{F}$ are pairwise disjoint. Two isotopic transversal lines induce the same geometric permutation [**387**], but the converse does not hold in general (although it does hold in the plane) [**386**].

Isotopy. The *combinatorial complexity* of $T^3(\mathcal{F})$ is defined as the total number of topological faces, of all dimensions, on the boundary of $T^3(\mathcal{F})$. Any upper bound on the maximal *combinatorial complexity* of $T^3(\mathcal{F})$ serves as a natural upper bound on the maximal number of connected components of $T^3(\mathcal{F})$.

In the planar case, the complexity of $T^2(\mathcal{F})$, when the elements of $\mathcal{F}$ are pairwise disjoint, is $O(n)$. In the 3-dimensional case, the complexity of $T^3(\mathcal{F})$ depends on the description complexity of the sets in $\mathcal{F}$. When the sets in $\mathcal{F}$ are semi-algebraic of *constant description complexity* (for example, if the sets in $\mathcal{F}$ are balls), the complexity of $T^3(\mathcal{F})$ is $O(n^{3+\varepsilon})$, as follows from the general result of Koltun and Sharir [**495**] on sandwich regions of trivariate functions. Indeed, similar to the analysis in Section 14.2 of Chapter 2, fix an input object $s \in \mathcal{F}$. We consider the representation of lines in space by equations of the form $y = ax + b$, $z = cx + d$, and parametrize such a line ℓ by the quadruple (a, b, c, d). When we translate ℓ up and down, a, b, and c remain fixed and d varies. We translate ℓ in this way until it becomes tangent to s, and denote the resulting upper and lower tangent lines by $(a, b, c, U_s(a, b, c))$ and $(a, b, c, L_s(a, b, c))$, respectively. (In general, the functions U_s and L_s are only partially defined, with a common domain of definition.) Then

ℓ intersects s if and only if
$$L_s(a,b,c) \leq d \leq U_s(a,b,c).$$
Hence if ℓ is a transversal of $\mathcal{F}$, we must have
$$\max_{s \in \mathcal{F}} L_s(a,b,c) \leq d \leq \min_{s \in \mathcal{F}} U_s(a,b,c).$$
In other words, $T^3(\mathcal{F})$ is the sandwich region between the upper envelope of the trivariate functions L_s and the lower envelope of the functions U_s. Hence, by [**495**], the complexity of $T^3(\mathcal{F})$ is $O(n^{3+\varepsilon})$, for any $\varepsilon > 0$.

This general bound was preceded by several works that considered special cases, and obtained slightly improved bounds. Pellegrini and Shor [**623**] showed that if $\mathcal{F}$ is a set of triangles in $\mathbb{R}^3$, then the space of line transversals of $\mathcal{F}$ has $n^3 2^{O(\sqrt{\log n})}$ complexity. The bound was slightly improved by Agarwal [**12**] to $O(n^3 \log n)$. He reduced the problem to bounding the complexity of a family of cells in an arrangement of $O(n)$ hyperplanes in $\mathbb{R}^5$.

There are also almost matching lower bounds, showing that the complexity of $T^3(\mathcal{F})$ can be $\Omega(n^3)$, see [**21, 621**]. However, for the number of connected components the best known lower bounds are $\Omega(n^2)$, or $\Omega(n^{d-1})$ in the d-dimensional case [**692**]. Narrowing this gap, even for restricted families of objects, is an intriguing open problem.

In the d-dimensional case, $d \geq 4$, where the sets of $\mathcal{F}$ are semi-algebraic of constant description complexity, the complexity of $T^d(\mathcal{F})$, defined in an analogous manner to the 3-dimensional case, is $O(n^{2d-2+\varepsilon})$, for any $\varepsilon > 0$, which actually bounds the overall complexity of the arrangement of surface patches described above. Extending the result of Koltun and Sharir [**495**] to higher dimensions is an important open problem.

These are the best general known bounds on the number of connected components, but there are some improved bounds in restricted cases: Aronov and Smorodinsky [**110**] proved that when restricting the transversals to pass through the origin the transversal space has $\Theta(n^{d-1})$ components. Brönnimann et al. [**181**] gave a complete description of the transversal space of n segments in $\mathbb{R}^3$. In this case the transversal space consists of $\Theta(n)$ connected components. Kaplan et al. [**458**] have shown that if the objects in $\mathcal{F}$ are n convex polytopes with a total of s facets then the complexity of $T^3(\mathcal{F})$ is close to $O(ns^2)$. In fact, they establish the stronger result that, for the same input, if we only consider lines that pass through a fixed line ℓ_0, the complexity of the transversal space is only close to $O(ns)$, a bound which is sharp in the worst case. Applying this bound to lines that pass through a fixed edge of one of the polytopes, and repeating the analysis to each such edge, the general bound follows. See also Brönnimann et al. [**180**] for an earlier related study.

Geometric permutations. Recall that a geometric permutation is the order in which a transversal line ℓ meets the objects of $\mathcal{F}$ (assuming them to be pairwise disjoint, this order is well defined, up to a reversal if ℓ is not oriented). The study of geometric permutations was initiated by Katchalski et al. [**463**]. One of the most interesting questions in this field is to prove tight asymptotic bounds on the maximal number $g_d(n)$ of geometric permutations that a family of n arbitrary pairwise disjoint convex sets in $\mathbb{R}^3$ can admit. The known bounds on $g_d(n)$ in the general case are:

$g_2(n) = 2n - 2$; see [**304**].
$g_d(n) = O(n^{2d-2})$, $d \geq 3$; see [**744**].
$g_d(n) = \Omega(n^{d-1})$, $d \geq 3$; see [**462, 692**].

Improved bounds are known in several special cases: Smorodinsky et al. [**692**] give a tight bound $\Theta(n^{d-1})$ for the case of pairwise disjoint balls in $\mathbb{R}^d$, which has been extended by Katz and Varadarajan [**465**] for families of α-*fat* convex objects in $\mathbb{R}^d$, where the constant of proportionality depends on α. If the sets in $\mathcal{F}$ have constant description complexity, then in $\mathbb{R}^3$ they induce at most $O(n^{3+\varepsilon})$ geometric permutations, as follows from the preceding discussion concerning the complexity of $T^3(\mathcal{F})$. An offshoot of the analysis of Kaplan et al. [**458**] shows that the number of geometric permutations is $O(n^3)$ in the special case where one of the objects in $\mathcal{F}$ is a line.

The case of balls was further studied by Zhou and Suri [**757**], who considered families $\mathcal{B}$ of n pairwise disjoint balls in $\mathbb{R}^d$ with a bounded *radial ratio* γ, namely the ratio between the largest radius of a ball in the family and the smallest one. Zhou and Suri show that, in this case, $\mathcal{B}$ admits at most $O(\gamma^{\log \gamma})$ geometric permutations.

If the balls all have the same radius, the number of geometric permutations is at most two for $n \geq 9$, and at most three for $3 \leq n \leq 8$. This has been recently shown by Cheong et al. [**234**], following several papers that obtained weaker bounds [**112, 447, 464, 692**].

Isotopy and geometric permutations. As mentioned, a simple continuity argument shows that two isotopic transversal lines induce the same geometric permutation. The converse is true in the plane, but an example, due to Goodman et al. [**386**], shows that the same geometric permutation can be attained by arbitrarily many connected components of $T^d(\mathcal{F})$, for any $d \geq 3$.

It is therefore natural to seek conditions on the family $\mathcal{F}$ that ensure that these two notions of equivalence coincide. For example, do these equivalence relations coincide for families of pairwise disjoint balls in $\mathbb{R}^d$? in $\mathbb{R}^3$?

The study of $T^d(\mathcal{F})$ can be reduced to the study of the space of orientations of line transversals to $\mathcal{F}$, as a subset of $\mathbb{S}^{d-1}$. Holmsen et al. [**443**] and Greenstein [**391**] show that, for families $\mathcal{B}$ of balls, the number of components of $T^3(\mathcal{B})$ is equal to the number of components of the orientation space.

Holmsen et al. [**443**] have studied the case of pairwise disjoint *unit* balls, and have shown that, for any triple of pairwise disjoint unit balls B_1, B_2, B_3 in $\mathbb{R}^3$, the space $D^3(B_1, B_2, B_3) \subseteq \mathbb{S}^2$ of orientations of line transversals to the triple that meet them in the fixed order (B_1, B_2, B_3) is *spherically convex*. This implies, among other consequences, that each geometric permutation of any family $\mathcal{B}$ of pairwise disjoint unit balls in $\mathbb{R}^3$ induces a single connected component of $T^3(\mathcal{B})$, or, equivalently, all transversals with the same geometric permutation are isotopic to each other.

Greenstein [**391**] gives an example that shows that the key technical step in the analysis of [**443**] fails for arbitrary balls. This has been partially overcome in a recent paper by Cheong et al. [**233**], who have shown that this property holds for so-called *pairwise inflatable* balls in any dimension, where two balls, with radii r_1, r_2 and distance c between their centers, are called pairwise inflatable if $c^2 > 2(r_1^2 + r_2^2)$. This suffices to handle the case of pairwise disjoint unit balls in any dimension, and leads to Helly-type theorems for transversals of pairwise disjoint unit balls, in any dimension.

The work of Greenstein [**391**] sheds more light on the topological structure of the transversal space of a family $\mathcal{B}$ of n pairwise disjoint balls in $\mathbb{R}^3$, where the ratio between the radius of the largest ball in the family and the radius of the smallest ball is γ. Specifically, Greenstein shows that in this case $T^3(\mathcal{B})$ consists of $O(\gamma n \lambda_{12}(n) + \gamma^2 n^2)$ connected components (where, recalling Chapter 3, $\lambda_{12}(n)$ is the maximal length of a Davenport–Schinzel sequence of order 12 on n symbols). This work is related to a long standing conjecture of Danzer [**258**], whose 3-dimensional version has recently been settled in the affirmative by Holmsen et al. [**443**]. The analysis proceeds by fixing a pair of balls B_i, B_j, whose centers are "too close", and by bounding the number of common tangents to B_i, B_j and two other balls that correspond to (potentially) non-convex features of the orientation space. The latter task is accomplished by a careful analysis of a specific representation of the tangent lines to a fixed pair of balls.

7. Open Problems

Although the manifold of all non-oriented lines in 3-space has been well studied [**442**], less seems to be known about the manifold of oriented lines that we have used in this chapter, and which seems to be computationally of significant advantage. It is known that this manifold is topologically equivalent to the oriented Grassmannian manifold $\tilde{M}_{4,2}(\mathbb{R})$, which happens to be the same[4] as $S^2 \times S^2$.

In general, in spite of the progress reviewed above, many questions about lines in space are still open. These questions can be gathered from the reviews of the various topics given above. Here is a short list of some of these questions. The reader can easily add many other open problems to the list.

PROBLEM 7.1.
   (a) *Tighten the bound on the number of joints in a line arrangement in 3-space.*
   (b) *Is there a general algorithm for ray shooting amid n triangles in space that uses near-linear storage and answers queries in $O(n^\gamma)$ time, for $\gamma < 3/4$? Similarly, is there a general algorithm that answers queries in logarithmic (or polylogarithmic) time but uses $O(n^\gamma)$ storage, with $\gamma < 4$?*
   (c) *Is there a subquadratic bound on the number of cuts that eliminate all cycles in a set of lines in general position?*
   (d) *Tighten the bound on the maximum number of geometric permutations in a set of n arbitrary pairwise disjoint convex objects in $\mathbb{R}^3$. Obtaining a (nearly) cubic upper bound would already constitute a major development.*
   (e) *Is there a constant upper bound on the number of connected components of the space of line transversals to pairwise disjoint balls in space (of arbitrary radii)?*
   (f) *Here is a nice (and probably very hard) visibility problem that has been around for nearly a decade: Given a collection $\mathcal{T}$ of n pairwise disjoint triangles in 3-space, report all pairs of mutually visible triangles in $\mathcal{T}$; a*

---

[4] A geometric proof can be given by associating to every pair of unit vectors u, v (placed at the origin of 3-space) the oriented line ℓ that passes through the tip of the vector $(u \times v)/(1 + u \cdot v)$ and has the direction of the vector $u + v$. When $u = -v$ the line ℓ is by definition the line at infinity on the plane normal to v and oriented relative to v according to the right-hand rule. It is easy to check that this mapping is continuous, one-to-one, and generates all oriented lines of 3-space.

pair (Δ_1, Δ_2) is mutually visible if there exists a segment that connects a point on Δ_1 to a point on Δ_2 and does not meet any other triangle of $\mathcal{T}$. The goal is to solve this problem in $O(n^\gamma)$ time, for some $\gamma < 4$.

CHAPTER 8

Geometric Coloring Problems: Sphere Packings and Frequency Allocation

1. Multiple Packings and Coverings

The notion of multiple packings and coverings was introduced independently by Davenport and László Fejes Tóth. Given a system $\mathcal{S}$ of subsets of an underlying set X, we say that they form a *k-fold packing (covering)* if every point of X belongs to *at most (at least)* k members of $\mathcal{S}$. A 1-fold packing (covering) is simply called a *packing (covering)*. Clearly, the union of k packings (coverings) is always a k-fold packing (covering). Today there is a vast literature on this subject [352, 354].

Many results are concerned with the determination of the maximum density $\delta^k(C)$ of a k-fold packing (minimum density $\theta^k(C)$ of a k-fold covering) with congruent copies of a fixed convex body C. The same question was studied for multiple *lattice packings (coverings)*, where only isothetic copies of C can be used, such that their relative displacement vectors form a lattice, giving rise to the parameter $\delta_L^k(C)$ ($\theta_L^k(C)$). In this section, it is usually assumed that the geometric arrangements, packings, and coverings under consideration are *locally finite*, that is, any bounded region intersects only finitely many members of the arrangement.

Because of the strongly combinatorial flavor of the definitions, it is not surprising that combinatorial methods have played an important role in these investigations. For instance, Erdős and Rogers [336] used the "probabilistic method" to show that $\mathbb{R}^d$ can be covered with congruent copies (actually, with translates) of a convex body so that no point is covered more than $e(d \ln d + \ln \ln d + 4d)$ times (see [585], and [372] for another combinatorial proof based on Lovász' Local Lemma). Note that this easily implies that

$$k\theta_d \leq \theta^k(C) \leq k\theta(C)$$

for a constant $\theta_d > 0$ depending on d (where $\theta(C)$ is shorthand for $\theta^1(C)$).

To establish almost tight density bounds, at least for lattice arrangements, it would be sufficient to show that any k-fold packing (covering) splits into roughly k packings (coverings), or into about k/l disjoint l-fold packings (coverings) for some $l < k$. The initial results were promising. Blundon [155] and Heppes [436] proved that for unit disks $C = B^2$, we have

$$\theta_L^2(C) = 2\theta_L(C), \quad \delta_L^k(C) = k\delta_L(C) \text{ for } k \leq 4,$$

and these results were extended to arbitrary centrally symmetric convex bodies in the plane by Dumir and Hans-Gill [278] and by G. Fejes Tóth [351, 353]. In fact, there was a simple reason for this phenomenon: It turned out that every 3-fold lattice packing of the plane can be decomposed into 3 packings, and every 4-fold lattice packing into *two* 2-fold ones. This simple scheme breaks down for larger values of k.

The situation becomes slightly more complicated if we do not restrict our attention to *lattice* arrangements. It is easy to see that any 2-fold packing of homothetic copies of a plane convex body splits into 4 packings [**579**]. Furthermore, any k-fold packing $\mathcal{C}$ with not too "elongated" convex sets splits into at most $9\lambda k$ packings, where
$$\lambda := \max_{C \in \mathcal{C}} \frac{(\text{circumradius}(C))^2 \pi}{\text{area}(C)}.$$
(Here the constant factor 9λ can be easily improved.)

One would expect that similar results hold for coverings rather than packings. However, in this respect we face considerable difficulties. For any k, it is easy to construct a k-fold covering of the plane with not too elongated convex sets (of different shapes but of roughly the same size) that cannot be decomposed even into *two* coverings [**579**]. The problem is far from being trivial even for coverings with congruent disks. In an unpublished manuscript, Mani-Levitska and Pach have shown that every 33-fold covering of the plane with congruent disks splits into two coverings [**523**]. Another positive result was established in [**580**].

THEOREM 1.1 (Pach [**580**]). *For any centrally symmetric convex polygon P, there exists a constant $k = k(P)$ such that every k-fold covering of the plane with translates of P can be decomposed into two coverings.*

At first glance, one may believe that approximating a disk by centrally symmetric polygons, the last theorem implies that any sufficiently thick covering with congruent disks is decomposable. The trouble is that, as we approximate a disk with polygons P, the value $k(P)$ tends to infinity. Nevertheless, it follows from Theorem 1.1 that if $k = k(\varepsilon)$ is sufficiently large, then any k-fold covering with disks of radius 1 splits into a covering and an "almost covering" in the sense that it becomes a covering if we replace each of its members by a concentric disk whose radius is $1 + \varepsilon$.

Recently, Tardos and Tóth [**719**] have managed to extend Theorem 1.1 to triangles and it seems likely that their method will go through for any (not necessarily centrally symmetric) convex polygon P. Here the assumption that P is convex cannot be dropped.

Surprisingly, the analogous decomposition result is false for multiple coverings with balls in *three* and higher dimensions.

THEOREM 1.2 (Mani and Pach [**523**]). *For any k, there exists a k-fold covering of $\mathbb{R}^3$ with unit balls that cannot be decomposed into two coverings.*

Somewhat paradoxically, it is the very heavily covered points that create problems. Pach [**579**] (see also [**77**, p. 68.]) noticed that by the Lovász Local Lemma we obtain

THEOREM 1.3 (Pach [**579**]). *Any k-fold covering of $\mathbb{R}^3$ with unit balls, no $c2^{(k-1)/3}$ of which have a point in common, can be decomposed into two coverings. (Here c is a positive constant.)*

Similar theorems hold in $\mathbb{R}^d$ ($d > 3$), except that the value $2^{(k-1)/3}$ must be replaced by $2^{(k-1)/d}$.

2. Cover-Decomposable Families and Hypergraph Colorings

These questions can be reformulated in a slightly more general combinatorial setting. A family $\mathcal{F}$ of sets in $\mathbb{R}^d$ is called *cover-decomposable* if there exists a positive integer $k = k(\mathcal{F})$ such that any k-fold covering of $\mathbb{R}^d$ with members from $\mathcal{F}$ can be decomposed into two coverings.

In particular, Theorem 1.1 above can be rephrased as follows. The family consisting of all translates of a given centrally symmetric convex polygon P in the plane is cover-decomposable. (P is considered to be an *open* region.)

Note that Theorem 1.1 has an equivalent "dual" form. Given a system $\mathcal{S}$ of translates of P, let $C(\mathcal{S})$ denote the set of centers of all members of $\mathcal{S}$. Clearly, $\mathcal{S}$ forms a k-fold covering of the plane if and only if every translate of P contains at least k elements of $C(\mathcal{S})$. The fact that the family of translates of P is cover-decomposable can be expressed by saying that there exists a positive integer k satisfying the following condition: any set C of points in the plane such that $|P' \cap C| \geq k$ for all translates P' of P can be partitioned into two disjoint subsets C_1 and C_2 with

$$|C_1 \cap P'| \neq \emptyset \text{ and } |C_2 \cap P'| \neq \emptyset \text{ for every translate } P' \text{ of } P.$$

We can think of C_1 and C_2 as "color classes."

This latter condition, in turn, can be reformulated as follows. Let $H(C)$ denote the (infinite) hypergraph whose vertex set is C and whose (hyper)edges are precisely those subsets of C that can be obtained by taking the intersection of C and a translate of P. By assumption, every hyperedge of $H(C)$ is of size at least k. The fact that C can be split into two color classes C_1 and C_2 with the above properties is equivalent to saying that $H(C)$ is *two-colorable*.

DEFINITION 2.1. *A hypergraph is* two-colorable *if its vertices can be colored by two colors such that no edge is monochromatic.*

A hypergraph is called two-edge-colorable *if its edges can be colored by two colors such that every vertex is contained in edges of both colors.*

Obviously, a hypergraph H is two-edge colorable if and only if its *dual hypergraph* H^* is two-colorable. (By definition, the vertex set and the edge set of H^* are the edge set and the vertex set of H, respectively, with the containment relation reversed.)

Summarizing, Theorem 1.1 can be rephrased in two equivalent forms. For any centrally symmetric convex polygon P in the plane, there is a $k = k(P)$ such that

(1) any k-fold covering of $\mathbb{R}^2$ with translates of P (regarded as an infinite hypergraph on the vertex set $\mathbb{R}^2$) is two-edge-colorable;

(2) for any set of points $C \subset \mathbb{R}^2$ with the property that each translate of P covers at least k elements of C, the hypergraph $H(C)$ whose edges are the intersections of C with all translates of P is two-colorable.

Here we outline a geometric construction showing that certain families of sets in the plane are not cover-decomposable.

Let T_k denote a rooted k-ary tree of depth $k-1$. That is, T_k has $1 + k + k^2 + k^3 + \ldots + k^{k-1} = \frac{k^k - 1}{k-1}$ vertices. The only vertex at level 0 is the root v_0. For $0 \leq i < k-1$, each vertex at level i has precisely k children. The k^{k-1} vertices at level $k-1$ are all leaves.

DEFINITION 2.2. *For any rooted tree T, let $H(T)$ denote the hypergraph on the vertex set $V(T)$, whose hyperedges are all sets of the following two types:*
1. **Sibling** *hyperedges: for each vertex $v \in V(T)$ that is not a leaf, take the set of all children of v;*
2. **Descendent** *hyperedges: for each leaf $v \in V(T)$, take all vertices along the unique path from the root to v.*

Obviously, $H_k := H(T_k)$ is a k-uniform hypergraph with the following property. No matter how we color the vertices of H_k by two colors, red and blue, say, at least one of the edges will be monochromatic. In other words, H_k is not two-colorable. Indeed, assume without loss of generality that the root v_0 is red. The children of the root form a sibling hyperedge $S(v_0)$. If all points of $S(v_0)$ are blue, we are done. Otherwise, pick a red point $v_1 \in S(v_0)$. Similarly, there is nothing to prove if all points of $S(v_1)$ are blue. Otherwise, there is a red point $v_2 \in S(v_1)$. Proceeding like this, we either find a sibling hyperedge $S(v_i)$, all of whose elements are blue, or we construct a red descendent hyperedge $\{v_0, v_1, \ldots, v_{k-1}\}$.

DEFINITION 2.3. *Given any hypergraph H, a* planar realization *of H is defined as a pair $(P, \mathcal{S})$, where P is a set of points in the plane and $\mathcal{S}$ is a system of sets in the plane such that the hypergraph obtained by taking the intersections of the members of $\mathcal{S}$ with P is isomorphic to H.*
A realization of the dual hypergraph of H is called a dual realization *of H.*

It was shown by Pach *et al.* [608] that for any rooted tree T, the hypergraph $H(T)$ defined above has both a planar and a dual realization, in which the members of $\mathcal{S}$ are open strips. In particular, the hypergraph $H_k = H(T_k)$ permits such realizations for every positive k. These results easily imply the following:

THEOREM 2.4 (Pach *et al.* [608]). *The family of open strips in the plane is not cover-decomposable.*

Indeed, fix a positive integer k, and assume that we have shown that $H_k = H(T_k)$ has a dual realization with strips. This means that the set of vertices of T_k can be represented by a collection $\mathcal{S}$ of strips, and the set of (sibling and descendent) hyperedges by a point set $P \subset \mathbf{R}^2$ whose every element is covered exactly by the corresponding k strips. Recall that H_k is not two-colorable, hence its dual hypergraph H_k^* is not two-edge-colorable. In other words, no matter how we color the strips in $\mathcal{S}$ with two colors, at least one point in P will be covered only by strips of the same color. Add now to $\mathcal{S}$ all open strips that do not contain any element of P. Clearly, the resulting (infinite) family of strips, $\mathcal{S}'$ (and hence a locally finite subfamily of $\mathcal{S}'$), forms a k-fold covering of the plane, and it does not split into two coverings. This proves Theorem 2.4.

In fact, a "degenerate" version of Theorem 2.4 is also true, in which strips are replaced by straight-lines (that is, by "strips of width zero").

THEOREM 2.5 (Pach *et al.* [608] and Pesant [626]). *The family of straight lines in the plane is not cover-decomposable.*

The last result implies the following generalization of Theorem 2.4: The family of open strips of *unit width* in the plane is not cover-decomposable.

In the unpublished manuscript [**523**] as well as in [**608**] it was shown that the hypergraph $H_k = H(T_k)$ permits a planar realization. In fact, in [**523**] a somewhat stronger form of this statement was used to establish that, for any $d \geq 3$, the family of open unit balls in $\mathbf{R}^d$ is not cover-decomposable, for any $d \geq 3$ (Theorem 1.2).

It is also true that the hypergraph $H_k = H(T_k)$ permits a dual realization in the plane with axis-parallel rectangles, for every positive k. This implies, in exactly the same way as outlined in the paragraph below Theorem 2.4, that the following is true.

THEOREM 2.6 ([**608**]). *The family of open axis-parallel rectangles in the plane is not cover-decomposable.*

We cannot decide whether H_k permits a planar realization. However, it can be shown [**232**] that a sufficiently large randomly and uniformly selected point set P in the unit square with large probability has the following property. No matter how we color the points of P with two (or with any fixed number of) colors, there is an axis-parallel rectangle containing at least k elements of P, all of the same color.

Recall that the family of translates of any triangle or any centrally symmetric convex polygon Q is cover-decomposable (see [**719**] and Theorem 1.1). The next result shows that this certainly does not hold for some *concave* polygons Q.

THEOREM 2.7 ([**608**]). *The family of all translates of a given (open) concave quadrilateral is not cover-decomposable.*

We conjecture that the same is true for every nonconvex polygon.

The proofs also yield that Theorems 2.4, 2.6, and 2.7 remain true for *closed* strips, rectangles, and quadrilaterals. Most arguments follow the same general inductional scheme, but the subtleties require separate treatment.

For illustration, we prove only Theorem 2.4; the other proofs are slightly trickier. As we have shown, Theorem 2.4 is an easy corollary of:

LEMMA 2.8. *For any rooted tree T, the hypergraph $H(T)$ permits a dual realization with strips.*

PROOF. Recall that a dual realization of a hypergraph H is a planar realization of its dual H^*. That is, given a tree T, a dual realization of $H(T)$ is a pair $(P, \mathcal{S})$, where P is a set of points in the plane representing the (sibling and descendent) hyperedges of $H(T)$, and $\mathcal{S}$ is a system of regions representing the vertices of T such that a region $S \in \mathcal{S}$ covers a point $p \in P$ if and only if the vertex corresponding to S is contained in the hyperedge corresponding to p.

We prove Lemma 2.8 by induction on the number of vertices of T. The statement is trivial if T has only one vertex. Suppose that T has n vertices and that the statement has been proved for all rooted trees with fewer than n vertices. Let v_0 be the root of T, and let $v_0 v_1 \cdots v_m$ be a path of maximum length starting at v_0. Let $U = \{u_1, u_2, \ldots, u_k\}$ denote the set of children of v_{m-1}. Clearly, each element of U is a leaf of T, one of them is v_m, and U is a sibling hyperedge of $H(T)$. Let T' denote the tree obtained by deleting from T all elements of U. The vertex v_{m-1} is then a leaf of T'.

By the induction hypothesis, $H(T')$ permits a dual realization $(P, \mathcal{S})$ with (open) strips. We can assume without loss of generality that no element of P

lies on the boundary of any strip in $\mathcal{S}$, otherwise, we could slightly decrease the widths of some strips without changing the containment relation.

Let $p \in P$ be the point corresponding to the descendent hyperedge $\{v_0, v_1, \ldots, v_{m-1}\}$ of $H(T')$. Take a short segment σ whose one endpoint is p and which does not intersect the boundary of any strip in $\mathcal{S}$. Then, for any point $x \in \sigma$ and for any strip $S \in \mathcal{S}$, we have $x \in S$ if and only if $p \in S$. Let $\sigma_1, \sigma_2, \ldots, \sigma_k$ be pairwise disjoint open subsegments of σ.

For any $1 \leq i \leq k$, choose a (very thin) open strip S^i, almost parallel to σ, such that

(1) $S^i \cap \sigma = \sigma_i$,
(2) $S^i \cap P = \emptyset$,
(3) all strips S^i have a point q in common.

Add $S^1, S^2, \ldots, S^k$ to $\mathcal{S}$. These strips will represent the vertices $u_1, u_2, \ldots, u_k \in V(T)$, respectively. For any $1 \leq i \leq k$, take a point $p_i \in \sigma_i$, these points will represent the descendent hyperedges of $H(T)$, corresponding to the paths connecting v_0 to $u_1, u_2, \ldots, u_k$, respectively. Finally, delete p and add q to the set P; the latter point represents the sibling hyperedge $U = \{u_1, u_2, \ldots, u_k\}$. We have obtained a dual realization of $H(T)$, so we are done. $\square$

3. Frequency Allocation and Conflict-Free Coloring

Motivated by a frequency assignment problem in cellular telephone networks, Even et al. [**339**] studied the following question. Given a set P of n points in general position in the plane, what is the smallest number of colors in a coloring of the elements of P with the property that any closed disk D with $D \cap P \neq \emptyset$ has an element whose color is not assigned to any other element of $D \cap P$. We refer to such a coloring as a *conflict-free* coloring of P with respect to disks.

In the specific application, the points correspond to *base stations* interconnected by a fixed backbone network. Each *client* continuously scans frequencies in search of a base station within its (circular) range with good reception. Once such a base station is found, the client establishes a radio link with it, using a frequency not shared by any other station within her range. Therefore, a conflict-free coloring of the points corresponds to an assignment of frequencies to the base stations, which enables every client to connect to a base station without interfering with the others.

THEOREM 3.1 (Even et al. [**339**]). *Any set of n points in the plane has a conflict-free coloring with respect to disks, using $O(\log n)$ colors. This bound is asymptotically tight in the worst case.*

PROOF. Assume for simplicity (as in the beginning of this section) that the points are in general position, that is, no four lie on a circle. Define the *Delaunay graph* $G(P)$ of the point set P by connecting a pair of its elements with a straight-line segment if there is a disk containing both of them but no other point in P. It is well known that these edges define a triangulation of the convex hull of P. In particular, $G(P)$ is a planar graph and so it is four-colorable. Take an at least $\frac{n}{4}$-element independent set $I_1 \subset P$ in this graph, and color each point of I_1 by color 1. Repeat the same procedure for the set $P_1 := P \setminus I_1$. That is, take an independent set $I_2 \subset P_1$ of at least $\frac{|P_1|}{4}$ vertices in the Delaunay graph $G(P_1)$, and color all of its elements with color 2. Set $P_2 := P_1 \setminus I_2$, and continue until no vertices are left.

Since in each step we color at least a positive fraction (one quarter) of the uncolored vertices, the algorithm will terminate in $O(\log n)$ steps.

It remains to verify that the resulting coloring is conflict-free. Take any disk D that contains at least one element of P. Pick a point in $P \cap D$ whose color j is the largest. We claim that the color j occurs only once in D. Suppose, in order to obtain a contradiction, that there are at least two points in D with color j. By continuously shrinking D, we can obtain a disk $D' \subset D$ that contains precisely two points of color j. These points must have been connected by an edge in the Delaunay graph $G(P_{j-1})$. Therefore, it is impossible that both of them belonged to the independent set of vertices I_j in this graph. Hence, at least one of them was not colored j, a contradiction.

This algorithm can easily be implemented in polynomial time. □

The logarithmic bound in Theorem 3.1 cannot be improved. To see this, consider a set P of n points on a line. Clearly, any disk contains an interval of this line and vice versa, for any interval there is a disk whose intersection with the line is precisely this interval. Consider an interval I_1 (or a disk) that contains all elements of P. If we start with a conflict-free coloring, I_1 must contain a point $p_1 \in P$ of unique color. This point cuts I_1 into two smaller intervals at least one of which contains at least $\lceil \frac{n-1}{2} \rceil$ points. Denote this interval by I_2, and choose a point of unique color in it. We can repeat this procedure at least $\log_2 n$ times, and in each step we discover a point whose color has not been encountered before. (The points of P are not in general position, but the argument just given continues to hold for any sufficiently small perturbation of the points of P.)

In fact, a stronger statement is true.

THEOREM 3.2 (Pach and Tóth [612]). *Any set of n points in general position in the plane requires at least $\Omega(\log n)$ colors for a conflict-free coloring with respect to disks.*

The above Delaunay-based argument suggests a general method for constructing conflict-free colorings of point sets with respect to various families of geometric figures (see, e.g., [691]). For instance, let P be a set of points in general position in three-dimensional space, and suppose that we want to color them with as few colors as possible so that in every *half-space* H that contains at least one element of P, there is a point whose color is not assigned to any other element of $P \cap H$. Notice that, using an extra color if necessary, it is sufficient to prove this statement in the special case when P is the vertex set of a convex polytope. (Otherwise, assign this additional color to all points lying in the interior of the convex hull of P.) As before, we can define a "Delaunay-type" graph $G(P)$ by connecting two points of P with an edge if there is a half-space containing both of them but no other element of P. If all points of P are vertices of its convex hull $\text{conv}(P)$, the edges of $G(P)$ are precisely the edges of the convex polytope $\text{conv}(P)$. In particular, they form a planar graph, and the rest of the argument can be repeated *verbatim*.

COROLLARY 3.3 (Even *et al.* [339]). *Any set of n points in general position in $\mathbb{R}^3$ has a conflict-free coloring with respect to half-spaces, using $O(\log n)$ colors. This bound is asymptotically tight in the worst case.*

The last result has an immediate corollary via duality, which is also interesting from the point of view of applications for cellular networks. Let now the base stations be represented by disks (by their reception ranges). Each base station uses a fixed frequency. If a client falls into the reception range of several stations, to avoid interference, she wants to establish connection with a station whose frequency is not used by the others. In this model, we wish to assign frequencies to the ranges, that is, we wish to color the disks rather than the points.

COROLLARY 3.4 (Even et al. [**339**]). *Any system $\mathcal{D}$ of n disks in the plane can be colored by $O(\log n)$ colors with the property that for each point p of the plane that is covered by at least one member of $\mathcal{D}$, there is a disk $D_p \in \mathcal{D}$ whose color is "unique", i.e., whose color differs from the color of any other disk that contains p.*

PROOF. To each point $p = (a,b)$ in $\mathbb{R}^2$, assign the plane p^* in $\mathbb{R}^3$, whose equation is $z = -2ax - 2by + a^2 + b^2$. To each disk D of radius r, centered at $(x,y) \in \mathbb{R}^2$ assign the point $D^* = (x, y, r^2 - x^2 - y^2) \in \mathbb{R}^3$. It is easy to verify that a point p lies in D if and only if the point D^* lies in the half-space above p^*. Thus, Corollary 3.4 follows directly from Corollary 3.3. □

Har-Peled and Smorodinsky [**428**] generalized Corollary 3.4 to any system of *pseudo-disks*, that is, simply connected regions whose boundary curves have no tangencies and have at most two crossings per pair. They use the probabilistic approach of Clarkson and Shor [**246**] combined with a lemma of Kedem et al. [**471**] (see Theorem 9.1 of Chapter 2), according to which the boundary of the union of any system $\mathcal{D}$ of n pseudo-disks in the plane consists of at most $O(n)$ "simple" arcs, where an arc is called simple if it belongs to the boundary of a member of $\mathcal{D}$. Smorodinsky [**691**] designed a quadratic-time algorithm for constructing a coloring of n pseudo-disks with $O(\log n)$ colors satisfying the properties in Corollary 3.4.

It is an interesting open problem to decide whether Corollary 3.4 can be strengthened under the additional condition that no point of the plane is covered by more than t disks in $\mathcal{D}$. It can be conjectured that in this case $O(\log t)$ colors suffice. An initial step in this direction was taken in [**76**].

As we have seen before, the independent sets in the Delaunay graph associated with a family of geometric figures play an important role in constructing conflict-free colorings. This raises some interesting questions for *axis-parallel* rectangles in the plane [**339**].

Given a set P of n points in general position in the plane, define their *Delaunay graph* $G_{\text{rect}}(P)$ with respect to axis-parallel rectangles, as a graph with vertex set P, in which two elements $p, q \in P$ are connected by an edge if and only there is an axis-parallel rectangle containing p and q that does not contain any other element of P. It was shown by Ajwani et al. [**62**] that $G_{\text{rect}}(P)$ always has an independent set of size at least $\Omega(n^{0.617})$. On the other hand, it was shown by Chen et al. [**232**] that this bound cannot be improved to linear. They have constructed n-element sets P such that the size of the largest independent set in $G_{\text{rect}}(P)$ is $O(n \log^2 \log n / \log n)$. In fact, they have proved that a randomly and uniformly selected set of n points in a square will meet the requirements with probability tending to 1, as $n \to \infty$.

COROLLARY 3.5 (Ajwani et al. [**62**]). *Any set of n points in general position in the plane permits a conflict-free coloring using $O(n^{0.383})$ colors, with respect to the family of all axis-parallel rectangles.*

4. Online Conflict-Free Coloring

Several recent papers have addressed *online conflict-free coloring*. Here we maintain a set P of input points. Initially, P is empty, and we repeatedly insert points into P, one point at a time. We denote by $P(t)$ the set P after the t-th point has been inserted. Each time we insert a point p, we need to assign a color $c(p)$ to it, which is a positive integer. Once the color has been assigned to p, it cannot be changed in the future. The coloring should remain conflict-free at all times, with respect to a given set of ranges. That is, as in the static case, for any range I that contains points of $P(t)$, there is a color that appears exactly once in I, at any given time.

As it turns out, the online version is considerably harder to analyze, even for points on a line, where the ranges are *intervals*. It follows from the static case that we need at least $\Omega(\log n)$ colors, where n is the final size of P. A matching upper bound, for points on a line, can be obtained with a randomized coloring algorithm, so that the expected number of colors is $O(\log n)$, but a deterministic solution with such a worst-case bound is not known.

Here is a sketch of the randomized algorithm. Imagine that we have an infinite sequence of available colors, which we identify with the integers $\mathbb{N}$. Each color i is independently assigned a random *label* λ_i from among three fixed labels a, b, c. For each point q in the current set P, let $c(q)$ denote its color. For each integer $j \leq c(q)$ we assign to q a label $L_j(q) \in \{a, b, c\}$, in a manner to be described shortly. These labels maintain the following invariant: For each i, let $P_{\geq i}$ denote the subset of all the points currently in P which have been colored by colors $\geq i$. Then, for each color (integer) j, and for each pair of consecutive points $q_1, q_2 \in P_{\geq j}$, we have $L_j(q_1) \neq L_j(q_2)$. That is, the j-th labels are a valid 3-coloring of the path graph that connects all the points that have colors $\geq j$ in their x-order.

When a new point p is inserted, we check, for each color $i \in \mathbb{N}$ in increasing order, whether p can be given color i, and color p with the first legal color. Suppose we have reached color j, and wish to determine whether p can be colored with j. We first give p its j-th label $L_j(p)$, which is a label different from those of its two neighbors in $P_{\geq j}$ (if p is the leftmost or rightmost current point of $P_{\geq j}$, there is more than one choice of label, and any of them will do). If $L_j(p) = \lambda_j$, we give p color j and stop. Otherwise, we proceed to the next potential color $j + 1$ and repeat the process. Note that the process terminates with probability 1.

It is easy to verify the *correctness* of this algorithm: Consider any step during the insertion process, and any interval I that contains points of the current set. Let j be the highest color assigned to points of the current set in I. We claim that I contains only one point with color j. Indeed, suppose to the contrary that I contains more than one such point, and let p be the last inserted point from among these points. Then, at the time of its insertion, p has at least one neighbor q in $P_{\geq j}$ that lies in I. By construction, $L_j(q) \neq L_j(p) = \lambda_j$. But this is a contradiction, because only points whose j-th label is equal to λ_j receive color j, so q has failed this test and must have been given a larger color, contrary to assumption.

To check the *efficiency* of the algorithm is not difficult either: We claim that, for each j, among the points of $P_{\geq j}$ at least a third receive color j, in expectation. Indeed, consider a newly inserted point p that reaches $P_{\geq j}$. Since the random choice of λ_j is independent of the insertion order (see a comment below), the probability that $L_j(p) = \lambda_j$ is exactly $1/3$, so, conditioned on having reached $P_{\geq j}$, p gets color

j with probability $1/3$. This is easily seen to imply that the expected number of points that get color j is $\frac{1}{3}\left(\frac{2}{3}\right)^{j-1} n$. From this, using Markov's inequality, it follows that the number of colors used is $O(\log n)$ with high probability.

We comment that this algorithm assumes an *oblivious adversary*, meaning that the insertion order does not depend in any way on the random choices of the labels λ_j. A non-oblivious adversary can force the algorithm to generate significantly more colors.

This algorithm follows recent work by Bar-Noy *et al.* [**123**] (who present a more general method). The first paper to address online conflict-free coloring is by Chen *et al.* [**230**]. Subsequent work on this problem also appear in Chen [**229**] and Chen *et al.* [**231**]. For a discussion of various other models of the adversary, see [**124**].

CHAPTER 9

From Sam Loyd to László Fejes Tóth: The 15 Puzzle and Motion Planning

1. Sam Loyd and the Fifteen Puzzle

Sam Loyd (1841–1911) was one of the greatest puzzle designers of all times. Was he a mathematician? Certainly not, but he could have become a great one. At the age of fourteen, with two of his brothers he joined a chess club in New York and soon he became obsessed with the game. His first chess problem was published in the New York Saturday Courier on 14 April, 1855. One year later, one of his problems won the first prize in a competition run by the New York Clipper, and next year he became the problem editor of Chess Monthly. He began to study engineering, but he abandoned his studies after he discovered that he can make a living by writing newspapers columns, inventing and selling games and puzzles, working as a plumbing contractor and running music stores. His 1878 collection "Chess Strategy," based on his columns in Scientific American, is a classic. He invented many surprising new themes and methods, e.g., the so-called *retrograde analysis*, which clearly required not only intuition but disciplined sharp mathematical thinking. He produced over ten thousand puzzles, many of which involved sophisticated mathematical ideas. After Sam Loyd's death, his son published "Sam Loyd's Cyclopaedia of 5000 Puzzles, Tricks and Conundrums," another classic, which is now freely available on the internet [**375**].

However, Loyd's most famous puzzle was the Fifteen Puzzle (formerly known as "Fourteen-Fifteen Puzzle") that conquered the world in two waves just like Rubik's cube did a hundred years later. The puzzle consists of fifteen moveable unit square blocks, numbered from 1 to 15, arranged in a four-by-four square box. The goal is to bring the squares in the standard position where they are numbered consecutively. "People became infatuated with the puzzle and ludicrous tales are told of shopkeepers who neglected to open their stores; of a distinguished clergyman who stood under a street lamp all through a wintry night trying to recall the way he had performed the feat. The mysterious feature of the puzzle is that none seem able to remember the sequence of moves whereby they feel sure they have succeeded in solving the puzzle. Pilots are said to have wrecked their ships, and engineers rush their trains past stations. A famous Baltimore editor tells how he went for his noon lunch and was discovered by frantic staff long past midnight pushing little pieces of pie around on a plate! Farmers are known to have deserted their ploughs..." Who wrote this? Sam Loyd, of course, who was also great advertising expert and self-promoter! But the truth is that at many places playing the puzzle during office hours was strictly prohibited, just like today using e-mail for personal purposes.

Loyd offered a thousand dollars for anyone who can solve the puzzle starting from the position that can be obtained from the standard one by swapping the last two squares. Of course, he knew that his money was safe, and the proof clearly required a precise mathematical argument. In 1978, Herstein and Kaplansky [**440**] wrote that "No really easy proof seems to be known," but this is not quite true! (Come up with such a proof! [**86, 455**].)

In his obituary in The Times, it was written that "Loyd had a real gift—such as that shown in the 'Curiosa Mathematica' of the Rev. C. L. Dodgson ... for the fantastic in mathematical science, and had he devoted himself to making use of it, might have earned fame as an investigator in the vast and poetic region of pure mathematics, a worthy follower of Cayley and Sylvester."

In 1984, Kornhauser, Miller, and Spirakis [**500**] generalized the Fifteen Puzzle to arbitrary graphs. Consider a graph G of n vertices, and put $k < n$ numbered "coins" ("chips" or "pebbles") on distinct vertices. A *move* consists of shifting a coin from one vertex to a neighboring, presently unoccupied, one. Obviously, one can ask two general algorithmic problems.

(1) Given an *initial* and a *target position* of the k coins, is it possible to reach the second position from the first one?
(2) If the answer to the first question is yes, design an algorithm for finding the shortest sequence of moves (or a reasonably short one).

According to Kornhauser *et al.*, these problems are relevant to memory management in totally distributed computing systems, where we want to coordinate the transfer of indivisible packets of data between the devices without ever exceeding the capacity of any device. In the above model, each device has unit capacity and each packet occupies a unit memory. The problem can also be studied in the framework of general motion planning problems or "Piano Movers' problems" [**660, 661**].

Kornhauser *et al.* proved that reachability is tractable, that is, the answer to the first question can be found, in polynomial time. If the answer is yes, then there is sequence of $O(n^3)$ moves that solves the problem, and this bound is asymptotically best possible.

However, if we want to find the *shortest* number of moves, that is, we want to answer the stronger version of question 2, the problem becomes *NP-hard*. Moreover, as was shown by Ratner and Warmuth [**644**], it remains NP-hard even if we restrict the problem to the case when G is the $\sqrt{n} \times \sqrt{n}$ grid and $k = n - 1$, which is the direct generalization of the Fifteen Puzzle.

Papadimitriou *et al.* [**616**] have further generalized the question by introducing unnumbered movable *obstacles* that can also reside on the vertices. The coins ("robots") are not allowed to collide with one another and with any of the obstacles. In any step we may move either a robot or an obstacle across a single edge. The problem is already interesting in the case when there is a *single* robot and several obstacles. Our goal is to bring the robot from a starting position (vertex) to a target one, while the final position of obstacles is irrelevant. It was shown that to decide whether this can be achieved by making at most M moves is NP-complete, even when restricted to a planar graph [**616**].

Auletta *et al.* [**114**] found a linear-time algorithm for solving the reachability problem on a tree with k robots and no obstacles.

In certain applications, objects are indistinguishable, therefore the chips (coins) are unlabeled; for instance, a modular robotic system consists of a number of identical modules (robots), each of which having identical capabilities [**280, 282**]. In this variant, the problem is always easier and feasible (in every nontrivial setting).

In some other variants, there is no reason to restrict the movement of the chips to a graph; any collision-free movement in the plane or in a region is permitted. See [**279**], for a survey.

2. Unlabeled Coins in Graphs and Grids

In this section, we use a different notion of *moves*, which is borrowed from the motion planning model. Given a connected graph G, a move from a vertex v_1 to a vertex v_2 is defined as shifting a chip from v_1 to v_2 along a *path* in G so that no intermediate vertices are occupied.

Let V and V' be two n-element subsets of $V(G)$. Imagine that we place a chip at each element of V and we want to move them into the positions of V' (V and V' may have common elements). A move is called a *target move* if it moves a chip to a final target position belonging to V'. Otherwise it is a *non-target* move.

THEOREM 2.1. *In G one can get from any n-element initial configuration V to any n-element final configuration V' using at most n moves, so that no chip moves twice. Moreover, for the case of a tree T with r vertices, there is a $O(r)$-time algorithm which performs the optimal (minimum) number of moves.*

PROOF. It is sufficient to prove the theorem for trees. We argue by induction on the number of chips. Take the smallest tree T containing V and V', and consider an arbitrary leaf l of T. Assume first that the leaf l belongs to V: say $l = v$. If v also belongs to V', the result trivially follows by induction, so assume that this is not the case. Choose a path P in T, connecting v to an element v' of V' such that no internal point of P belongs to V'. Apply the induction hypothesis to $V \setminus \{v\}$ and $V' \setminus \{v'\}$ to obtain a sequence of at most $n-1$ moves, and add a final (unobstructed) move from v to v'.

The remaining case when the leaf l belongs to V' is symmetric: say $l = v'$; choose a path P in T, connecting v' to an element v of V such that no internal point of P belongs to V. Move first v to v' and append the sequence of at most $n-1$ moves obtained from the induction hypothesis applied to $V \setminus \{v\}$ and $V' \setminus \{v'\}$.

We further refine this algorithm so as to minimize the number of moves. We call a vertex that is both a start and target position an *obstacle*. We have four types of vertices: free vertices, chip-only vertices, target-only vertices, and obstacles. Denote by c (resp., t) the number of chip-only (resp., target-only) vertices, and by o the number of obstacles. We have $c+o = o+t = n$. We call a tree *balanced* if it contains an equal number of chip-only and target-only vertices. Clearly, the initial tree T is balanced. If there exists an obstacle whose removal from T breaks T into balanced subtrees, we keep this obstacle fixed and proceed recursively (by induction) on the subtrees. If no obstacle removal breaks T into balanced subtrees, then all obstacles must move (each at least once), hence the number of moves necessary is at least $o+c = n$, and the algorithm in the first part of our proof can be used to obtain an optimal schedule.

The observation above, together with postorder traversal, keeping additional information for every node, is the basis of the linear time algorithm, whose details are left to the reader. □

COROLLARY 2.2 ([**190**]). *In the infinite rectangular grid, we can get from any starting position to any ending position of the same size n in at most n moves.*

Using the intractability of the *Set Cover* problem, it is not hard to see that to determine the *smallest* number of moves necessary to solve the above *Graph Reconfiguration* (GR) problem is NP-hard.

THEOREM 2.3 (Călinescu et al. [**190**]). *The Graph Reconfiguration problem is NP-complete. Moreover, assuming $P \neq NP$, there is an absolute constant $\varepsilon_1 > 0$ such that no polynomial-time algorithm has approximation guarantee at most $1+\varepsilon_1$. That is, the problem is APX-hard.*

PROOF. The decision version of the problem is clearly in NP, so we only have to prove its NP-hardness. We reduce the *Set Cover* problem SC to GR. An instance of the set cover problem consists of a family $\mathcal{F}$ of subsets of a finite set U. The problem is to decide whether there is a set cover of size k for $\mathcal{F}$, i.e., a subset $\mathcal{F}' \subset \mathcal{F}$, with $|\mathcal{F}'| \leq k$, such that every element in U belongs to at least one member of $\mathcal{F}'$. SC is known to be NP-complete [**376**].

Consider an instance of SC represented by a bipartite graph $(B \cup C, E)$, where $U = C$, $\mathcal{F} = B$, and edges describe the membership relation. Construct the undirected graph G shown in Figure 1, with $|A| = |C|$. At the beginning, the chips

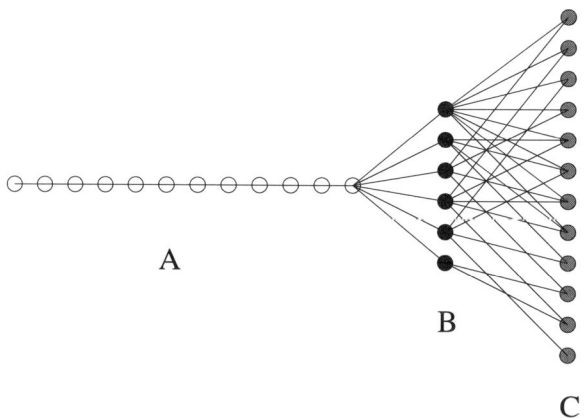

FIGURE 1. The "broom" graph G corresponding to a set cover instance with $|U| = 12$, and $|\mathcal{F}| = 6$. The vertices occupied only by chips are white, those occupied by both chips and targets are black, and those occupied only by targets are shaded. An optimal reconfiguration takes 15 moves (an optimal set cover has size 3).

occupy the positions in $S = A \cup B$, and the set of targets is $T = B \cup C$. Clearly, G can be constructed in polynomial time. The reduction is complete once we establish the following claim, whose proof is left to the reader.

CLAIM 2.4. *There is a set cover consisting of at most q sets if and only if reconfiguration in G can be done using at most $|A| + q$ moves.*

To prove the approximation hardness, we use the same reduction and the fact that 3-SC, the set cover problem in which the size of each set in $\mathcal{F}$ is bounded from above by 3 is APX-hard [**69**, **617**]. □

THEOREM 2.5 (Călinescu et al. [**190**]). *There is a 3-approximation algorithm, which runs in polynomial-time, for estimating the minimum number of moves needed to solve the Graph Reconfiguration problem.*

The algorithm is based on Bar-Yehuda's local ratio algorithm [**125**].

It is interesting to note that, if we count as a move every *edge* traversed by a chip (as we did in Section 1), we can easily minimize the number of moves in *polynomial time*, as follows. Construct a complete weighted bipartite graph $B = (V \cup V', F)$ with bipartition V: the vertices containing chips and V': the vertices containing targets (with obstacles in both sides of the bipartition). The weight of an edge in F is equal to the length of the shortest path connecting the endpoints of the edge in G. Now apply an algorithm for Minimum Weight Perfect Matching in B, and move accordingly: if the path a chip c_1 would take to reach its destination has another chip c_2 on it, have the two chips switch destinations and continue moving c_2. One can check that the number of moves does not exceed the weight of the perfect matching. On the other hand, the optimum solution must move chips to targets and cannot do better than the total length of the shortest paths in a minimum matching.

3. László Fejes Tóth and Sliding Coins

Consider a set (system) of n pairwise disjoint, unlabeled, congruent disks ("coins") in the plane that need to be brought from a given start (initial) configuration S into a desired goal (target) configuration T. In one *move*, we are allowed to take one coin and slide it to another position without colliding with the other coins.

It is easy to see that for the class of congruent disks this problem is always feasible. More generally, it is also feasible for the class of all convex objects.

This follows from an old result (Theorem 3.1) that appears in the work of Fejes Tóth and Heppes [**356**], but it can be traced back to de Bruijn [**183**]; the algorithmic aspects of the problem have been studied by Guibas and Yao [**404**]. We refer to this set of motion rules (moves) as the *sliding model*.

THEOREM 3.1. *Any set of n convex objects in the plane can be separated via translations all parallel to any given fixed direction, with each object moving once only. If the topmost and bottommost points of each object are given (or can be computed in $O(n \log n)$ time), an ordering of the moves can be computed in $O(n \log n)$ time.*

The following simple *universal* algorithm that can be adapted to any set of n convex objects performs $2n$ moves for reconfiguration of n disks. In the first step (n moves), in decreasing order of the x-coordinates of their centers, slide the disks initially along a horizontal direction, one by one to the far right. Note that no collisions can occur. In the second step (n moves), bring the disks "back" to target positions in increasing order of the x-coordinates of their centers. (General convex objects need rotations and translations in the second step). Already for the class of all (not necessarily congruent) disks, one cannot do much better in terms of the number of moves. For the class of congruent segments (as objects), it is easy to construct examples that require $2n - 1$ moves for reconfiguration.

As in the previous section, we call a move a *target move* if it slides a disk to a final target position. Otherwise, it is a *non-target* move. Our *lower* bounds use the following argument: if no target disk coincides with a start disk (so each disk must move), a schedule with x non-target moves consists of at least $n + x$ moves.

In order to obtain an *upper* bound for the number of moves required for the reconfiguration in the worst case, we need to prove a geometric lemma of independent interest: there always exists a line bisecting the set of centers of the start disks into two almost equal parts such that the strip of width 6 around it contains only a small number of disks. A slightly weaker statement guaranteeing the existence of a bisecting line that cuts through few disks was given by Alon *et al.* [**74**]. We include an almost identical proof, for completeness.

We say that a line ℓ bisects an n-element point set into two *almost equal* parts if the number of points on one side of ℓ differs from the number of points on the side by at most one.

LEMMA 3.2. *Let S be a set of n pairwise disjoint unit (radius) disks in the plane. Then there exists a line ℓ bisecting the set of centers of the disks into two almost equal parts such that the parallel strip of width 6 around ℓ contains entirely at most $O(\sqrt{n \log n})$ disks.*

PROOF. Set $m = c_2 \sqrt{n \log n}$ where $c_2 > 0$ is a suitable large constant to be chosen later. Assume for contradiction that the strip of width $w = 6$ around each line bisecting the set of centers of S into two almost equal parts contains at least m disks. Set $k = \lceil \sqrt{n/\log n} \rceil$ and consider k bisecting lines that form angles $i\theta$ with the positive direction of the x-axis (in counterclockwise order), where $i = 0, \ldots, k-1$, and $\theta = \pi/k$.

Let A_i be the set of disks contained (entirely) in the i-strip of width $w = 6$ around the ith bisecting line, $i = 0, \ldots, k-1$. Clearly, we have

$$(3.18) \qquad n \geq |A_0 \cup \ldots \cup A_{k-1}| \geq \sum_{i=0}^{k-1} |A_i| - \sum_{0 \leq i < j \leq k-1} |A_i \cap A_j|$$

by the inclusion-exclusion formula. By our assumption, $\sum_{i=0}^{k-1} |A_i| \geq km$. The summand $|A_i \cap A_j|$ counts the number of disks contained in the intersection of the strips i and j. This intersection is a rhombus whose area is

$$F_{ij} = \frac{w^2}{\sin(j-i)\theta}.$$

Since the disks are pairwise disjoint,

$$|A_i \cap A_j| \leq \frac{F_{ij}}{\pi}.$$

We thus have

$$\sum_{0 \leq i < j \leq k-1} |A_i \cap A_j| = O\left(\sum_{0 \leq i < j \leq k-1} \frac{1}{\sin(j-i)\theta} \right).$$

The identity $\sin \alpha = \sin(\pi - \alpha)$ yields

$$\sum_{0 \leq i < j \leq k-1} \frac{1}{\sin (j-i)\theta} \leq k \sum_{i=1}^{\lfloor k/2 \rfloor} \frac{1}{\sin i\theta}.$$

For $1 \leq i \leq k/2$

$$\frac{1}{\sin i\theta} = \frac{1}{\sin \frac{i\pi}{k}} = O\left(\frac{k}{i}\right).$$

Consequently the second sum in Equation (3.18) is bounded as follows:

$$\sum_{0 \leq i < j \leq k-1} |A_i \cap A_j| = O\left(k^2 \sum_{i=1}^{\lfloor k/2 \rfloor} \frac{1}{i}\right) = O(k^2 \log k).$$

Let $c_1 > 0$ be an absolute constant such that $\sum_{0 \leq i < j \leq k-1} |A_i \cap A_j| \leq c_1 k^2 \log k$. Since $\log k \leq (\log n)/2$ for $n \geq 16$, and using the above estimates, Equation (3.18) can be rewritten as

$$n \geq mk - c_1 k^2 \log k \geq c_2 \sqrt{n \log n} \sqrt{\frac{n}{\log n}} - 2c_1 \frac{n}{\log n} \frac{\log n}{2} = (c_2 - c_1)n.$$

Take now $c_2 = c_1 + 2$, and obtain $n \geq 2n$ which is a contradiction. $\square$

At first glance one would conjecture that Lemma 3.2 holds with $O(\sqrt{n})$ in the place of $O(\sqrt{n \log n})$, but surprisingly this is not the case!

THEOREM 3.3 (Bereg et al. [**138**]). *Given a pair of start and target configurations S and T, consisting of n congruent disks each, $\frac{3n}{2} + O(\sqrt{n \log n})$ moves always suffice for transforming the start configuration into the target configuration. The entire motion can be computed in $O(n^{3/2}/(\log n)^{1/2})$ time. On the other hand, there exist pairs of configurations that require $\left(1 + \frac{1}{15}\right)n - O(\sqrt{n})$ moves for this task.*

PROOF. We start with the upper bound. Let S' and T' be the centers of the start disks and target disks, respectively, and let ℓ be the line guaranteed by Lemma 3.2. Without loss of generality we can assume that ℓ is vertical. Denote by $s_1 = \lfloor n/2 \rfloor$ and $s_2 = \lceil n/2 \rceil$ the number of centers of start disks to the left and to the right of ℓ, respectively. Let $m = O(\sqrt{n \log n})$ be the number of start disks contained in the vertical strip of width 6 around ℓ. Denote by t_1 and t_2 the number of centers of target disks to the left and to the right of ℓ, respectively. By symmetry we can assume that $t_1 \leq n/2 \leq t_2$.

Let R be a region containing all start and target disks (e.g., the smallest axis-aligned rectangle that contains all disks). The algorithm has three steps. All moves in the region R are taken along horizontal lines, i.e., perpendicularly to the line ℓ.

STEP 1: Slide to the far right all start disks whose centers are to the right of ℓ and the (other) start disks in the strip, one by one, in decreasing order of their x-coordinates (with ties broken arbitrarily). At this point all $t_2 \geq n/2$ target disks whose centers are right of ℓ are free.

STEP 2: Using all the $s'_1 \leq n/2$ remaining disks whose centers are to the left of ℓ, in increasing order of their x-coordinates, we fill free target positions to the right of ℓ, in increasing order of their x-coordinates: each disk slides first to the left, then to the right on a wide arc and to the left again in the end. Note that $s'_1 \leq n/2 \leq t_2$. Now all the target positions whose centers are to the left of ℓ are free.

STEP 3: Move to place the far away disks: first continue to fill target positions whose centers are to the right of ℓ, in increasing order of their x-coordinates. When we are done, we fill target positions whose centers are to the left of ℓ, in decreasing order of their x-coordinates. Note that at this point all target positions to the left of ℓ are "free".

The only non-target moves are those done in STEP 1 and their number is $n/2 + O(\sqrt{n \log n})$, so the total number of moves is $3n/2 + O(\sqrt{n \log n})$.

Algorithm. A trivial implementation examines all $k = \lceil \sqrt{n/\log n} \rceil$ strip directions each in $O(n)$ time, in order to find a suitable one, as described in the proof of Lemma 3.2. After that, $O(n \log n)$ time is spent for this direction for sorting and performing the moves. The resulting time complexity is $O(n^{3/2}(\log n)^{-1/2})$.

Lower bound. The target configuration consists of a set of n densely packed unit (radius) disks contained, for example, in a square of side length $\approx 2\sqrt{n}$. The disks in the start configuration enclose the target positions in a ring-like structure with long "legs". Its design is more complicated and uses "rigidity" considerations as described below.

Following László Fejes Tóth, we say that a packing $\mathcal{C}$ of unit (radius) disks in the plane is *stable* if each disk is kept fixed by its neighbors [**174**]. More precisely, $\mathcal{C}$ is stable if none of its elements can be translated by any small distance in any direction without colliding with the others. It is easy to see that any stable system of (unit) disks in the plane has infinitely many elements. Refuting a conjecture of Fejes Tóth, K. Böröczky [**163**] showed that there exist stable systems of unit disks with arbitrarily small density.

The main building block used in Böröczky's construction is a one-way infinite "bridge" made up of disks, which can be defined as follows. In Figure 2, the initial section of such a one-way infinite bridge appears on the left of the vertical line ℓ. Fix a rectilinear xy-coordinate system in the plane. Let us start with five unit disks

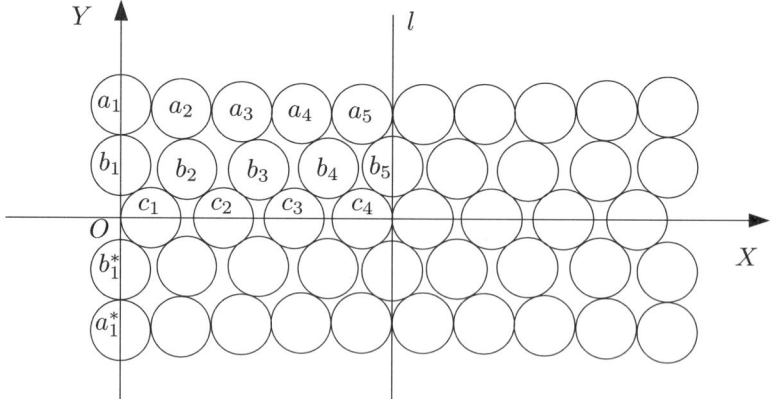

FIGURE 2. A double bridge and its vertical line of symmetry ℓ. The part left of ℓ forms the initial section of a one-way infinite bridge.

centered at
$$a_1 = (0, 2+\sqrt{3}), \quad b_1 = (0, \sqrt{3}), \quad c_1 = (1,0), \quad b_1^* = -b_1, \quad a_1^* = -a_1,$$
that serve as an "abutment." The bridge will be symmetric about the x-axis, so it is sufficient to describe the part of the packing in the upper half-plane. The set of centers of the disks is denoted by C.

Take a strictly convex function $f(x)$ defined for all $x \geq 0$ such that $f(0) = 2+\sqrt{3}$ and $\lim_{x \to \infty} f(x) = 2\sqrt{3}$. Starting with a_1, choose a series of points $a_2, a_3, a_4, \ldots$ belonging to the graph of f such that the distance between any two consecutive points satisfies
$$|a_i - a_{i+1}| = 2 \quad (i = 1, 2, 3, \ldots).$$
All unit disks around these points belong to the packing, so that $a_i \in C$ for every i. These points will uniquely determine all other elements of C, according to the following rules.

Let b_2 be the point at distance 2 from both c_1 and a_2, which lies to the right of the line $c_1 a_2$. Once b_2 is defined, let c_2 be the point on the x-axis, different from c_1, whose distance from b_2 is 2. In general, if b_i and c_i have already been defined, let b_{i+1} denote the point at distance 2 from both c_i and a_{i+1}, lying on the right-hand side of their connecting line, and let $c_{i+1} \neq c_i$ be the (other) point of the x-axis at distance 2 from b_{i+1}. Let C, the set of centers of the disks forming the bridge, consist of all points a_i, b_i, c_i ($i = 1, 2, 3, \ldots$) and their reflections about the x-axis. Note that the points $c_i \in C$ lie on the x-axis, so they are identical with their reflections.

We need four properties of this construction, whose simple trigonometric proofs can be found in [**163**]:

(1) the distance between any two points in C is at least 2;
(2) all unit disks around a_i, b_i, c_i ($i = 2, 3, 4, \ldots$) are kept fixed by their neighbors;
(3) all points $b_2, b_3, b_4, \ldots$ lie strictly below the line $y = \sqrt{3}$;
(4) the x-coordinate of c_i is smaller than that of a_{i+1} ($i = 1, 2, 3, \ldots$).

It is not hard to see that the difference between the x-coordinates of c_i and a_{i+1} tends to zero as i tends to infinity.

Next, we slightly modify the above construction. Take a small positive ε and replace $f(x)$ by the strictly convex function
$$f_\varepsilon(x) := (1+\varepsilon)f(x) - \varepsilon f(0)$$
whose asymptote is the line $y = 2\sqrt{3} - (2-\sqrt{3})\varepsilon$. Clearly, $f_\varepsilon(0) = f(0)$. If we carry out the same construction as above, nothing changes before we first find a point a_i that lies below the line $y = 2\sqrt{3}$. However, if ε is sufficiently large, sooner or later we get stuck: the construction cannot be continued forever without violating any of the conditions listed above. Let k be the first integer for which such an event occurs, involving a_k, a_{k+1}, b_k, or c_k. By varying $\varepsilon > 0$, it can be shown by a simple case analysis that the construction can be realized up to level k so that the difference between the x-coordinates of b_k and a_k is 1. It follows that the disk around a_k is tangent to the vertical line ℓ passing through b_k. Remove the rightmost disk centered at c_k from the set. Thus from the above condition, by taking the union of the part of C built so far together with its reflection about ℓ, we obtain the following:

LEMMA 3.4. *There exist arbitrarily long finite packings ("double-bridges") consisting of five rows of unit disks, symmetric about the coordinate axes, in which all but eight disks are kept fixed by their neighbors. These eight exceptional disks are at the two abutments of the double-bridge and their y-coordinates are $\pm\sqrt{3}, \pm(2+\sqrt{3})$.*

Notice that three such bridges can be connected at a "junction" depicted in Figure 3 so that the angles between their "long" half-axes of symmetries (corresponding to the positive x-axis) are $\frac{2\pi}{3}$. Consequently, using six double-bridges connected by six junctions one can enclose an arbitrarily large hexagonal region H. Let us attach a one-way infinite bridge to each of the unused sides of the junctions. As Böröczky pointed out, the resulting packing is stable.

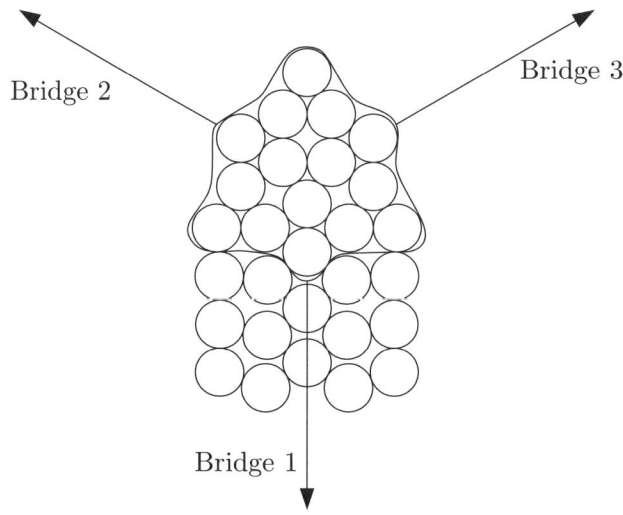

FIGURE 3. Junction of type 1.

Let us refer to the disks in the start (resp., target) configuration as white (resp., black) disks. Now fix a large n, and take n white disks. Use $O(\sqrt{n})$ of them to build six junctions connected by six double-bridges (as described above) to enclose a hexagonal region that can accommodate the n nonoverlapping black disks. See also Figure 4. Divide the remaining white disks into six roughly equal groups, each of size $\frac{n}{6} - O(\sqrt{n})$, and rearrange each group to form the initial section of a one-way infinite bridge attached to the unused sides ("ports") of the junctions. Notice that the number of necessary moves is at least $\left(1 + \frac{1}{30}\right)n - O(\sqrt{n})$. To see this, it is enough to observe, that in order to fill the first target, we have to break up the hexagonal ring around the black disks. That is, we have to move at least one element of the six double-bridges enclosing H. However, with the exception of the at most $6 \times 5 = 30$ white disks at the far ends of the truncated one-way infinite bridges, every white disk is fixed by its neighbors. Each of these bridges consists of five rows of disks of "length" roughly $\frac{n}{30}$, where the length of a bridge is the number of disks along its side. Therefore, before we could move any element of the ring around H, we must start at a far end and move a sequence of roughly $\frac{n}{30}$ white adjacent disks.

Instead of enclosing the n black disks by a hexagon, we can construct a triangular ring T around them, consisting of three double-bridges (see Figure 4). To

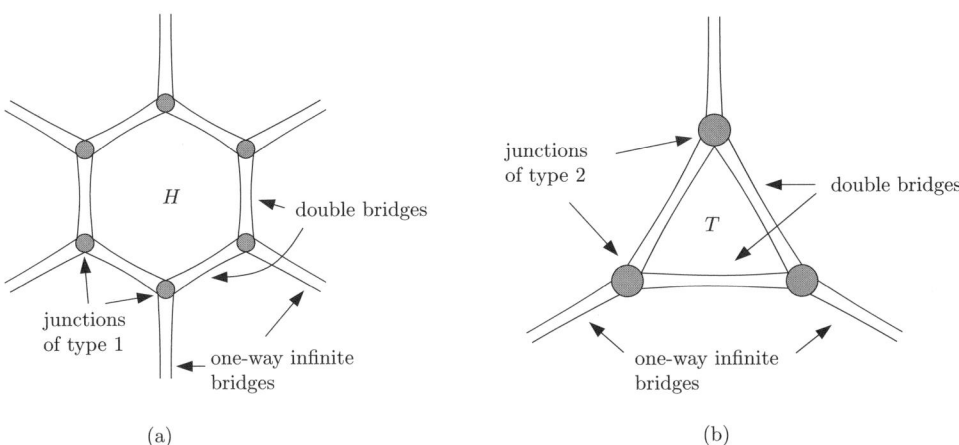

FIGURE 4. Two start configurations based on hexagonal and triangular rings.

achieve this, we have to build a junction of three sides establishing a connection between the abutments of three bridges such that the angles between their half-axes of symmetry are $\frac{5\pi}{6}, \frac{5\pi}{6}$, and $\frac{\pi}{3}$. Such a junction is shown on Figure 5. The convex

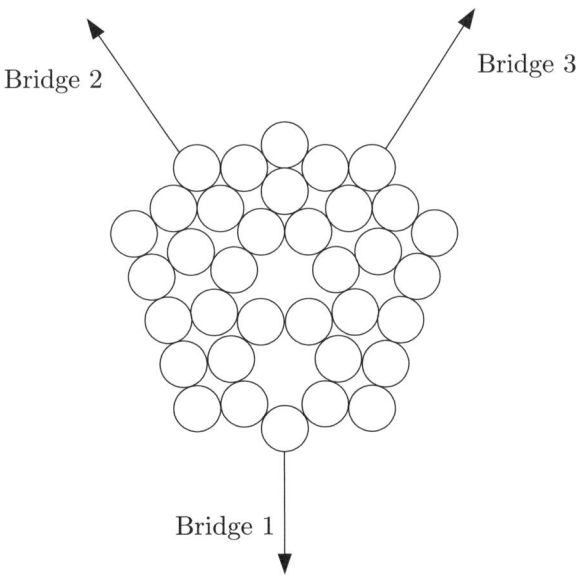

FIGURE 5. Junction of type 2.

hull of the disk centers (for the disks in the junction) is a pentagon symmetric with respect to a vertical line passing through the top vertex. Four out of the five centers along each of the three sides of the pentagon connected to bridges are collinear. The disk centers on the other two sides form two slightly concave chains. The number of necessary moves is at least $\left(1 + \frac{1}{15}\right) n - O(\sqrt{n})$ for this second construction. This completes the proof of Theorem 3.3. □

László Fejes Tóth's (1915–2005) name appeared and his ideas were used at two crucial points in this argument. In fact, it is not a coincidence! He played a pioneering role in creating the field that is called today *Discrete and Computational Geometry*. In the 1940's he started to explore the structural properties of 2- and 3-dimensional packings and coverings with disks, spheres, and other convex sets. Apart from some scattered results by Thue and Kershner, all previous work in this area concerned lattice packings and lattice coverings. Lattices play a fundamental role in number theory and in the geometry of numbers, but restricting investigations to lattice arrangements appeared to be an artificial exercise from a geometric point of view. Fejes Tóth's idea was that the solution of many extremal problems would remain the same if this restriction were dropped. For example, he showed that a densest packing of congruent copies of a centrally symmetric convex set in the plane is necessarily latticelike. In a similar spirit he proved that most regular polytopes can be obtained as the unique solution to a suitable "natural" optimization problem, in the same way that the regular hexagonal honeycomb is, in a certain well defined sense, the most economical structure to house bees. This area of geometry can be regarded as the "genetics"[1] of symmetric objects.

In 1953, Fejes Tóth summarized his results in the seminal book, *Lagerungen in der Ebene auf der Kugel und im Raum* [**355**] that appeared in the prestigious series of Springer-Verlag: Die Grundlehren der Mathematischen Wissenschaften. It is hard to overestimate the impact of this monograph. As C. A. Rogers wrote, "until recently, the theory of packing and covering was not sufficiently well developed to justify the publication of a book devoted exclusively to it. After the publication of L. Fejes Tóth's book in 1953, there would be no need for a second work on the subject..."

László Fejes Tóth was a high school teacher for many years. Perhaps that is why he loved simply stated problems that have an aesthetic appeal and can be explained to a layman. In fact, he is a master of raising such questions, and he knows exactly who the right people are to tell them to. Because of these special abilities, his straightforward, modest style, and his warm personality, he became one of the most influential discrete geometers of the second half of the twentieth century. He passed away in 2005, at the age of 90.

4. Pushing Squares Around

A modular *metamorphic* system consists of a number of identical modules that can connect to, disconnect from, and relocate relative to adjacent modules (see, for example, [**237, 566, 615, 656, 754**]). While individual modules are not capable of moving by themselves, the entire system may be able to reconfigure or move to a new position, when its members repeatedly change their positions relative to their neighbors, by rotating or sliding around other modules [**188, 566, 753**], or by expansion and contraction [**656**]. It is usually assumed that the entire system must remain connected during reconfiguration.

The *motion planning* problem for such a system is that of computing a sequence of module motions that brings the system from a given initial configuration I into a desired goal configuration F. Depending on the existence of such a sequence of motions, we say that the problem is *feasible* or respectively, *infeasible*.

---

[1]This expression was coined by László Fejes Tóth, see, e.g., [**725**].

For instance, in [**237, 615**] upper and lower bounds on the number of moves needed to change I to F are discussed for a system of hexagonal modules. In [**568**] it is shown that in such a system any two connected configurations are mutually reachable as long as they do not contain a certain prohibited pattern. Demaine *et al.* [**261**] have considered a family of one-player games, involving the movement of coins from one configuration to another. Moves are restricted so that a coin can only be placed in a free position adjacent to at least two other coins, in contrast to our motion rules that are required to maintain overall connectedness throughout the reconfiguration process.

Consider a plane that is partitioned into a rectangular integer grid of square cells indexed by their center coordinates in the underlying xy-coordinate system. Of the eight *adjacent* cells of cell $c = c_{x,y}$ in the E $(+x)$, W $(-x)$, N $(+y)$, S $(-y)$, NE, SE, NW and SW directions, the four in the E, W, N and S directions are said to be *side-adjacent* to c, while the other four in the NE, SE, NW and SW directions are said to be *corner-adjacent* to c. We denote by $N(c)$ (resp., $NE(c)$) the cell side-adjacent to c in the N direction (resp., the cell corner-adjacent to c in the NE direction). Similar notation is used to denote the cells side-adjacent or corner-adjacent to c in the other axis and diagonal directions.

At any time each cell may be empty or occupied by a module. The *reconfiguration* of a metamorphic system consisting of n modules is a sequence of configurations (distributions) of the modules in the grid at discrete time steps $t = 0, 1, 2, \ldots$, see below. Let V_t be the configuration of the modules at time t, where we often identify V_t with the set of cells occupied by the modules or with the set of their centers. We are only interested in configurations that are connected, i.e., for each t, the graph $G_t = (V_t, E_t)$ must be connected, where for any t, E_t is the set of edges connecting pairs of cells in V_t that are side-adjacent. V_t yields V_{t+1} when one module m moves from its current location to new location in step t. Here we restrict ourselves to sequential reconfiguration, in which only one module moves at each discrete time step, as explained above. Note that, according to the above definition, the pattern (set of cells or set of integer points) V_t uniquely determines the edge set E_t so that the graph G_t can be characterized by its vertex set V_t. The union of all closed squares belonging to V_t is a connected point set of area $|V_t|$, which will be denoted by $S(V_t)$.

The following two generic motion rules (Figure 6) define the *rectangular model*. These are to be understood as possible in all-axis parallel orientations, in fact generating 16 rules, eight for rotation and eight for sliding.

- *Rotation*: A module m side-adjacent to a stationary module f rotates through an angle of $90°$ around f either clockwise or counterclockwise. Figure 6(a) shows a clockwise NE rotation. For rotation to take place, both the target cell and the cell at the corresponding corner of f that m passes through (NW in the given example) have to be empty.
- *Sliding*: Let f_1 and f_2 be stationary cells that are side-adjacent. A module m that is side-adjacent to f_1 and adjacent to f_2 slides along the sides of f_1 and f_2 into the cell that is adjacent to f_1 and side-adjacent to f_2. Figure 6(b) shows a sliding move in the E direction. For sliding to take place, the target cell has to be empty.

In order to ensure motion precision, each move is guided by one or two modules that are stationary during the same step. The two motion rules of this model also appear in [**282, 283**]. A somewhat similar model is presented in [**188**].

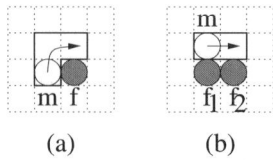

(a) (b)

FIGURE 6. (a) Clockwise NE rotation and (b) sliding in the E direction. Fixed modules are shaded. The cells in which the moves take place are outlined in the figure.

THEOREM 4.1 (Dumitrescu and Pach [**280**]). *The set of motion rules of the rectangular model guarantees the feasibility of motion planning for any pair of connected configurations V and V' having the same number of modules. That is, following the above rules, V and V' can always be transformed into each other so that all intermediate configurations are connected.*

A set of modules that form a straight-line chain in the grid is called a *straight chain*. It is easy to construct examples so that neither sliding nor rotation alone can reconfigure them to straight chains. However, it was proved in [**280**] that the motion rules of the rectangular model (rotation and sliding, Figure 6) are sufficient to guarantee reachability, while maintaining the system connected at each discrete time step. Moreover, this can be achieved by a $O(n^3)$-time algorithm. We conjecture that there exists a quadratic-time algorithm for this task.

At this point we cannot resist mentioning a similar question [**576**]. A configuration consisting of unit cubes of integer coordinates in d-space is called an *animal* if the boundary of their union is homeomorphic to a $(d-1)$-sphere. It is easy to see that in the plane, any animal can be transformed into any other by adding or removing one square at a time so that all intermediate configurations are animals. The corresponding statement in higher dimensions is not known to be true. There exist, however, relatively small animals in 3-space with the property that no cube can be removed from them without violating the condition. This is in sharp contrast to the situation in the plane.

Bibliography

[1] Bernardo M. Ábrego, György Elekes, and Silvia Fernández-Merchant, Structural results for planar sets with many similar subsets, *Combinatorica* **24** (2004), no. 4, 541–554.

[2] Bernardo M. Ábrego and Silvia Fernández-Merchant, On the maximum number of equilateral triangles. I, *Discrete Comput. Geom.* **23** (2000), no. 1, 129–135.

[3] _____, Convex polyhedra in R^3 spanning $\Omega(n^{4/3})$ congruent triangles, *J. Combin. Theory Ser. A* **98** (2002), no. 2, 406–409.

[4] Eyal Ackerman, On the maximum number of edges in topological graphs with no four pairwise crossing edges, *SCG '06: Proc. Twenty-Second Annual Symposium on Computational Geometry* (Sedona, Arizona, USA), ACM, 2006, pp. 259–263.

[5] Eyal Ackerman and Gábor Tardos, On the maximum number of edges in quasi-planar graphs, *J. Combin. Theory Ser. A* **114** (2007), no. 3, 563–571.

[6] Wilhelm Ackermann, Zum Hilbertschen Aufbau der reellen Zahlen, *Math. Ann.* **99** (1928), no. 1, 118–133.

[7] Radek Adamec, Martin Klazar, and Pavel Valtr, Generalized Davenport-Schinzel sequences with linear upper bound, *Discrete Math.* **108** (1992), no. 1-3, 219–229, *Topological, algebraical and combinatorial structures. Frolík's memorial volume.*

[8] Pankaj K. Agarwal, Partitioning arrangements of lines. I. An efficient deterministic algorithm, *Discrete Comput. Geom.* **5** (1990), no. 5, 449–483.

[9] _____, Partitioning arrangements of lines. II. Applications, *Discrete Comput. Geom.* **5** (1990), no. 6, 533–573.

[10] _____, Geometric partitioning and its applications, *Discrete and Computational Geometry*, DIMACS Ser. Discrete Math. Theoret. Comput. Sci., vol. 6, Amer. Math. Soc., Providence, RI, 1991, pp. 1–37.

[11] _____, Ray shooting and other applications of spanning trees with low stabbing number, *SIAM J. Comput.* **21** (1992), no. 3, 540–570.

[12] _____, On stabbing lines for convex polyhedra in 3D, *Comput. Geom.* **4** (1994), no. 4, 177–189.

[13] _____, Range searching, in Goodman and O'Rourke [381], pp. 809–837.

[14] Pankaj K. Agarwal, Nina Amenta, and Micha Sharir, Largest placement of one convex polygon inside another, *Discrete Comput. Geom.* **19** (1998), no. 1, 95–104.

[15] Pankaj K. Agarwal, Roel Apfelbaum, George Purdy, and Micha Sharir, Similar simplices in a d-dimensional point set, *SCG '07: Proc. 23-rd Annual Symposium on Computational Geometry* (Gyeongju, South Korea), ACM, 2007, pp. 232–238.

[16] Pankaj K. Agarwal, Lars Arge, Jeff Erickson, Paulo G. Franciosa, and Jeffrey Scott Vitter, Efficient searching with linear constraints, *J. Comput. System Sci.* **61** (2000), no. 2, 194–216, *Special issue on the Seventeenth ACM SIGACT-SIGMOD-SIGART Symposium on Principles of Database Systems* (Seattle, WA, 1998).

[17] Pankaj K. Agarwal and Boris Aronov, Counting facets and incidences, *Discrete Comput. Geom.* **7** (1992), no. 4, 359–369.

[18] Pankaj K. Agarwal, Boris Aronov, Timothy M. Chan, and Micha Sharir, On levels in arrangements of lines, segments, planes, and triangles, *Discrete Comput. Geom.* **19** (1998), no. 3, 315–331.

[19] Pankaj K. Agarwal, Boris Aronov, János Pach, Richard Pollack, and Micha Sharir, Quasi-planar graphs have a linear number of edges, *Combinatorica* **17** (1997), no. 1, 1–9.

[20] Pankaj K. Agarwal, Boris Aronov, and Micha Sharir, Computing envelopes in four dimensions with applications, *SIAM J. Comput.* **26** (1997), no. 6, 1714–1732.

[21] _____, Line transversals of balls and smallest enclosing cylinders in three dimensions, *Discrete Comput. Geom.* **21** (1999), no. 3, 373–388.

[22] _____, Motion planning for a convex polygon in a polygonal environment, *Discrete Comput. Geom.* **22** (1999), no. 2, 201–221.

[23] _____, On the complexity of many faces in arrangements of pseudo-segments and circles, *Discrete and Computational Geometry, The Goodman–Pollack Festsschrift* (S. Basu, B. Aronov, J. Pach, and M. Sharir, eds.), *Algorithms Combin.*, vol. 25, Springer, Berlin, 2003, pp. 1–24.

[24] Pankaj K. Agarwal, Boris Aronov, Micha Sharir, and Subhash Suri, Selecting distances in the plane, *Algorithmica* **9** (1993), no. 5, 495–514.

[25] Pankaj K. Agarwal, Sergey Bereg, Ovidiu Daescu, Haim Kaplan, Simeon Ntafos, and Binhai Zhu, Guarding a terrain by two watchtowers, *SCG '05: Proc. Twenty-First Annual Symposium on Computational Geometry* (Pisa, Italy), ACM, 2005, pp. 346–355.

[26] Pankaj K. Agarwal, Mark T. de Berg, Jiří Matoušek, and Otfried Schwarzkopf, Constructing levels in arrangements and higher order Voronoi diagrams, *SIAM J. Comput.* **27** (1998), no. 3, 654–667.

[27] Pankaj K. Agarwal, Sergio Cabello, J. Antoni Sellarès, and Micha Sharir, Computing a center-transversal line, *FSTTCS* (2006), *Lecture Notes in Comput. Sci.*, vol. 4337, Springer, 2006, 93–104. Also *Comput. Geom.*, submitted.

[28] Pankaj K. Agarwal, Danny Z. Chen, S. K. Ganjugunte, E. Misiolek, Micha Sharir, and K. Tang, Stabbing convex polygons with a segment or a polygon, *Proceedings European Symposium on Algorithms* (2008), to appear.

[29] Pankaj K. Agarwal, Alon Efrat, and Micha Sharir, Vertical decomposition of shallow levels in 3-dimensional arrangements and its applications, *SIAM J. Comput.* **29** (2000), no. 3, 912–953.

[30] Pankaj K. Agarwal, Alon Efrat, Micha Sharir, and Sivan Toledo, Computing a segment center for a planar point set, *J. Algorithms* **15** (1993), no. 2, 314–323.

[31] Pankaj K. Agarwal and Jeff Erickson, Geometric range searching and its relatives, *Advances in discrete and computational geometry*, C. Chazelle, J. E.Goodman, and R. Pollack (eds.), *Contemp. Math.*, vol. 223, Amer. Math. Soc., Providence, RI, 1999, pp. 1–56.

[32] Pankaj K. Agarwal, Ido Grabinsky, and Micha Sharir, Online range searching with disks, work in progress.

[33] Pankaj K. Agarwal, Sariel Har-Peled, and Kasturi R. Varadarajan, Approximating extent measures of points, *J. ACM* **51** (2004), no. 4, 606–635.

[34] _____, Geometric approximation via coresets, *Combinatorial and Computational Geometry*, *Math. Sci. Res. Inst. Publ.*, vol. 52, Cambridge Univ. Press, Cambridge, 2005, pp. 1–30.

[35] Pankaj K. Agarwal, Haim Kaplan, and Micha Sharir, Computing the volume of the union of cubes, *SCG '07: Proc. Twenty-Third Annual Symposium on Computational Geometry* (Gyeongju, South Korea), ACM, 2007, pp. 294–301.

[36] Pankaj K. Agarwal, Marc van Kreveld, and Mark H. Overmars, Intersection queries in curved objects, *J. Algorithms* **15** (1993), no. 2, 229–266.

[37] Pankaj K. Agarwal and Jiří Matoušek, Ray shooting and parametric search, *SIAM J. Comput.* **22** (1993), no. 4, 794–806.

[38] _____, On range searching with semialgebraic sets, *Discrete Comput. Geom.* **11** (1994), no. 4, 393–418.

[39] _____, Dynamic half-space range reporting and its applications, *Algorithmica* **13** (1995), no. 4, 325–345.

[40] Pankaj K. Agarwal, Jiří Matoušek, and Otfried Schwarzkopf, Computing many faces in arrangements of lines and segments, *SIAM J. Comput.* **27** (1998), no. 2, 491–505.

[41] Pankaj K. Agarwal, Eran Nevo, János Pach, Rom Pinchasi, Micha Sharir, and Shakhar Smorodinsky, Lenses in arrangements of pseudo-circles and their applications, *J. ACM* **51** (2004), no. 2, 139–186.

[42] Pankaj K. Agarwal, János Pach, and Micha Sharir, State of the union (of geometric objects), *Proc. 2006 AMS/IMS/SIAM Joint Summer Research Conference on Discrete and Computational Geometry: 20 Years Later* (J. Goodman, J. Pach, and R. Pollack, eds.), *Contemp. Math.*, vol. 453, Amer. Math. Soc., Providence, RI, 2008, pp. 9–48.

[43] Pankaj K. Agarwal, Otfried Schwarzkopf, and Micha Sharir, The overlay of lower envelopes and its applications, *Discrete Comput. Geom.* **15** (1996), no. 1, 1–13.

[44] Pankaj K. Agarwal and Sandeep Sen, Randomized algorithms for geometric optimization problems, *Handbook of Randomized Computing, Vol. I*, (S. Rajesekaran, P.M. Pardalos, J.H. Reif, J. Rolim, eds.), *Comb. Optim.*, vol. 9, Kluwer Acad. Publ., Dordrecht, 2001, pp. 151–202.

[45] Pankaj K. Agarwal and Micha Sharir, Red-blue intersection detection algorithms, with applications to motion planning and collision detection, *SIAM J. Comput.* **19** (1990), no. 2, 297–321.

[46] _____, Applications of a new space-partitioning technique, *Discrete Comput. Geom.* **9** (1993), no. 1, 11–38.

[47] _____, On the number of views of polyhedral terrains, *Discrete Comput. Geom.* **12** (1994), no. 2, 177–182.

[48] _____, Efficient randomized algorithms for some geometric optimization problems, *Discrete Comput. Geom.* **16** (1996), no. 4, 317–337.

[49] _____, Ray shooting amidst convex polyhedra and polyhedral terrains in three dimensions, *SIAM J. Comput.* **25** (1996), no. 1, 100–116.

[50] _____, Efficient algorithms for geometric optimization, *ACM Computing Surveys* **30** (1998), no. 4, 412–458.

[51] _____, Arrangements and their applications, in Sack and Urrutia [**657**], pp. 49–119.

[52] _____, Davenport-Schinzel sequences and their geometric applications, in Sack and Urrutia [**657**], pp. 1–47.

[53] _____, Pipes, cigars, and kreplach: The union of Minkowski sums in three dimensions, *Discrete Comput. Geom.* **24** (2000), no. 4, 645–685.

[54] _____, The number of congruent simplices in a point set, *Discrete Comput. Geom.* **28** (2002), no. 2, 123–150.

[55] _____, Pseudo-line arrangements: Duality, algorithms, and applications, *SIAM J. Comput.* **34** (2005), no. 3, 526–552.

[56] Pankaj K. Agarwal, Micha Sharir, and Peter W. Shor, Sharp upper and lower bounds on the length of general Davenport-Schinzel sequences, *J. Combin. Theory Ser. A* **52** (1989), no. 2, 228–274.

[57] Pankaj K. Agarwal, Micha Sharir, and Sivan Toledo, Applications of parametric searching in geometric optimization, *J. Algorithms* **17** (1994), no. 3, 292–318.

[58] Pankaj K. Agarwal, Micha Sharir, and Emo Welzl, Algorithms for center and Tverberg points, *ACM Trans. Algorithms*, to appear.

[59] Hee-Kap Ahn, Mark T. de Berg, Prosenjit Bose, Siu-Wing Cheng, Dan Halperin, Jiří Matoušek, and Otfried Schwarzkopf, Separating an object from its cast, *Computer-Aided Design* **34** (2002), no. 8, 547–559.

[60] Martin Aigner and Günter M. Ziegler, *Proofs from The Book*, third ed., Springer, Berlin, 2004, Including illustrations by Karl H. Hofmann.

[61] Miklós Ajtai, Václav Chvátal, Monroe M. Newborn, and Endre Szemerédi, Crossing-free subgraphs, *Annals Discrete Math.* **12** (1982), 9–12.

[62] Deepak Ajwani, Khaled Elbassioni, Sathish Govindarajan, and Saurabh Ray, Conflict-free coloring for rectangle ranges using $o(n^{.382})$ colors, *SPAA '07: Proc. Nineteenth Annual ACM Symposium on Parallel Algorithms and Architectures* (San Diego, California, USA), ACM, 2007, pp. 181–187.

[63] Tatsuya Akutsu, Hisao Tamaki, and Takeshi Tokuyama, Distribution of distances and triangles in a point set and algorithms for computing the largest common point sets, *Discrete Comput. Geom.* **20** (1998), no. 3, 307–331.

[64] Panagiotis Alevizos, Jean-Daniel Boissonnat, and Franco P. Preparata, An optimal algorithm for the boundary of a cell in a union of rays, *Algorithmica* **5** (1990), no. 4, 573–590.

[65] _____, Corrigendum: "An optimal algorithm for the boundary of a cell in a union of rays" [Algorithmica **5** (1990), no. 4, 573–590; MR1072808 (91g:68154)], *Algorithmica* **6** (1991), no. 2, 292–293.

[66] Gerald L. Alexanderson and John E. Wetzel, Dissections of a plane oval, *Amer. Math. Monthly* **84** (1977), no. 6, 442–449.

[67] _____, Simple partitions of space, *Math. Mag.* **51** (1978), no. 4, 220–225.

[68] _____, Arrangements of planes in space, *Discrete Math.* **34** (1981), no. 3, 219–240, Erratum in **45** (1983), no. 1, 140.

[69] Paola Alimonti and Viggo Kann, Hardness of approximating problems on cubic graphs, *Algorithms and complexity (Rome, 1997)*, Lecture Notes in Comput. Sci., vol. 1203, Springer, Berlin, 1997, pp. 288–298.

[70] Noga Alon, Tools from higher algebra, *Handbook of Combinatorics, Vol. 2*, Elsevier, Amsterdam, 1995, pp. 1749–1783.

[71] Noga Alon, Imre Bárány, Zoltán Füredi, and Daniel J. Kleitman, Point selections and weak ε-nets for convex hulls, *Combin. Probab. Comput.* **1** (1992), no. 3, 189–200.

[72] Noga Alon and Paul Erdős, Disjoint edges in geometric graphs, *Discrete Comput. Geom.* **4** (1989), no. 4, 287–290.

[73] Noga Alon and Ervin Győri, The number of small semispaces of a finite set of points in the plane, *J. Combin. Theory Ser. A* **41** (1986), no. 1, 154–157.

[74] Noga Alon, Meir Katchalski, and William R. Pulleyblank, Cutting disjoint disks by straight lines, *Discrete Comput. Geom.* **4** (1989), no. 3, 239–243.

[75] Noga Alon, Hagit Last, Rom Pinchasi, and Micha Sharir, On the complexity of arrangements of circles in the plane, *Discrete Comput. Geom.* **26** (2001), no. 4, 465–492.

[76] Noga Alon and Shakhar Smorodinsky, Conflict-free colorings of shallow discs, *SCG '06: Proc. Twenty-second Annual Symposium on Computational Geometry* (Sedona, Arizona, USA), ACM, 2006, pp. 41–43.

[77] Noga Alon and Joel H. Spencer, *The Probabilistic Method*, second ed., Wiley-Interscience Series in Discrete Mathematics and Optimization, Wiley-Interscience [John Wiley & Sons], New York, 2000, With an appendix on the life and work of Paul Erdős.

[78] Helmut Alt, Stefan Felsner, Ferran Hurtado, Marc Noy, and Emo Welzl, A class of point-sets with few k-sets, *Comput. Geom.* **16** (2000), no. 2, 95–101.

[79] Helmut Alt, Rudolf Fleischer, Michael Kaufmann, Kurt Mehlhorn, Stefan Näher, Stefan Schirra, and Christian Uhrig, Approximate motion planning and the complexity of the boundary of the union of simple geometric figures, *Algorithmica* **8** (1992), no. 5-6, 391–406.

[80] Nancy M. Amato, O. Burchan Bayazit, Lucia K. Dale, Christopher Jones, and Daniel Vallejo, OBPRM: an obstacle-based PRM for 3D workspaces, *WAFR '98: Proceedings of the Third Workshop on the Algorithmic Foundations of Robotics: the Algorithmic Perspective* (Houston, Texas, United States) (Pankaj K. Agarwal, Lydia. E. Kavraki, and Matt. Mason, eds.), A. K. Peters,, Natick, MA, 1998, pp. 155–168.

[81] Nancy M. Amato, Michael T. Goodrich, and Edgar A. Ramos, Computing faces in segment and simplex arrangements, *STOC'95: Proc. Twenty-seventh Annual ACM Symposium on Theory of Computing* (Las Vegas, Nevada, United States), ACM, 1995, pp. 672–682.

[82] Efthymios G. Anagnostou, Leonidas J. Guibas, and Vassilios G. Polimenis, Topological sweeping in three dimensions, *Algorithms (Tokyo, 1990)*, Lecture Notes in Comput. Sci., vol. 450, Springer, Berlin, 1990, pp. 310–317.

[83] Artur Andrzejak, Boris Aronov, Sariel Har-Peled, Raimund Seidel, and Emo Welzl, Results on k-sets and j-facets via continuous motion, *SCG '98: Proc. Fourteenth Annual Symposium on Computational Geometry* (Minneapolis, Minnesota, United States), ACM, 1998, pp. 192–199.

[84] Roel Apfelbaum and Micha Sharir, Repeated angles in three and four dimensions, *SIAM J. Discrete Math.*, **19** (2005), 294–300.

[85] ———, Large complete bipartite subgraphs in incidence graphs of points and hyperplanes, *SIAM J. Discrete Math.* **21** (2007), no. 3, 707–725.

[86] Aaron F. Archer, A modern treatment of the 15 puzzle, *Amer. Math. Monthly* **106** (1999), no. 9, 793–799.

[87] Esther M. Arkin, Dan Halperin, Klara Kedem, Joseph S. B. Mitchell, and Nir Naor, Arrangements of segments that share endpoints: single face results, *Discrete Comput. Geom.* **13** (1995), no. 3–4, 257–270.

[88] Dennis S. Arnon, George E. Collins, and Scott McCallum, Cylindrical algebraic decomposition. II. An adjacency algorithm for the plane, *SIAM J. Comput.* **13** (1984), no. 4, 878–889.

[89] Boris Aronov, A lower bound on Voronoi diagram complexity, *Inform. Process. Lett.* **83** (2002), no. 4, 183–185.

[90] Boris Aronov, Marshall Bern, and David Eppstein, Arrangements of convex polytopes, Unpublished manuscript, 1995.

[91] Boris Aronov, Bernard Chazelle, Herbert Edelsbrunner, Leonidas J. Guibas, Micha Sharir, and Rephael Wenger, Points and triangles in the plane and halving planes in space, *Discrete Comput. Geom.* **6** (1991), no. 5, 435–442.

[92] Boris Aronov, Herbert Edelsbrunner, Leonidas J. Guibas, and Micha Sharir, The number of edges of many faces in a line segment arrangement, *Combinatorica* **12** (1992), no. 3, 261–274.

[93] Boris Aronov, Alon Efrat, Dan Halperin, and Micha Sharir, On the number of regular vertices of the union of Jordan regions, *Discrete Comput. Geom.* **25** (2001), no. 2, 203–220.

[94] Boris Aronov, Alon Efrat, Vladlen Koltun, and Micha Sharir, On the union of κ-round objects in three and four dimensions, *Discrete Comput. Geom.* **36** (2006), no. 4, 511–526.

[95] Boris Aronov, Paul Erdős, W. Goddard, Daniel J. Kleitman, M. Klugerman, János Pach, and L. J. Schulman, Crossing families, *Combinatorica* **14** (1994), no. 2, 127–134.

[96] Boris Aronov, Vladlen Koltun, and Micha Sharir, Cutting triangular cycles of lines in space, *Discrete Comput. Geom.* **33** (2005), no. 2, 231–247.

[97] _____, Incidences between points and circles in three and higher dimensions, *Discrete Comput. Geom.* **33** (2005), no. 2, 185–206.

[98] Boris Aronov, Jiří Matoušek, and Micha Sharir, On the sum of squares of cell complexities in hyperplane arrangements, *J. Combin. Theory Ser. A* **65** (1994), no. 2, 311–321.

[99] Boris Aronov, János Pach, Micha Sharir, and Gábor Tardos, Distinct distances in three and higher dimensions, *Combin. Probab. Comput.* **13** (2004), no. 3, 283–293.

[100] Boris Aronov, Marco Pellegrini, and Micha Sharir, On the zone of a surface in a hyperplane arrangement, *Discrete Comput. Geom.* **9** (1993), no. 2, 177–186.

[101] Boris Aronov, Robert Schiffenbauer, and Micha Sharir, On the number of views of translates of a cube and related problems, *Comput. Geom.* **27** (2004), no. 2, 179–192.

[102] Boris Aronov and Micha Sharir, Triangles in space or building (and analyzing) castles in the air, *Combinatorica* **10** (1990), no. 2, 137–173.

[103] _____, The union of convex polyhedra in three dimensions, *Proc. 34th Annual Symposium on Foundations of Computer Science* (Palo Alto, CA, USA), IEEE Comput. Soc. Press, 1993, pp. 518–527.

[104] _____, Castles in the air revisited, *Discrete Comput. Geom.* **12** (1994), no. 2, 119–150.

[105] _____, The common exterior of convex polygons in the plane, *Comput. Geom.* **8** (1997), no. 3, 139–149.

[106] _____, On translational motion planning of a convex polyhedron in 3-space, *SIAM J. Comput.* **26** (1997), no. 6, 1785–1803.

[107] _____, Cutting circles into pseudo-segments and improved bounds for incidences, *Discrete Comput. Geom.* **28** (2002), no. 4, 475–490.

[108] _____, Cell complexities in hyperplane arrangements, *Discrete Comput. Geom.* **32** (2004), no. 1, 107–115.

[109] Boris Aronov, Micha Sharir, and Boaz Tagansky, The union of convex polyhedra in three dimensions, *SIAM J. Comput.* **26** (1997), no. 6, 1670–1688.

[110] Boris Aronov and Shakhar Smorodinsky, On geometric permutations induced by lines transversal through a fixed point, *Proc. Sixteenth Annual ACM-SIAM Symposium on Discrete Algorithms* (New York, NY), ACM, 2005, pp. 251–256.

[111] Tetsuo Asano, Leonidas J. Guibas, and Takeshi Tokuyama, Walking in an arrangement topologically, *Internat. J. Comput. Geom. Appl.* **4** (1994), no. 2, 123–151.

[112] Andrei Asinowski, *Geometric permutations for planar families of disjoint translates of a convex set*, Master's thesis, Technion–Israel Institute of Technology, Haifa, 1999.

[113] Mikhail J. Atallah, Some dynamic computational geometry problems, *Comput. Math. Appl.* **11** (1985), no. 12, 1171–1181.

[114] V. Auletta, A. Monti, M. Parente, and P. Persiano, A linear-time algorithm for the feasibility of pebble motion on trees, *Algorithmica* **23** (1999), no. 3, 223–245.

[115] Franz Aurenhammer, Voronoi diagrams—a survey of a fundamental geometric data structure, *ACM Comput. Surv.* **23** (1991), no. 3, 345–405.

[116] David Avis, David Bremner, and Raimund Seidel, How good are convex hull algorithms?, *Comput. Geom.* **7** (1997), no. 5-6, 265–301.

[117] David Avis and Komei Fukuda, A pivoting algorithm for convex hulls and vertex enumeration of arrangements and polyhedra, *Discrete Comput. Geom.* **8** (1992), no. 3, 295–313.

[118] _____, Reverse search for enumeration, *Discrete Appl. Math.* **65** (1996), no. 1-3, 21–46.

[119] S. Avital and Haim Hanani (Chaim Chojnacki), Graphs, *Gilyonot leMatematika* **3** (1966), 2–8. (Hebrew)
[120] M. Badent, E. Di Giacomo, and G. Liotta, Drawing colored graphs on colored points, *10th Workshop on Algorithms and Data Structures* (Berlin/Heidelberg), Lecture Notes in Computer Science, vol. 4619, Springer, 2007, pp. 102–113.
[121] Ivan J. Balaban, An optimal algorithm for finding segment intersections, *SCG '95: Proc. Eleventh Annual Symposium on Computational Geometry* (Vancouver, British Columbia, Canada), ACM, 1995, pp. 211–219.
[122] József Balogh, Oded Regev, Clifford Smyth, William L. Steiger, and Mario Szegedy, Long monotone paths in line arrangements, *Discrete Comput. Geom.* **32** (2004), no. 2, 167–176.
[123] Amotz Bar-Noy, Panagiotis Cheilaris, Svetlana Olonetsky, and Shakhar Smorodinsky, Conflict-free colorings for hypergraphs, *Automata, Languages, and Programming (Wroclaw, 2007)*, Lecture Notes in Comput. Sci., vol. 4596, Springer, Berlin/Heidelberg, 2007, pp. 219–230.
[124] Amotz Bar-Noy, Panagiotis Cheilaris, and Shakhar Smorodinsky, Conflict-free coloring for intervals: from offline to online, *SPAA '06: Proc. Eighteenth Annual ACM Symposium on Parallelism in Algorithms and Architectures* (Cambridge, Massachusetts, USA), ACM, 2006, pp. 128–137.
[125] Reuven Bar-Yehuda, One for the price of two: a unified approach for approximating covering problems, *Algorithmica* **27** (2000), no. 2, 131–144.
[126] Reuven Bar-Yehuda and S. Fogel, Good splitters with applications to ray shooting, *Proc. Second Canadian Conference on Computational Geometry*, 1990, pp. 81–84.
[127] Imre Bárány, Zoltan Füredi, and László Lovász, On the number of halving planes, *Combinatorica* **10** (1990), no. 2, 175–183.
[128] Imre Bárány and William L. Steiger, On the expected number of k-sets, *Discrete Comput. Geom.* **11** (1994), no. 3, 243–263.
[129] János Barát, Jiří Matoušek, and David R. Wood, Bounded-degree graphs have arbitrarily large geometric thickness, *Electron. J. Combin.* **13** (2006), no. 1, Research Paper 3, 14 pages.
[130] Jérôme Barraquand and Jean-Claude Latombe, Robot motion planning: A distributed representation approach, *Internat. J. Robotics Research* **10** (1991), no. 6, 628–649.
[131] Saugata Basu, The combinatorial and topological complexity of a single cell, *Discrete Comput. Geom.* **29** (2003), no. 1, 41–59.
[132] Saugata Basu, Richard Pollack, and Marie-Françoise Roy, On the number of cells defined by a family of polynomials on a variety, *Mathematika* **43** (1996), no. 1, 120–126.
[133] _____, Computing roadmaps of semi-algebraic sets on a variety, *J. Amer. Math. Soc.* **13** (2000), no. 1, 55–82.
[134] _____, Computing the first Betti number and the connected components of semi-algebraic sets, *STOC'05: Proc. 37th Annual ACM Symposium on Theory of Computing* (New York), ACM, 2005, pp. 304–312.
[135] _____, *Algorithms in Real Algebraic Geometry*, Second Edition, Algorithms and Computation in Mathematics, vol. 10, Springer, Berlin, 2006.
[136] József Beck, On the lattice property of the plane and some problems of Dirac, Motzkin and Erdős in combinatorial geometry, *Combinatorica* **3** (1983), no. 3–4, 281–297.
[137] J. L. Bentley and Thomas A. Ottmann, Algorithms for reporting and counting geometric intersections, *IEEE Trans. Comput.* **28** (1979), no. 9, 643–647.
[138] Sergey Bereg, Adrian Dumitrescu, and János Pach, Sliding disks in the plane, *Discrete and computational geometry*, Lecture Notes in Comput. Sci., vol. 3742, Springer, Berlin, 2005, pp. 37–47.
[139] Mark T. de Berg, *Ray Shooting, Depth Orders and Hidden Surface Removal*, Lecture Notes in Comput. Sci., vol. 703, Springer, New York, 1993.
[140] _____, Improved bounds on the union complexity of fat objects, *FSTTCS 2005: Foundations of Software Technology and Theoretical Computer Science*, Lecture Notes in Comput. Sci., vol. 3821, Springer, Berlin, 2005, pp. 116–127.
[141] Mark T. de Berg, K. Dobrindt, and Otfried Schwarzkopf, On lazy randomized incremental construction, *Discrete Comput. Geom.* **14** (1995), no. 3, 261–286.
[142] Mark T. de Berg, Leonidas J. Guibas, and Dan Halperin, Vertical decompositions for triangles in 3-space, *Discrete Comput. Geom.* **15** (1996), no. 1, 35–61.

[143] Mark T. de Berg, Leonidas J. Guibas, Dan Halperin, Mark H. Overmars, Micha Sharir, Otfried Schwarzkopf, and Monique Teillaud, Reaching a goal with directional uncertainty, *Theoret. Comput. Sci.* **140** (1995), no. 2, 301–317.

[144] Mark T. de Berg, Dan Halperin, Mark H. Overmars, and Marc van Kreveld, Sparse arrangements and the number of views of polyhedral scenes, *Internat. J. Comput. Geom. Appl.* **7** (1997), no. 3, 175–195.

[145] Mark T. de Berg, Dan Halperin, Mark H. Overmars, Jack Snoeyink, and Marc van Kreveld, Efficient ray shooting and hidden surface removal, *Algorithmica* **12** (1994), no. 1, 30–53.

[146] Mark T. de Berg, Marc van Kreveld, Mark H. Overmars, and Otfried Schwarzkopf, *Computational geometry. Algorithms and applications*, Revised Edition, Springer, Berlin, 2000.

[147] Mark T. de Berg, Marc van Kreveld, Otfried Schwarzkopf, and Jack Snoeyink, Point location in zones of k-flats in arrangements, *Comput. Geom.* **6** (1996), no. 3, 131–143.

[148] Marshall Bern, David Eppstein, Paul Plassmann, and F. Frances Yao, Horizon theorems for lines and polygons, *Discrete and Computational Geometry*, DIMACS Ser. Discrete Math. Theoret. Comput. Sci., vol. 6, Amer. Math. Soc., Providence, RI, 1991, pp. 45–66.

[149] Sergey Bespamyatnikh, An efficient algorithm for the three-dimensional diameter problem, *Discrete Comput. Geom.* **25** (2001), no. 2, 235–255.

[150] Lilach Bien, *Incidences between points and curves in the plane*, Master's thesis, School of Computer Science, Tel Aviv University, Tel Aviv, Israel, 2006.

[151] Daniel Bienstock and Nathaniel Dean, Bounds for rectilinear crossing numbers, *J. Graph Theory* **17** (1993), no. 3, 333–348.

[152] Anders Björner, Michel Las Vergnas, Bernd Sturmfels, Neil White, and Günter M. Ziegler, *Oriented Matroids*, second ed., Encyclopedia of Mathematics and its Applications, vol. 46, Cambridge University Press, Cambridge, 1999.

[153] Anders Björner and Günter M. Ziegler, Combinatorial stratification of complex arrangements, *J. Amer. Math. Soc.* **5** (1992), no. 1, 105–149.

[154] Aart Blokhuis and Ákos Seress, The number of directions determined by points in the three-dimensional Euclidean space, *Discrete Comput. Geom.* **28** (2002), no. 4, 491–494.

[155] W. J. Blundon, Multiple covering of the plane by circles, *Mathematika* **4** (1957), 7–16.

[156] Jacek Bochnak, Michel Coste, and Marie-Françoise Roy, *Real Algebraic Geometry*, Ergebnisse der Mathematik und ihrer Grenzgebiete (3), vol. 36, Springer, Berlin, 1998, Translated from the 1987 French original, Revised by the authors.

[157] Karl-Friedrich Böhringer, Bruce Donald, and Dan Halperin, On the area bisectors of a polygon, *Discrete Comput. Geom.* **22** (1999), no. 2, 269–285.

[158] Jean-Daniel Boissonnat and Katrin T. G. Dobrindt, Randomized construction of the upper envelope of triangles in R^3, Proc. Fourth Canadian Conference on Computational Geometry, 1992, pp. 311–315.

[159] _____, On-line construction of the upper envelope of triangles and surface patches in three dimensions, *Comput. Geom.* **5** (1996), no. 6, 303–320.

[160] Jean-Daniel Boissonnat, Micha Sharir, Boaz Tagansky, and Mariette Yvinec, Voronoi diagrams in higher dimensions under certain polyhedral distance functions, *Discrete Comput. Geom.* **19** (1998), no. 4, 485–519.

[161] Jean-Daniel Boissonnat and Monique Teillaud (eds.), *Effective Computational Geometry of Curves and Surfaces*, Springer, Berlin/Heidelberg, 2006.

[162] Jean-Daniel Boissonnat and Mariette Yvinec, *Algorithmic Geometry*, Cambridge University Press, Cambridge, 1998, Translated from the 1995 French original by Hervé Brönnimann.

[163] Károly Böröczky, Sr., Über stabile Kreis- und Kugelsysteme, *Ann. Univ. Sci. Budapest. Eötvös Sect. Math.* **7** (1964), 79–82.

[164] Peter B. Borwein, Sylvester's problem and Motzkin's theorem for countable and compact sets, *Proc. Amer. Math. Soc.* **90** (1984), no. 4, 580–584.

[165] Peter B. Borwein and Michael Edelstein, Unsolved problems: A conjecture related to Sylvester's problem, *Amer. Math. Monthly* **90** (1983), no. 6, 389–390.

[166] Peter B. Borwein and William O. J. Moser, A survey of Sylvester's problem and its generalizations, *Aequationes Math.* **40** (1990), no. 2-3, 111–135.

[167] O. Bottema and B. Roth, *Theoretical kinematics*, Dover Publications Inc., New York, 1990, Corrected reprint of the 1979 edition.

[168] Kevin W. Bowyer and Charles R. Dyer, Aspect graphs: An introduction and survey of recent results, *Internat. J. Imaging Systems Technologies* **2** (1990), 315–328.

[169] K. A. Brakke, Some new values for Sylvester's function for n noncollinear points, *J. Undergrad. Math.* **4** (1972), 11–14.

[170] Peter Brass, On the maximum number of unit distances among n points in dimension four, *Intuitive Geometry (Budapest, 1995)*, Bolyai Soc. Math. Stud., vol. 6, János Bolyai Math. Soc., Budapest, 1997, pp. 277–290.

[171] _____, Exact point pattern matching and the number of congruent triangles in a three-dimensional pointset, *Algorithms—ESA 2000 (Saarbrücken)*, Lecture Notes in Comput. Sci., vol. 1879, Springer, Berlin, 2000, pp. 112–119.

[172] _____, Combinatorial geometry problems in pattern recognition, *Discrete Comput. Geom.* **28** (2002), no. 4, 495–510.

[173] Peter Brass and Christian Knauer, On counting point-hyperplane incidences, *Comput. Geom.* **25** (2003), no. 1-2, 13–20.

[174] Peter Brass, William O. J. Moser, and János Pach, *Research Problems in Discrete Geometry*, Springer, New York, 2005.

[175] Peter Brass and János Pach, Problems and results on geometric patterns, *Graph Theory and Combinatorial Optimization*, GERAD 25th Anniv. Ser., vol. 8, Springer, New York, 2005, pp. 17–36.

[176] David Bremner, Komei Fukuda, and Ambros Marzetta, Primal-dual methods for vertex and facet enumeration, *Discrete Comput. Geom.* **20** (1998), no. 3, 333–357.

[177] Erik Brisson, Representing geometric structures in d dimensions: Topology and order, *Discrete Comput. Geom.* **9** (1993), no. 4, 387–426.

[178] Hervé Brönnimann and Bernard Chazelle, Optimal slope selection via cuttings, *Comput. Geom.* **10** (1998), no. 1, 23–29.

[179] Hervé Brönnimann, Bernard Chazelle, and Jiří Matoušek, Product range spaces, sensitive sampling, and derandomization, *SIAM J. Comput.* **28** (1999), no. 5, 1552–1575.

[180] Hervé Brönnimann, Olivier Devillers, Vida Dujmovic, Hazel Everett, Marc Glisse, Xavier Goaoc, Sylvain Lazard, Hyeon-Suk Na, and Sue Whitesides, Lines and free line segments tangent to arbitrary three-dimensional convex polyhedra, *SIAM J. Comput.* **37** (2007), no. 2, 522–551.

[181] Hervé Brönnimann, H. Everett, S. Lazard, F. Sottile, and S. Whitesides, Transversals to line segments in three-dimensional space, *Discrete Comput. Geom.* **34** (2005), no. 3, 381–390.

[182] U. A. Brousseau, A mathematician's progress, *The Mathematics Teacher* **59** (1966), 722–727.

[183] Nicolaas Govert de Bruijn, Aufgaben 17 and 18, *Nieuw Archief voor Wiskunde* **2** (1954), 67. (Dutch)

[184] Nicolaas Govert de Bruijn and Paul Erdős, On a combinatorial problem, *Nederl. Akad. Wetensch., Proc.* **51** (1948), 1277–1279 = *Indag. Math.* **10** (1948), 421–423.

[185] R. C. Buck, Partition of space, *Amer. Math. Monthly* **50** (1943), 541–544.

[186] Christoph Burnikel, Kurt Mehlhorn, and Stefan Schirra, On degeneracy in geometric computations, *Proc. Fifth Annual ACM-SIAM Symposium on Discrete Algorithms* (Arlington, VA), ACM, 1994, pp. 16–23.

[187] G. R. Burton and G. B. Purdy, The directions determined by n points in the plane, *J. London Math. Soc. (2)* **20** (1979), no. 1, 109–114.

[188] Z. Butler, K. Kotay, D. Rus, and K. Tomita, Generic decentralized control for a class of self-reconfigurable robots, *Proc. ICRA'02: IEEE International Conf. on Robotics and Automation*, vol. 1, 2002, pp. 809–816.

[189] G. Cairns and Y. Nikolayevsky, Bounds for generalized thrackles, *Discrete Comput. Geom.* **23** (2000), no. 2, 191–206.

[190] Gruia Călinescu, Adrian Dumitrescu, and János Pach, Reconfigurations in graphs and grids, *LATIN 2006: Theoretical informatics*, Lecture Notes in Comput. Sci., vol. 3887, Springer, Berlin, 2006, pp. 262–273.

[191] R. J. Canham, A theorem on arrangements of lines in the plane, *Israel J. Math.* **7** (1969), 393–397.

[192] John F. Canny, *The Complexity of Robot Motion Planning*, ACM Doctoral Dissertation Awards, vol. 1987, MIT Press, Cambridge, MA, 1988.

[193] _____, Some algebraic and geometric computations in PSPACE, *STOC'88: Proc. Twentieth Annual ACM Symposium on Theory of Computing* (Chicago, IL), ACM, 1988, pp. 460–469.

[194] _____, Computing roadmaps of general semi-algebraic sets, *Comput. J.* **36** (1993), no. 5, 504–514.

[195] John F. Canny, Dima Yu. Grigor'ev, and Nicolai N. Vorobjov, Jr., Finding connected components of a semialgebraic set in subexponential time, *Appl. Algebra Engrg. Comm. Comput.* **2** (1992), no. 4, 217–238.

[196] Edwin E. Catmull, *A Subdivision Algorithm for Computer Display of Curved Surfaces*, Ph.D. thesis, University of Utah, Salt Lake City, 1974.

[197] Jakub Černý, Geometric graphs with no three disjoint edges, *Discrete Comput. Geom.* **34** (2005), no. 4, 679–695.

[198] G. D. Chakerian, Sylvester's problem on collinear points and a relative, *Amer. Math. Monthly* **77** (1970), 164–167.

[199] Timothy M. Chan, Optimal output-sensitive convex hull algorithms in two and three dimensions, *Discrete Comput. Geom.* **16** (1996), no. 4, 361–368.

[200] _____, Output-sensitive results on convex hulls, extreme points, and related problems, *Discrete Comput. Geom.* **16** (1996), no. 4, 369–387.

[201] _____, Geometric applications of a randomized optimization technique, *Discrete Comput. Geom.* **22** (1999), no. 4, 547–567.

[202] _____, Remarks on k-level algorithms in the plane, Unpublished manuscript, 1999.

[203] _____, Approximating the diameter, width, smallest enclosing cylinder, and minimum-width annulus, *Internat. J. Comput. Geom. Appl.* **12** (2002), no. 1-2, 67–85.

[204] _____, On levels in arrangements of curves, *Discrete Comput. Geom.* **29** (2003), no. 3, 375–393.

[205] _____, On levels in arrangements of curves. II. A simple inequality and its consequences, *Discrete Comput. Geom.* **34** (2005), no. 1, 11–24.

[206] _____, On levels in arrangements of curves. III. Further developments, *SCG '08: Proc. Twenty Fourth Annual Symposium on Computational Geometry* (College Park, MD, USA), ACM, 2008, pp. 85–93.

[207] _____, On levels in arrangements of surfaces in three dimensions, *Proc. Sixteenth Annual ACM-SIAM Symposium on Discrete Algorithms* (New York), ACM, 2005, pp. 232–240.

[208] Timothy M. Chan, Jack Snoeyink, and Chee-Keng Yap, Primal dividing and dual pruning: Output-sensitive construction of four-dimensional polytopes and three-dimensional Voronoi diagrams, *Discrete Comput. Geom.* **18** (1997), no. 4, 433–454.

[209] Bernard Chazelle, An optimal convex hull algorithm and new results on cuttings, *Proc. 32nd Annual Symposium on Foundations of Computer Science* (San Juan, Puerto Rico), IEEE Comput. Soc. Press, 1991, pp. 29–38.

[210] _____, Cutting hyperplanes for divide-and-conquer, *Discrete Comput. Geom.* **9** (1993), no. 2, 145–158.

[211] _____, An optimal convex hull algorithm in any fixed dimension, *Discrete Comput. Geom.* **10** (1993), no. 4, 377–409.

[212] Bernard Chazelle and Herbert Edelsbrunner, An optimal algorithm for intersecting line segments in the plane, *J. Assoc. Comput. Mach.* **39** (1992), no. 1, 1–54.

[213] Bernard Chazelle, Herbert Edelsbrunner, Michelangelo Grigni, Leonidas J. Guibas, John E. Hershberger, Micha Sharir, and Jack Snoeyink, Ray shooting in polygons using geodesic triangulations, *Algorithmica* **12** (1994), no. 1, 54–68.

[214] Bernard Chazelle, Herbert Edelsbrunner, Leonidas J. Guibas, Richard Pollack, Raimund Seidel, Micha Sharir, and Jack Snoeyink, Counting and cutting cycles of lines and rods in space, *Comput. Geom.* **1** (1992), no. 6, 305–323.

[215] Bernard Chazelle, Herbert Edelsbrunner, Leonidas J. Guibas, and Micha Sharir, A singly-exponential stratification scheme for real semi-algebraic varieties and its applications, *Automata, Languages and Programming (Stresa, 1989)*, Lecture Notes in Comput. Sci., vol. 372, Springer, Berlin, 1989, pp. 179–193.

[216] _____, A singly exponential stratification scheme for real semi-algebraic varieties and its applications, *Theor. Comput. Sci.* **84** (1991), no. 1, 77–105.

[217] _____, Diameter, width, closest line pair, and parametric searching, *Discrete Comput. Geom.* **10** (1993), no. 2, 183–196.

[218] _____, Algorithms for bichromatic line-segment problems and polyhedral terrains, *Algorithmica* **11** (1994), no. 2, 116–132.

[219] Bernard Chazelle, Herbert Edelsbrunner, Leonidas J. Guibas, Micha Sharir, and Jack Snoeyink, Computing a face in an arrangement of line segments and related problems, *SIAM J. Comput.* **22** (1993), no. 6, 1286–1302.

[220] Bernard Chazelle, Herbert Edelsbrunner, Leonidas J. Guibas, Micha Sharir, and J. Stolfi, Lines in space: combinatorics and algorithms, *Algorithmica* **15** (1996), no. 5, 428–447.

[221] Bernard Chazelle and Joel Friedman, A deterministic view of random sampling and its use in geometry, *Combinatorica* **10** (1990), no. 3, 229–249.

[222] Bernard Chazelle and Joel Friedman, Point location among hyperplanes and unidirectional ray-shooting, *Comput. Geom.* **4** (1994), no. 2, 53–62.

[223] Bernard Chazelle and Leonidas J. Guibas, Visibility and intersection problems in plane geometry, *Discrete Comput. Geom.* **4** (1989), no. 6, 551–581.

[224] Bernard Chazelle, Leonidas J. Guibas, and Der-Tsai Lee, The power of geometric duality, *BIT* **25** (1985), no. 1, 76–90.

[225] Bernard Chazelle and Jiří Matoušek, Derandomizing an output-sensitive convex hull algorithm in three dimensions, *Comput. Geom.* **5** (1995), no. 1, 27–32.

[226] Bernard Chazelle and Franco P. Preparata, Halfspace range search: An algorithmic application of k-sets, *Discrete Comput. Geom.* **1** (1986), no. 1, 83–93.

[227] Bernard Chazelle, Micha Sharir, and Emo Welzl, Quasi-optimal upper bounds for simplex range searching and new zone theorems, *Algorithmica* **8** (1992), no. 5-6, 407–429.

[228] Bernard M. Chazelle and Der-Tsai Lee, On a circle placement problem, *Computing* **36** (1986), no. 1-2, 1–16.

[229] Ke Chen, How to play a coloring game against a color-blind adversary, *SCG '06: Proc. Twenty-second Annual Symposium on Computational Geometry* (Sedona, Arizona, USA), ACM, 2006, pp. 44–51.

[230] Ke Chen, Amos Fiat, Haim Kaplan, Meital Levy, Jiří Matoušek, Elchanan Mossel, János Pach, Micha Sharir, Shakhar Smorodinsky, Uli Wagner, and Emo Welzl, Online conflict-free coloring for intervals, *SIAM J. Comput.* **36** (2006/07), no. 5, 1342–1359.

[231] Ke Chen, Haim Kaplan, and Micha Sharir, Online CF coloring for halfplanes, congruent disks, and axis-parallel rectangles, *Trans. Algorithms*, to appear.

[232] Xiaomin Chen, János Pach, Mario Szegedy, and Gábor Tardos, Delaunay graphs of point sets in the plane with respect to axis-parallel rectangles, *SODA '08: Proc. Nineteenth Annual ACM-SIAM Symposium on Discrete Algorithms* (San Francisco, California), SIAM, 2008, pp. 94–101.

[233] Otfried Cheong, Xavier Goaoc, Andreas Holmsen and Sylvain Petitjean, Helly-type theorems for line transversals to disjoint unit balls, *Discrete Comput. Geom.* **39** (2008), 194–212.

[234] Otfried Cheong, Xavier Goaoc, and Hyeon-Suk Na, Geometric permutations of disjoint unit spheres, *Comput. Geom.* **30** (2005), no. 3, 253–270.

[235] L. Paul Chew, Near-quadratic bounds for the L_1 Voronoi diagram of moving points, *Comput. Geom.* **7** (1997), no. 1-2, 73–80.

[236] L. Paul Chew, Klara Kedem, Micha Sharir, Boaz Tagansky, and Emo Welzl, Voronoi diagrams of lines in 3-space under polyhedral convex distance functions, *J. Algorithms* **29** (1998), no. 2, 238–255.

[237] Ghain Chirikjian, Amit Pamecha, and Imme Ebert-Uphoff, Evaluating efficiency of self-reconfiguration in a class of modular robots, *J. Robotic Systems* **13** (1996), no. 5, 317–328.

[238] Chaim Chojnacki (Haim Hanani), Über wesentlich unplättbare Kurven im dreidimensionalen Raume, *Fund. Math.* **23** (1934), 135–142.

[239] Fan R. K. Chung, The number of different distances determined by n points in the plane, *J. Combin. Theory Ser. A* **36** (1984), no. 3, 342–354.

[240] Fan R. K. Chung, Endre Szemerédi, and William T. Trotter, Jr., The number of different distances determined by a set of points in the Euclidean plane, *Discrete Comput. Geom.* **7** (1992), no. 1, 1–11.

[241] Kenneth L. Clarkson, New applications of random sampling in computational geometry, *Discrete Comput. Geom.* **2** (1987), no. 2, 195–222.

[242] _____, A randomized algorithm for closest-point queries, *SIAM J. Comput.* **17** (1988), no. 4, 830–847.

[243] _____, Computing a single face in arrangements of segments, Unpublished manuscript, 1990.

[244] _____, Randomized geometric algorithms, *Computing in Euclidean Geometry* (D.-Z. Du and F. K. Hwang, eds.), *Lecture Notes Ser. Comput.*, vol. 1, World Sci. Publ., River Edge, NJ, 1992, pp. 117–162.

[245] Kenneth L. Clarkson, Herbert Edelsbrunner, Leonidas J. Guibas, Micha Sharir, and Emo Welzl, Combinatorial complexity bounds for arrangements of curves and spheres, *Discrete Comput. Geom.* **5** (1990), no. 2, 99–160.

[246] Kenneth L. Clarkson and Peter W. Shor, Applications of random sampling in computational geometry. II, *Discrete Comput. Geom.* **4** (1989), no. 5, 387–421.

[247] Richard Cole, Jeffrey S. Salowe, William L. Steiger, and Endre Szemerédi, An optimal-time algorithm for slope selection, *SIAM J. Comput.* **18** (1989), no. 4, 792–810.

[248] Richard Cole and Micha Sharir, Visibility problems for polyhedral terrains, *J. Symbolic Comput.* **7** (1989), no. 1, 11–30.

[249] Richard Cole, Micha Sharir, and Chee-Keng Yap, On k-hulls and related problems, *SIAM J. Comput.* **16** (1987), no. 1, 61–77.

[250] George E. Collins, Quantifier elimination for real closed fields by cylindrical algebraic decomposition, *Automata Theory and Formal Languages (Second GI Conf., Kaiserslautern, 1975)*, Lecture Notes in Comput. Sci., vol. 33, Springer, Berlin, 1975, pp. 134–183.

[251] M. L. Connolly, Analytical molecular surface calculation, *J. Applied Crystallography* **16** (1983), no. 5, 548–558.

[252] Raul Cordovil, The directions determined by n points in the plane, a matroidal generalization, *Discrete Mathematics* **43** (1983), no. 2–3, 131–332.

[253] H. S. M. Coxeter, A problem of collinear points, *Amer. Math. Monthly* **55** (1948), 26–28.

[254] D. W. Crowe and T. A. McKee, Sylvester's problem on collinear points, *Math. Mag.* **41** (1968), 30–34.

[255] J. Csima and E. T. Sawyer, There exist $6n/13$ ordinary points, *Discrete Comput. Geom.* **9** (1993), no. 2, 187–202.

[256] _____, The $6n/13$ theorem revisited, *Graph Theory, Combinatorics, and Algorithms, Vol. 1, 2 (Kalamazoo, MI, 1992)*, Wiley, New York, 1995, pp. 235–249.

[257] Ilda P. F. Da Silva and Komei Fukuda, Isolating points by lines in the plane, *J. Geom.* **62** (1998), no. 1-2, 48–65.

[258] L. Danzer, Über ein Problem aus der kombinatorischen Geometrie, *Arch. Math.* **8** (1957), 347–351.

[259] H. Davenport, A combinatorial problem connected with differential equations. II, *Acta Arith.* **17** (1970/1971), 363–372.

[260] H. Davenport and A. Schinzel, A combinatorial problem connected with differential equations, *Amer. J. Math.* **87** (1965), 684–694.

[261] Erik D. Demaine, Martin L. Demaine, and Helena A. Verrill, Coin-moving puzzles, *More Games of no Chance (Berkeley, CA, 2000)*, Math. Sci. Res. Inst. Publ., vol. 42, Cambridge Univ. Press, Cambridge, 2002, pp. 405–431.

[262] Erik D. Demaine, Joseph S. B. Mitchell, and Joseph O'Rourke, The open problems project, http://www.cs.smith.edu/~orourke/TOPP/.

[263] Tamal K. Dey, Improved bounds on planar k-sets and k-levels, *Proc. 38th IEEE Symposium on Foundations of Computer Science* (Los Alamitos, CA, USA), IEEE Computer Society, 1997, pp. 156–161.

[264] _____, Improved bounds for planar k-sets and related problems, *Discrete Comput. Geom.* **19** (1998), no. 3, 373–382.

[265] Tamal K. Dey and Herbert Edelsbrunner, Counting triangle crossings and halving planes, *Discrete Comput. Geom.* **12** (1994), no. 3, 281–289.

[266] Tamal K. Dey and János Pach, Extremal problems for geometric hypergraphs, *Discrete Comput. Geom.* **19** (1998), no. 4, 473–484.

[267] Reinhard Diestel, *Graph Theory*, third ed., Graduate Texts in Mathematics, vol. 173, Springer, Berlin, 2005.

[268] Gabriel A. Dirac, Collinearity properties of sets of points, *Quart. J. Math., Oxford Ser. (2)* **2** (1951), 221–227.

[269] David Dobkin, Leonidas J. Guibas, John E. Hershberger, and Jack Snoeyink, An efficient algorithm for finding the CSG representation of a simple polygon, *SIGGRAPH '88: Proceedings of the 15th Annual Conference on Computer Graphics and Interactive Techniques*, ACM, 1988, pp. 31–40.

[270] David Dobkin and Seth Teller, Computer graphics, ch. 49, in Goodman and O'Rourke [**381**], pp. 1095–1116.

[271] David P. Dobkin and David G. Kirkpatrick, Fast detection of polyhedral intersection, *Theoret. Comput. Sci.* **27** (1983), no. 3, 241–253.

[272] _____, Determining the separation of preprocessed polyhedra—a unified approach, *Automata, Languages and Programming (Coventry, 1990), Lecture Notes in Comput. Sci.*, vol. 443, Springer, New York, 1990, pp. 400–413.

[273] David P. Dobkin and Michael J. Laszlo, Primitives for the manipulation of three-dimensional subdivisions, *Algorithmica* **4** (1989), no. 1, 3–32.

[274] Annette J. Dobson and Sheila O. Macdonald, Lower bounds for the lengths of Davenport-Schinzel sequences, *Utilitas Math.* **6** (1974), 251–257.

[275] Susan E. Dorward, A survey of object-space hidden surface removal, *Internat. J. Comput. Geom. Appl.* **4** (1994), no. 3, 325–362.

[276] Vida Dujmović, Matthew Suderman, and David R. Wood, Really straight graph drawings, *Graph Drawing, Lecture Notes in Comput. Sci.*, vol. 3383, Springer, Berlin/Heidelberg, 2005, pp. 122–132.

[277] _____, Graph drawings with few slopes, *Comput. Geom.* **38** (2007), no. 3, 181–193.

[278] V. C. Dumir and R. J. Hans-Gill, Lattice double packings in the plane, *Indian J. Pure Appl. Math.* **3** (1972), no. 3, 481–487.

[279] Adrian Dumitrescu, Motion planning and reconfiguration for systems of multiple objects, *Mobile Robots: Perception & Navigation* (Sascha Kolski, ed.), pro literatur Verlag / ARS, Germany, 2007, pp. 523–542.

[280] Adrian Dumitrescu and János Pach, Pushing squares around, *Graphs Combin.* **22** (2006), no. 1, 37–50.

[281] Adrian Dumitrescu, Micha Sharir, and Csaba D. Tóth, Extremal problems on triangle areas in two and three dimensions, *J. Combinat. Theory, Ser. A* (2007), submitted; *Proc. 24th ACM Symposium on Computational Geoemtry* (College Park, MD, USA) ACM, 2008, pp. 208–217.

[282] Adrian Dumitrescu, Ichiro Suzuki, and Masafumi Yamashita, Formations for fast locomotion of metamorphic robotic systems, *Internat. J. Robotics Research* **23** (2004), 583–593.

[283] _____, Motion planning for metamorphic systems: feasibility, decidability and distributed reconfiguration, *IEEE Trans. Robotics and Automation* **20** (2004), no. 3, 409–418.

[284] Adrian Dumitrescu and Csaba D. Tóth, Distinct triangle areas in a planar point set, *Integer Programming and Combinatorial Optimization, Lecture Notes in Comput. Sci.*, vol. 4513, Springer, Berlin/Heidelberg, 2007, pp. 119–129.

[285] _____, On the number of tetrahedra with minimum, unit, and distinct volumes in three-space, *SODA '07: Proc. Eighteenth Annual ACM-SIAM Symposium on Discrete Algorithms* (New Orleans, Louisiana), SIAM, 2007, pp. 1114–1123.

[286] Rex A. Dwyer, Voronoi diagrams of random lines and flats, *Discrete Comput. Geom.* **17** (1997), no. 2, 123–136.

[287] Herbert Edelsbrunner, Edge-skeletons in arrangements with applications, *Algorithmica* **1** (1986), no. 1, 93–109.

[288] _____, *Algorithms in Combinatorial Geometry*, EATCS Monographs on Theoretical Computer Science, vol. 10, Springer, Berlin, 1987.

[289] _____, The upper envelope of piecewise linear functions: Tight bounds on the number of faces, *Discrete Comput. Geom.* **4** (1989), no. 4, 337–343.

[290] _____, The union of balls and its dual shape, *Discrete Comput. Geom.* **13** (1995), no. 3–4, 415–440.

[291] _____, Geometry for modeling biomolecules, *Robotics: the Algorithmic Perspective* (Houston, TX, 1998), A. K. Peters, Natick, MA, 1998, pp. 265–277.

[292] Herbert Edelsbrunner, Michael A. Facello, Ping Fu, and Jie Liang, Measuring proteins and voids in proteins, *HICSS '95: Proc. 28th Hawaii International Conference on System Sciences* (Los Alamitos, CA, USA), IEEE Computer Society, 1995, pp. 256–264.

[293] Herbert Edelsbrunner and Leonidas J. Guibas, Topologically sweeping an arrangement, *J. Comput. System Sci.* **38** (1989), no. 1, 165–194, Corrigendum in 42 (1991), no. 2, 249–251.

[294] Herbert Edelsbrunner, Leonidas J. Guibas, John E. Hershberger, János Pach, Richard Pollack, Raimund Seidel, Micha Sharir, and Jack Snoeyink, On arrangements of Jordan arcs with three intersections per pair, *Discrete Comput. Geom.* **4** (1989), no. 5, 523–539.

[295] Herbert Edelsbrunner, Leonidas J. Guibas, János Pach, Richard Pollack, Raimund Seidel, and Micha Sharir, Arrangements of curves in the plane—topology, combinatorics, and algorithms, *Theoret. Comput. Sci.* **92** (1992), no. 2, 319–336.

[296] Herbert Edelsbrunner, Leonidas J. Guibas, and Micha Sharir, The upper envelope of piecewise linear functions: Algorithms and applications, *Discrete Comput. Geom.* **4** (1989), no. 4, 311–336.

[297] _____, The complexity and construction of many faces in arrangements of lines and of segments, *Discrete Comput. Geom.* **5** (1990), no. 2, 161–196.

[298] _____, The complexity of many cells in arrangements of planes and related problems, *Discrete Comput. Geom.* **5** (1990), no. 2, 197–216.

[299] Herbert Edelsbrunner and David Haussler, The complexity of cells in three-dimensional arrangements, *Discrete Math.* **60** (1986), 139–146.

[300] Herbert Edelsbrunner and Ernst Peter Mücke, Simulation of simplicity: a technique to cope with degenerate cases in geometric algorithms, *Proc. Fourth Annual Symposium on Computational Geometry* (Urbana, IL), ACM, 1988, pp. 118–133.

[301] Herbert Edelsbrunner, Joseph O'Rourke, and Raimund Seidel, Constructing arrangements of lines and hyperplanes with applications, *SIAM J. Comput.* **15** (1986), no. 2, 341–363.

[302] Herbert Edelsbrunner and Raimund Seidel, Voronoï diagrams and arrangements, *Discrete Comput. Geom.* **1** (1986), no. 1, 25–44.

[303] Herbert Edelsbrunner, Raimund Seidel, and Micha Sharir, On the zone theorem for hyperplane arrangements, *SIAM J. Comput.* **22** (1993), no. 2, 418–429.

[304] Herbert Edelsbrunner and Micha Sharir, The maximum number of ways to stab n convex nonintersecting sets in the plane is $2n-2$, *Discrete Comput. Geom.* **5** (1990), no. 1, 35–42.

[305] _____, A hyperplane incidence problem with applications to counting distances, *Applied geometry and discrete mathematics*, DIMACS Ser. Discrete Math. Theoret. Comput. Sci., vol. 4, Amer. Math. Soc., Providence, RI, 1991, pp. 253–263.

[306] Herbert Edelsbrunner and Diane L. Souvaine, Computing least median of squares regression lines and guided topological sweep, *J. Amer. Statist. Assoc.* **85** (1990), 115–119.

[307] Herbert Edelsbrunner, Pavel Valtr, and Emo Welzl, Cutting dense point sets in half, *Discrete Comput. Geom.* **17** (1997), no. 3, 243–255.

[308] Herbert Edelsbrunner and Emo Welzl, On the number of line separations of a finite set in the plane, *J. Combin. Theory Ser. A* **38** (1985), no. 1, 15–29.

[309] _____, Constructing belts in two-dimensional arrangements with applications, *SIAM J. Comput.* **15** (1986), no. 1, 271–284.

[310] _____, On the maximal number of edges of many faces in an arrangement, *J. Combin. Theory Ser. A* **41** (1986), no. 2, 159–166.

[311] M. Edelstein, Generalizations of the Sylvester problem, *Math. Mag.* **43** (1970), 250–254.

[312] M. Edelstein, F. Herzog, and L. M. Kelly, A further theorem of the Sylvester type, *Proc. Amer. Math. Soc.* **14** (1963), 359–363.

[313] Alon Efrat, The complexity of the union of (α,β)-covered objects, *SIAM J. Comput.* **34** (2005), no. 4, 775–787.

[314] Alon Efrat and Matthew J. Katz, On the union of κ-curved objects, *Comput. Geom.* **14** (1999), no. 4, 241–254.

[315] Alon Efrat, Günter Rote, and Micha Sharir, On the union of fat wedges and separating a collection of segments by a line, *Comput. Geom.* **3** (1993), no. 5, 277–288.

[316] Alon Efrat and Micha Sharir, A near-linear algorithm for the planar segment-center problem, *Discrete Comput. Geom.* **16** (1996), no. 3, 239–257.

[317] _____, On the complexity of the union of fat convex objects in the plane, *Discrete Comput. Geom.* **23** (2000), no. 2, 171–189.

[318] György Elekes, SUMS versus PRODUCTS in number theory, algebra and Erdős geometry, *Paul Erdős and his mathematics, II (Budapest, 1999)*, Bolyai Soc. Math. Stud., vol. 11, János Bolyai Math. Soc., Budapest, 2002, pp. 241–290.

[319] György Elekes and Paul Erdős, Similar configurations and pseudo grids, *Intuitive Geometry (Szeged, 1991)*, Colloq. Math. Soc. János Bolyai, vol. 63, North-Holland, Amsterdam, 1994, pp. 85–104.

[320] Ioannis Z. Emiris and John F. Canny, A general approach to removing degeneracies, *SIAM J. Comput.* **24** (1995), no. 3, 650–664.

[321] Ioannis Z. Emiris, John F. Canny, and Raimund Seidel, Efficient perturbations for handling geometric degeneracies, *Algorithmica* **19** (1997), no. 1-2, 219–242.

[322] David Eppstein, Improved bounds for intersecting triangles and halving planes, *J. Combin. Theory Ser. A* **62** (1993), no. 1, 176–182.

[323] Paul Erdős, On sets of distances of n points, *Amer. Math. Monthly* **53** (1946), 248–250.

[324] _____, On sets of distances of n points in Euclidean space, *Magyar Tud. Akad. Mat. Kutató Int. Közl.* **5** (1960), 165–169.

[325] Paul Erdős, Richard Bellman, H. S. Wall, James Singer, and V. Thebault, Problems and Solutions: Advanced Problems: Problems for Solution: 4065-4069, *Amer. Math. Monthly* **50** (1943), no. 1, 65–66.

[326] Paul Erdős, Ronald L. Graham, P. Montgomery, Bruce L. Rothschild, Joel Spencer, and E. G. Straus, Euclidean Ramsey theorems. I, *J. Combin. Theory Ser. A* **14** (1973), 341–363.

[327] _____, Euclidean Ramsey theorems. III, *Infinite and finite sets (Colloq., Keszthely, 1973; dedicated to P. Erdős on his 60th birthday), Vol. I*, North-Holland, Amsterdam, 1975, pp. 559–583. Colloq. Math. Soc. János Bolyai, Vol. 10.

[328] Paul Erdős and R. K. Guy, Crossing number problems, *Amer. Math. Monthly* **80** (1973), 52–58.

[329] Paul Erdős, Dean Hickerson, and János Pach, A problem of Leo Moser about repeated distances on the sphere, *Amer. Math. Monthly* **96** (1989), no. 7, 569–575.

[330] Paul Erdős, László Lovász, A. Simmons, and E. G. Straus, Dissection graphs of planar point sets, *A survey of combinatorial theory (Proc. Internat. Sympos., Colorado State Univ., Fort Collins, Colo., 1971)*, North-Holland, Amsterdam, 1973, pp. 139–149.

[331] Paul Erdős and George Purdy, Some extremal problems in geometry, *J. Combin. Theory Ser. A* **10** (1971), 246–252.

[332] _____, Some extremal problems in geometry. III, *Proc. Sixth Southeastern Conference on Combinatorics, Graph Theory and Computing* (Florida Atlantic Univ., Boca Raton, FL), Utilitas Math., 1975, pp. 291–308. Congressus Numerantium, No. XIV.

[333] _____, Some extremal problems in geometry. IV, Utilitas Math., Louisiana State Univ., Baton Rouge, LA, 1976.

[334] _____, Some extremal problems in geometry. V, *Proc. Eighth Southeastern Conference on Combinatorics, Graph Theory and Computing* (Louisiana State Univ., Baton Rouge, LA), Utilitas Math., 1977, pp. 569–578. Congressus Numerantium, No. XIX.

[335] _____, Some combinatorial problems in the plane, *J. Combin. Theory Ser. A* **25** (1978), no. 2, 205–210.

[336] Paul Erdős and C. A. Rogers, Covering space with convex bodies, *Acta Arith.* **7** (1961/1962), 281–285.

[337] Paul Erdős and Robert Steinberg, Problems and Solutions: Advanced Problems: Solutions: 4065, *Amer. Math. Monthly* **51** (1944), no. 3, 169–171.

[338] Jeff Erickson, New lower bounds for Hopcroft's problem, *Discrete Comput. Geom.* **16** (1996), no. 4, 389–418.

[339] Guy Even, Zvi Lotker, Dana Ron, and Shakhar Smorodinsky, Conflict-free colorings of simple geometric regions with applications to frequency assignment in cellular networks, *SIAM J. Comput.* **33** (2003), no. 1, 94–136.

[340] Hazel Everett, Sylvain Lazard, Daniel Lazard, and Mohab Safey El Din, The Voronoi diagram of three lines, *SCG '07: Proc. Twenty-third Annual Symposium on Computational Geometry* (Gyeongju, South Korea), ACM, 2007, pp. 255–264.

[341] Hazel Everett, Jean-Marc Robert, and Marc van Kreveld, An optimal algorithm for computing ($\leq K$)-levels, with applications, *Internat. J. Comput. Geom. Appl.* **6** (1996), no. 3, 247–261.

[342] Esther Ezra, On the union of cylinders in three dimensions, *Proc. 49th IEEE Symp. on Foundations of Computer Science* (Philadelphia, PA, USA), 2008, to appear.

[343] Esther Ezra, János Pach, and Micha Sharir, On regular vertices on the union of planar objects, *SCG '07: Proc. Twenty-third Annual Symposium on Computational Geometry* (Gyeongju, South Korea), ACM, 2007, pp. 220–226.

[344] Esther Ezra and Micha Sharir, Lower envelopes of 3-intersecting surfaces, In preparation.

[345] _____, Output-sensitive construction of the union of triangles, *SIAM J. Comput.* **34** (2005), no. 6, 1331–1351.

[346] _____, Almost tight bound for the union of fat tetrahedra in three dimensions, *J. ACM* (2008), submitted. *Proc. 48th IEEE Symp. on Foundations of Computer Science* (2007), 525–535.

[347] Esther Ezra, Dan Halperin, and Micha Sharir, Speeding up the incremental construction of the union of geometric objects in practice, *Comput. Geom.* **27** (2004), no. 1, 63–85.

[348] Michael Falk and Richard Randell, On the homotopy theory of arrangements, *Complex analytic singularities*, Adv. Stud. Pure Math., vol. 8, North-Holland, Amsterdam, 1987, pp. 101–124.

[349] Gino Fano and Alessandro Terracini, *Lezioni di Geometria Analitica e Proiettiva*, G. B. Paravia & C., Torino, 1958.

[350] István Fáry, On straight line representation of planar graphs, *Acta Univ. Szeged. Sect. Sci. Math.* **11** (1948), 229–233.

[351] Gábor Fejes Tóth, A problem connected with multiple circle-packings and circle-coverings, *Studia Sci. Math. Hungar.* **12** (1977), no. 3–4, 447–456.

[352] _____, New results in the theory of packing and covering, *Convexity and its Applications*, Birkhäuser, Basel, 1983, pp. 318–359.

[353] _____, Multiple lattice packings of symmetric convex domains in the plane, *J. London Math. Soc.* **29** (1984), no. 3, 556–561.

[354] Gábor Fejes Tóth and Włodzimierz Kuperberg, A survey of recent results in the theory of packing and covering, *New Trends in Discrete and Computational Geometry* (J. Pach, ed.), Algorithms Combin., vol. 10, Springer, Berlin, 1993, pp. 251–279.

[355] László Fejes Tóth, *Lagerungen in der Ebene auf der Kugel und im Raum*, Zweite Auflage, Springer, Berlin, 1972.

[356] László Fejes Tóth and A. Heppes, Über stabile Körpersysteme, *Compositio Math.* **15** (1963), 119–126.

[357] Sharona Feldman and Micha Sharir, An improved bound for joints in arrangements of lines in space, *Discrete Comput. Geom.* **33** (2005), no. 2, 307–320.

[358] Stefan Felsner, On the number of arrangements of pseudolines, *Discrete Comput. Geom.* **18** (1997), no. 3, 257–267.

[359] Lukas Finschi and Komei Fukuda, Combinatorial generation of small point configurations and hyperplane arrangements, *Discrete and Computational Geometry, The Goodman–Pollack Festsschrift* (S. Basu, B. Aronov, J. Pach, and M. Sharir, eds.), Algorithms Combin., vol. 25, Springer, Berlin, 2003, pp. 425–440.

[360] M. Formann, T. Hagerup, J. Haralambides, M. Kaufmann, F. T. Leighton, A. Symvonis, E. Welzl, and G. Woeginger, Drawing graphs in the plane with high resolution, *SIAM J. Comput.* **22** (1993), no. 5, 1035–1052.

[361] Steven Fortune, Progress in computational geometry, *Directions in Geometric Computing* (Ralph R. Martin, ed.), Information Geometers, Winchester, 1993, pp. 81–128.

[362] Jacob Fox and János Pach, Coloring K_k-free intersection graphs of geometric objects in the plane, *SCG '08: The 24th Annual Symposium on Computational Geometry* (College Park, MD, USA), ACM, 2008, 346–354.

[363] Jacob Fox, János Pach, and Csaba D. Tóth, A bipartite strengthening of the crossing lemma, *Graph Drawing*, Lecture Notes in Comput. Sci., vol. 4875, Springer, Berlin/Heidelberg, 2008, pp. 13–24.

[364] Jacob Fox and Csaba D. Tóth, On the decay of crossing numbers, *J. Combin. Theory Ser. B* **98** (2008), no. 1, 33–42.

[365] Peter Frankl and Vojtěch Rödl, All triangles are Ramsey, *Trans. Amer. Math. Soc.* **297** (1986), no. 2, 777–779.

[366] Greg N. Frederickson and Donald B. Johnson, The complexity of selection and ranking in $X+Y$ and matrices with sorted columns, *J. Comput. System Sci.* **24** (1982), no. 2, 197–208.

[367] Jyh Jong Fu and R. C. T. Lee, Voronoï diagrams of moving points in the plane, *Internat. J. Comput. Geom. Appl.* **1** (1991), no. 1, 23–32.

[368] Henry Fuchs, Zvi M. Kedem, and Bruce F. Naylor, On visible surface generation by a priori tree structures, *SIGGRAPH Computer Graphics* **14** (1980), no. 3, 124–133.

[369] Komei Fukuda, Thomas M. Liebling, and François Margot, Analysis of backtrack algorithms for listing all vertices and all faces of a convex polyhedron, *Comput. Geom.* **8** (1997), no. 1, 1–12.

[370] Komei Fukuda, Shigemasa Saito, and Akihisa Tamura, Combinatorial face enumeration in arrangements and oriented matroids, *Discrete Appl. Math.* **31** (1991), no. 2, 141–149.
[371] Komei Fukuda, Shigemasa Saito, Akihisa Tamura, and Takeshi Tokuyama, Bounding the number of k-faces in arrangements of hyperplanes, *Discrete Appl. Math.* **31** (1991), no. 2, 151–165.
[372] Zoltan Füredi and Jeong-Hyun Kang, Covering the n-space by convex bodies and its chromatic number, *Discrete Math.* **308** (2008), no. 19, 4495–4500.
[373] A. Gabrièlov and N. Vorobjov, Complexity of stratifications of semi-Pfaffian sets, *Discrete Comput. Geom.* **14** (1995), no. 1, 71–91.
[374] Alexander A. Gaifullin, On isotopic weavings, *Arch. Math. (Basel)* **81** (2003), no. 5, 596–600.
[375] Martin Gardner (ed.), *Mathematical Puzzles of Sam Loyd*, Dover Publications Inc., New York, 1959.
[376] Michael R. Garey and David S. Johnson, *Computers and Intractability: A Guide to the Theory of NP-completeness*, W. H. Freeman and Co., San Francisco, Calif., 1979.
[377] _____, Crossing number is NP-complete, *SIAM J. Algebraic Discrete Methods* **4** (1983), no. 3, 312–316.
[378] O. Giering, Zum Problem von Sylvester Punktmengen mit k-Tripeln, *Österreich. Akad. Wiss. Math.-Natur. Kl. Sitzungsber. II* **204** (1995), 119–143.
[379] Wayne Goddard, Meir Katchalski, and Daniel J. Kleitman, Forcing disjoint segments in the plane, *European J. Combin.* **17** (1996), no. 4, 391–395.
[380] Jacob E. Goodman, Proof of a conjecture of Burr, Grünbaum, and Sloane, *Discrete Math.* **32** (1980), no. 1, 27–35.
[381] Jacob E. Goodman and Joseph O'Rourke (eds.), *Handbook of Discrete and Computational Geometry*, 2nd ed., Discrete Mathematics and its Applications, vol. 27, CRC Press LLC, Boca Raton, FL, 2004.
[382] Jacob E. Goodman and Richard Pollack, A combinatorial perspective on some problems in geometry, *Proc. Twelfth Southeastern Conference on Combinatorics, Graph Theory and Computing, Vol. I* (Baton Rouge, LA), vol. 32, 1981, pp. 383–394.
[383] _____, On the number of k-subsets of a set of n points in the plane, *J. Combin. Theory Ser. A* **36** (1984), no. 1, 101–104.
[384] _____, Semispaces of configurations, cell complexes of arrangements, *J. Combin. Theory Ser. A* **37** (1984), no. 3, 257–293.
[385] _____, Allowable sequences and order types in discrete and computational geometry, *New Trends in Siscrete and Computational Geometry* (J. Pach, ed.), Algorithms Combin., vol. 10, Springer, Berlin, 1993, pp. 103–134.
[386] Jacob E. Goodman, Richard Pollack, and Rephael Wenger, Geometric transversal theory, *New Trends in Discrete and Computational Geometry* (J. Pach, ed.), Algorithms Combin., vol. 10, Springer, Berlin, 1993, pp. 163–198.
[387] _____, On the connected components of the space of line transversals to a family of convex sets, *Discrete Comput. Geom.* **13** (1995), no. 3–4, 469–476.
[388] Michael T. Goodrich, Constructing arrangements optimally in parallel, *Discrete Comput. Geom.* **9** (1993), no. 4, 371–385.
[389] Mark Goresky and Robert MacPherson, *Stratified Morse Theory*, Springer, Berlin, 1988.
[390] L. Gournay and J.-J. Risler, Construction of roadmaps in semi-algebraic sets, *Appl. Algebra Engrg. Comm. Comput.* **4** (1993), no. 4, 239–252.
[391] Erez Greenstein, *Transversals of pairwise disjoint balls in R^3*, Master's thesis, Tel-Aviv University, Tel Aviv, 2005.
[392] Dima Yu. Grigor'ev and Nicolai N. Vorobjov, Jr., Counting connected components of a semialgebraic set in subexponential time, *Comput. Complexity* **2** (1992), no. 2, 133–186.
[393] Branko Grünbaum, Arrangements of hyperplanes, *Proc. Second Louisiana Conference on Combinatorics, Graph Theory and Computing* (Baton Rouge, LA), Louisiana State Univ., 1971, pp. 41–106.
[394] _____, *Arrangements and spreads*, American Mathematical Society Providence, R.I., 1972, Conference Board of the Mathematical Sciences Regional Conference Series in Mathematics, No. 10.
[395] _____, Arrangements of colored lines, abstract 720-50-5, *Notices of the Amer. Math. Soc.* **22** (1975), A–200.

[396] _____, Monochromatic intersection points in families of colored lines, *Geombinatorics* **9** (1999), no. 1, 3–9.

[397] Leonidas J. Guibas, Dan Halperin, Hirohisa Hirukawa, Jean-Claude Latombe, and Randall H. Wilson, Polyhedral assembly partitioning using maximally covered cells in arrangements of convex polytopes, *Internat. J. Comput. Geom. Appl.* **8** (1998), no. 2, 179–199.

[398] Leonidas J. Guibas, Dan Halperin, Jiří Matoušek, and Micha Sharir, Vertical decomposition of arrangements of hyperplanes in four dimensions, *Discrete Comput. Geom.* **14** (1995), no. 2, 113–122.

[399] Leonidas J. Guibas, Joseph S. B. Mitchell, and Thomas Roos, Voronoï diagrams of moving points in the plane, *Graph-theoretic concepts in computer science (Fischbachau, 1991)*, Lecture Notes in Comput. Sci., vol. 570, Springer, Berlin, 1992, pp. 113–125.

[400] Leonidas J. Guibas, Mark H. Overmars, and Jean-Marc Robert, The exact fitting problem in higher dimensions, *Comput. Geom.* **6** (1996), no. 4, 215–230.

[401] Leonidas J. Guibas and Micha Sharir, Combinatorics and algorithms of arrangements, *New Trends in Discrete and Computational Geometry* (J. Pach, ed.), Algorithms Combin., vol. 10, Springer, Berlin, 1993, pp. 9–36.

[402] Leonidas J. Guibas, Micha Sharir, and Shmuel Sifrony, On the general motion-planning problem with two degrees of freedom, *Discrete Comput. Geom.* **4** (1989), no. 5, 491–521.

[403] Leonidas J. Guibas and Jorge Stolfi, Primitives for the manipulation of general subdivisions and the computation of Voronoi diagrams, *ACM Trans. Graphics* **4** (1985), no. 2, 74–123.

[404] Leonidas J. Guibas and F. Frances Yao, On translating a set of rectangles, *Computational Geometry* (Franco P. Preparata, ed.), Advances in Computing Research, vol. 1, JAI Press, London, 1983, pp. 61–67.

[405] Dan Gusfield, Bounds for the parametric minimum spanning tree problem, *Proc. West Coast Conference on Combinatorics, Graph Theory and Computing (Humboldt State Univ., Arcata, Calif., 1979)*, Congress. Numer., XXVI, Utilitas Math., 1980, pp. 173–181.

[406] Richard K. Guy, The decline and fall of Zarankiewicz's theorem, *Proof Techniques in Graph Theory (Proc. Second Ann Arbor Graph Theory Conf., Ann Arbor, Mich., 1968)*, Academic Press, New York, 1969, pp. 63–69.

[407] Hugo Hadwiger, Ein Überdeckungssatz für den Euklidischen Raum, *Portugal. Math.* **4** (1944), 140–144.

[408] _____, Eulers Charakteristik und kombinatorische Geometrie, *J. Reine Angew. Math.* **194** (1955), 101–110.

[409] _____, Ungelöste Probleme No. 40, *Elemente d. Math.* **16** (1961), 103–104.

[410] Torben Hagerup, H. Jung, and Emo Welzl, Efficient parallel computation of arrangements of hyperplanes in d dimensions, *SPAA '90: Proceedings of the second annual ACM Symposium on Parallel Algorithms and Architectures* (Island of Crete, Greece), ACM, 1990, pp. 290–297.

[411] A. W. Hales and R. I. Jewett, Regularity and positional games, *Trans. Amer. Math. Soc.* **106** (1963), 222–229.

[412] Dan Halperin, *Algorithmic Motion Planning via Arrangements of Curves and of Surfaces*, Ph.D. thesis, Tel-Aviv University, Tel Aviv, 1992.

[413] _____, On the complexity of a single cell in certain arrangements of surfaces related to motion planning, *Discrete Comput. Geom.* **11** (1994), no. 1, 1–33.

[414] _____, Arrangements, in Goodman and O'Rourke [381], pp. 529–562.

[415] Dan Halperin, Lydia E. Kavraki, and Jean-Claude Latombe, Robotics, in Goodman and O'Rourke [381], pp. 1065–1094.

[416] Dan Halperin, Jean-Claude Latombe, and Randall H. Wilson, A general framework for assembly planning: the motion space approach, *Algorithmica* **26** (2000), no. 3–4, 577–601.

[417] Dan Halperin and Mark H. Overmars, Spheres, molecules, and hidden surface removal, *Comput. Geom.* **11** (1998), no. 2, 83–102.

[418] Dan Halperin and Micha Sharir, On disjoint concave chains in arrangements of (pseudo) lines, *Inform. Process. Lett.* **40** (1991), no. 4, 189–192, Corrigendum in **51** (1994), no. 1, 53–56.

[419] _____, New bounds for lower envelopes in three dimensions, with applications to visibility in terrains, *Discrete Comput. Geom.* **12** (1994), no. 3, 313–326.

[420] _____, Almost tight upper bounds for the single cell and zone problems in three dimensions, *Discrete Comput. Geom.* **14** (1995), no. 4, 385–410.

[421] _____, Arrangements and their applications in robotics: recent developments, *WAFR: Proceedings of the workshop on Algorithmic foundations of robotics* (San Francisco, California, United States), A. K. Peters, Natick, MA, 1995, pp. 495–511.

[422] _____, A near-quadratic algorithm for planning the motion of a polygon in a polygonal environment, *Discrete Comput. Geom.* **16** (1996), no. 2, 121–134.

[423] Dan Halperin and Christian R. Shelton, A perturbation scheme for spherical arrangements with application to molecular modeling, *Comput. Geom.* **10** (1998), no. 4, 273–287.

[424] Dan Halperin and Chee-Keng Yap, Combinatorial complexity of translating a box in polyhedral 3-space, *Comput. Geom.* **9** (1998), no. 3, 181–196.

[425] Sariel Har-Peled, *The Complexity of Many Faces in the Overlay of Arrangements*, Master's thesis, Tel-Aviv University, Tel Aviv, 1995.

[426] _____, Constructing planar cuttings in theory and practice, *SIAM J. Comput.* **29** (2000), no. 6, 2016–2039.

[427] Sariel Har-Peled and Micha Sharir, Online point location in planar arrangements and its applications, *Discrete Comput. Geom.* **26** (2001), no. 1, 19–40.

[428] Sariel Har-Peled and Shakhar Smorodinsky, Conflict-free coloring of points and simple regions in the plane, *Discrete Comput. Geom.* **34** (2005), no. 1, 47–70.

[429] Joe Harris, *Algebraic Geometry: A First Course*, Graduate Texts in Mathematics, vol. 133, Springer, New York, 1995, Corrected reprint of the 1992 original.

[430] Sergiu Hart and Micha Sharir, Nonlinearity of Davenport-Schinzel sequences and of generalized path compression schemes, *Combinatorica* **6** (1986), no. 2, 151–177.

[431] David Haussler and Emo Welzl, ε-nets and simplex range queries, *Discrete Comput. Geom.* **2** (1987), no. 2, 127–151.

[432] Joos Heintz, Tomas Recio, and Marie-Françoise Roy, Algorithms in real algebraic geometry and applications to computational geometry, *Discrete and computational geometry (New Brunswick, NJ, 1989/1990)*, DIMACS Ser. Discrete Math. Theoret. Comput. Sci., vol. 6, Amer. Math. Soc., Providence, RI, 1991, pp. 137–163.

[433] Joos Heintz, Marie-Françoise Roy, and Pablo Solernó, Description of the connected components of a semialgebraic set in single exponential time, *Discrete Comput. Geom.* **11** (1994), no. 2, 121–140.

[434] _____, Single exponential path finding in semi-algebraic sets. II. The general case, *Algebraic geometry and its applications (West Lafayette, IN, 1990)*, Springer, New York, 1994, pp. 449–465.

[435] Joos Heintz, Pablo Solernó, and Marie-Françoise Roy, On the complexity of semialgebraic sets, *IFIP Congress* (San Francisco), 1989, pp. 293–298.

[436] A. Heppes, Mehrfache gitterförmige Kreislagerungen in der Ebene, *Acta Math. Acad. Sci. Hungar.* **10** (1959), 141–148.

[437] John E. Hershberger, Finding the upper envelope of n line segments in $O(n \log n)$ time, *Inform. Process. Lett.* **33** (1989), no. 4, 169–174.

[438] John E. Hershberger and Jack S. Snoeyink, Erased arrangements of lines and convex decompositions of polyhedra, *Comput. Geom.* **9** (1998), no. 3, 129–143.

[439] John E. Hershberger and Subhash Suri, A pedestrian approach to ray shooting: shoot a ray, take a walk, *J. Algorithms* **18** (1995), no. 3, 403–431.

[440] I. N. Herstein and I. Kaplansky, *Matters mathematical*, Chelsea, New York, 1978.

[441] Fritz Herzog and L. M. Kelly, A generalization of the theorem of Sylvester, *Proc. Amer. Math. Soc.* **11** (1960), 327–331.

[442] W. V. D. Hodge and D. Pedoe, *Methods of Algebraic Geometry. Vol. II. Book III: General theory of algebraic varieties in projective space. Book IV: Quadrics and Grassmann varieties*, Cambridge, at the University Press, 1952.

[443] Andreas Holmsen, Meir Katchalski, and Ted Lewis, A Helly-type theorem for line transversals to disjoint unit balls, *Discrete Comput. Geom.* **29** (2003), no. 4, 595–602.

[444] L. S. Homem de Mello and A. C. Sanderson, A correct and complete algorithm for the generation of mechanical assembly sequences, *IEEE Trans. Robot. Autom.* **7** (1991), no. 2, 228–240.

[445] H. Hopf and E. Pannwitz, Aufgabe no. 167, *Jahresbericht Deutsche Mahtematiker-Vereinigung* **43** (1934), 114.

[446] David Hsu, Lydia E. Kavraki, Jean-Claude Latombe, Rajeev Motwani, and Stephen Sorkin, On finding narrow passages with probabilistic roadmap planners, *Robotics: the Algorithmic Perspective (Houston, TX, 1998)*, A. K. Peters, Natick, MA, 1998, pp. 141–153.

[447] Yingping Huang, Jinhui Xu, and Danny Z. Chen, Geometric permutations of high dimensional spheres, *Proc. Twelfth Annual ACM-SIAM Symposium on Discrete Algorithms (Washington, DC, 2001)*, SIAM, 2001, pp. 244–245.

[448] K. H. Hunt, *Kinematic Geometry of Mechanisms*, Oxford Engineering Science Series, vol. 7, The Clarendon Press Oxford University Press, New York, 1990, Corrected reprint of the 1978 original, Oxford Science Publications.

[449] Daniel P. Huttenlocher, Klara Kedem, and Jon M. Kleinberg, On dynamic Voronoi diagrams and the minimum Hausdorff distance for point sets under Euclidean motion in the plane, *SCG '92: Proceedings of the Eighth Annual Symposium on Computational Geometry* (Berlin, Germany), ACM, 1992, pp. 110–119.

[450] Daniel P. Huttenlocher, Klara Kedem, and Micha Sharir, The upper envelope of Voronoï surfaces and its applications, *Discrete Comput. Geom.* **9** (1993), no. 3, 267–291.

[451] S. Jadhav and A. Mukhopadhyay, Computing a centerpoint of a finite planar set of points in linear time, *Discrete Comput. Geom.* **12** (1994), no. 3, 291–312.

[452] Robert E. Jamison, A survey of the slope problem, *Discrete geometry and convexity (New York, 1982)*, Ann. New York Acad. Sci., vol. 440, New York Acad. Sci., New York, 1985, pp. 34–51.

[453] Robert E. Jamison and Dale Hill, A catalogue of sporadic slope-critical configurations, *Proc. Fourteenth Southeastern Conference on Combinatorics, Graph Theory and Computing (Boca Raton, Fla., 1983)*, vol. 40, 1983, pp. 101–125.

[454] C. M. Jessop, *A Treatise on the Line Complex*, Chelsea, New York, 1969.

[455] Wm. Woolsey Johnson and William E. Story, Notes on the "15" Puzzle, *Amer. J. Math.* **2** (1879), no. 4, 397–404.

[456] Lior Kapelushnik, *Computing the k-centrum hyperplane and related problems*, Master's thesis, School of Computer Science, Tel Aviv University, Tel Aviv, Israel, 2008.

[457] Haim Kaplan, Natan Rubin, and Micha Sharir, Linear data structures for fast ray-shooting amidst convex polyhedra, *Algorithms—ESA 2007 (Eilat, Israel)*, Lecture Notes in Comput. Sci., vol. 4698, Springer, Berlin, 2007, pp. 287–298.

[458] _____, Line transversals of convex polyhedra in $\mathbb{R}^3$, *SODA 2009*, submitted.

[459] Györgi Károlyi, János Pach, Gábor Tardos, and Géza Tóth, An algorithm for finding many disjoint monochromatic edges in a complete 2-colored geometric graph, *Intuitive geometry (Budapest, 1995)*, Bolyai Soc. Math. Stud., vol. 6, János Bolyai Math. Soc., Budapest, 1997, pp. 367–372.

[460] Györgi Károlyi, János Pach, and Géza Tóth, Ramsey-type results for geometric graphs. I, *Discrete Comput. Geom.* **18** (1997), no. 3, 247–255.

[461] Györgi Károlyi, János Pach, Géza Tóth, and Pavel Valtr, Ramsey-type results for geometric graphs. II, *Discrete Comput. Geom.* **20** (1998), no. 3, 375–388.

[462] Meir Katchalski, T. Lewis, and A. Liu, The different ways of stabbing disjoint convex sets, *Discrete Comput. Geom.* **7** (1992), no. 2, 197–206.

[463] Meir Katchalski, T. Lewis, and J. Zaks, Geometric permutations for convex sets, *Discrete Math.* **54** (1985), no. 3, 271–284.

[464] Meir Katchalski, Subhash Suri, and Yunhong Zhou, A constant bound for geometric permutations of disjoint unit balls, *Discrete Comput. Geom.* **29** (2003), no. 2, 161–173.

[465] Matthew J. Katz and Kasturi R. Varadarajan, A tight bound on the number of geometric permutations of convex fat objects in R^d, *Discrete Comput. Geom.* **26** (2001), no. 4, 543–548.

[466] Matthew J. Katz and Micha Sharir, Optimal slope selection via expanders, *Inform. Process. Lett.* **47** (1993), no. 3, 115–122.

[467] _____, An expander-based approach to geometric optimization, *SIAM J. Comput.* **26** (1997), no. 5, 1384–1408.

[468] Nets Hawk Katz and Gábor Tardos, A new entropy inequality for the Erdős distance problem, *Towards a Theory of Geometric Graphs* (J. Pach, ed.), Contemp. Math., vol. 342, Amer. Math. Soc., Providence, RI, 2004, pp. 119–126.

[469] Lydia E. Kavraki, Jean-Claude Latombe, Rajeev Motwani, and Prabhakar Raghavan, Randomized query processing in robot path planning, *J. Comput. System Sci.* **57** (1998), no. 1, 50–60.

[470] Lydia E. Kavraki, Petr Svestka, Jean-Claude Latombe, and Mark H. Overmars, Probabilistic roadmaps for path planning in high-dimensional configuration spaces, *IEEE Trans. Robot. Autom.* **12** (1996), no. 4, 566–580.

[471] Klara Kedem, Ron Livné, János Pach, and Micha Sharir, On the union of Jordan regions and collision-free translational motion amidst polygonal obstacles, *Discrete Comput. Geom.* **1** (1986), no. 1, 59–71.

[472] Klara Kedem and Micha Sharir, An efficient motion-planning algorithm for a convex polygonal object in two-dimensional polygonal space, *Discrete Comput. Geom.* **5** (1990), no. 1, 43–75.

[473] Klara Kedem, Micha Sharir, and Sivan Toledo, On critical orientations in the Kedem-Sharir motion planning algorithm, *Discrete Comput. Geom.* **17** (1997), no. 2, 227–239.

[474] L. M. Kelly and William O. J. Moser, On the number of ordinary lines determined by n points, *Canad. J. Math.* **1** (1958), 210–219.

[475] L. M. Kelly and R. Rottenberg, Simple points in pseudoline arrangements, *Pacific J. Math.* **40** (1972), 617–622.

[476] Balázs Keszegh, János Pach, Dömötör Pálvölgyi, and Géza Tóth, Drawing cubic graphs with at most five slopes, *Graph drawing, Lecture Notes in Comput. Sci.*, vol. 4372, Springer, Berlin/Heidelberg, 2007, pp. 114–125.

[477] Jussi Ketonen and Robert Solovay, Rapidly growing Ramsey functions, *Ann. of Math.* **113** (1981), no. 2, 267–314.

[478] Lutz Kettner, Designing a data structure for polyhedral surfaces, *SCG '98: Proceedings of the Fourteenth Annual Symposium on Computational Geometry* (Minneapolis, Minnesota, United States), ACM, 1998, pp. 146–154.

[479] A. G. Khovanskiĭ, *Fewnomials, Translations of Mathematical Monographs*, vol. 88, American Mathematical Society, Providence, RI, 1991, Translated from the Russian by Smilka Zdravkovska.

[480] David G. Kirkpatrick and Raimund Seidel, The ultimate planar convex hull algorithm?, *SIAM J. Comput.* **15** (1986), no. 1, 287–299.

[481] Maria M. Klawe, Superlinear bounds for matrix searching problems, *J. Algorithms* **13** (1992), no. 1, 55–78.

[482] Maria M. Klawe, Michael S. Paterson, and N. Pippenger, Inversions with $n2^{\Omega(\sqrt{\log n})}$ transpositions at the median, Unpublished manuscript, 1982.

[483] Martin Klazar, A general upper bound in extremal theory of sequences, *Comment. Math. Univ. Carolin.* **33** (1992), no. 4, 737–746.

[484] _____, Two results on a partial ordering of finite sequences, *Comment. Math. Univ. Carolin.* **34** (1993), no. 4, 697–705.

[485] _____, A linear upper bound in extremal theory of sequences, *J. Combin. Theory Ser. A* **68** (1994), no. 2, 454–464.

[486] _____, *Combinatorial Aspects of Davenport–Schinzel Sequences*, Ph.D. thesis, Charles University, Prague, Czech Republic, 1995.

[487] _____, On the maximum lengths of Davenport–Schinzel sequences, *Contemporary Trends in Discrete Mathematics (DIMACS Series in Discrete Mathematics and Theoretical Computer Science, Vol. 49)* (R. L. Graham, J. Kratochvíl, J. Nesteril, and F. S. Roberts, eds.), pp. 169–178, Amer. Math. Soc., Providence, RI, 1999.

[488] Martin Klazar and Pavel Valtr, Generalized Davenport–Schinzel sequences, *Combinatorica* **14** (1994), no. 4, 463–476.

[489] Victor Klee, On the complexity of d-dimensional Voronoi diagrams, *Arch. Math. (Basel)* **34** (1980), no. 1, 75–80.

[490] Daniel J. Kleitman, The crossing number of $K_{5,n}$, *J. Combin. Theory* **9** (1970), 315–323.

[491] Petr Kolman and Jiří Matoušek, Crossing number, pair-crossing number, and expansion, *J. Combin. Theory Ser. B* **92** (2004), no. 1, 99–113.

[492] Vladlen Koltun, Almost tight upper bounds for vertical decompositions in four dimensions, *J. ACM* **51** (2004), no. 5, 699–730.

[493] _____, Sharp bounds for vertical decompositions of linear arrangements in four dimensions, *Discrete Comput. Geom.* **31** (2004), no. 3, 435–460.

[494] Vladlen Koltun and Micha Sharir, 3-dimensional Euclidean Voronoi diagrams of lines with a fixed number of orientations, *SIAM J. Comput.* **32** (2003), no. 3, 616–642.

[495] _____, The partition technique for overlays of envelopes, *SIAM J. Comput.* **32** (2003), no. 4, 841–863.

[496] _____, Polyhedral Voronoi diagrams of polyhedra in three dimensions, *Discrete Comput. Geom.* **31** (2004), no. 1, 83–124.

[497] _____, On overlays and minimization diagrams, *SCG '06: Proceedings of the twenty-second annual symposium on Computational geometry* (Sedona, Arizona, USA), ACM, 2006, pp. 395–401.

[498] Vladlen Koltun and Carola Wenk, Matching polyhedral terrains using overlays of envelopes, *Algorithmica* **41** (2005), no. 3, 159–183.

[499] Péter Komjáth, A simplified construction of nonlinear Davenport-Schinzel sequences, *J. Combin. Theory Ser. A* **49** (1988), no. 2, 262–267.

[500] Daniel Kornhauser, Gary Miller, and Paul Spirakis, Coordinating pebble motion on graphs, the diameter of permutation groups, and applications, *Proc. 25th Annual Symposium on Foundations of Computer Science* (West Palm Beach, FL, USA), IEEE, 1984, pp. 241–250.

[501] Marc J. van Kreveld and Mark T. de Berg, Finding squares and rectangles in sets of points, *BIT* **31** (1991), no. 2, 202–219.

[502] Yaakov S. Kupitz, On a generalization of the Gallai-Sylvester theorem, *Discrete Comput. Geom.* **7** (1992), no. 1, 87–103.

[503] _____, *Extremal Problems in Combinatorial Geometry*, Lecture Notes Series, vol. 53, Aarhus Univ. Mat. Institut, Aarhus, 1979.

[504] M. Laczkovich and I. Z. Ruzsa, The number of homothetic subsets, *The Mathematics of Paul Erdős, II*, Algorithms Combin., vol. 14, Springer, Berlin, 1997, pp. 294–302.

[505] D. W. Lang, The dual of a well-known theorem, *The Mathematical Gazette* **39** (1955), no. 330, 314.

[506] Michel Las Vergnas, Convexity in oriented matroids, *J. Combin. Theory Ser. B* **29** (1980), no. 2, 231–243.

[507] Jean-Claude Latombe, *Robot Motion Planning*, Kluwer International Series in Engineering and Computer Science, vol. 124, Kluwer Academic Publishers, Norwell, MA, USA, 1991.

[508] B. Lee and F. M. Richards, The interpretation of protein structures: Estimation of static accessibility, *J. Molecular Biology* **55** (1971), no. 3, 379–380.

[509] Frank Thomson Leighton, *Complexity Issues in VLSI: Optimal Layouts for the Shuffle-Exchange Graph and Other Networks*, MIT Press, Cambridge, MA, 1983.

[510] Jonathan Lenchner, Wedges in Euclidean arrangements, *Discrete and Computational Geometry*, Lecture Notes in Comput. Sci., vol. 3742, Springer, Berlin, 2005, pp. 131–142.

[511] Thomas Lengauer, Algorithmic research problems in molecular bioinformatics, *ISTCS'93: Proc. Second Israel Symposium on the Theory of Computing Systems* (Natanya, Israel), IEEE Computer Society, 1993, pp. 177–192.

[512] Daniel Leven and Micha Sharir, On the number of critical free contacts of a convex polygonal object moving in two-dimensional polygonal space, *Discrete Comput. Geom.* **2** (1987), no. 3, 255–270.

[513] Pascal Lienhardt, Topological models for boundary representation: a comparison with n-dimensional generalized maps, *Computer Aided Design* **23** (1991), no. 1, 59–82.

[514] _____, n-dimensional generalized combinatorial maps and cellular quasi-manifolds, *Internat. J. Comput. Geom. Appl.* **4** (1994), no. 3, 275–324.

[515] X.-B. Lin, Another brief proof of the Sylvester theorem, *Amer. Math. Monthly* **95** (1988), no. 10, 932–933.

[516] Richard J. Lipton and Robert Endre Tarjan, A separator theorem for planar graphs, *SIAM J. Appl. Math.* **36** (1979), no. 2, 177–189.

[517] Chi-Yuan Lo, Jiří Matoušek, and William L. Steiger, Algorithms for ham-sandwich cuts, *Discrete Comput. Geom.* **11** (1994), no. 4, 433–452.

[518] László Lovász, On the number of halving lines, *Ann. Univ. Sci. Budapest. Eötvös Sect. Math.* **14** (1971), 107–108.

[519] _____, On the ratio of optimal integral and fractional covers, *Discrete Math.* **13** (1975), no. 4, 383–390.

[520] László Lovász, János Pach, and Mario Szegedy, On Conway's thrackle conjecture, *Discrete Comput. Geom.* **18** (1997), no. 4, 369–376.

[521] Desmond MacHale, *Comic Sections: The Book of Mathematical Jokes, Humour, Wit and Wisdom*, Boole Press, Dublin, 1993.
[522] Seth Malitz and Achilleas Papakostas, On the angular resolution of planar graphs, *SIAM J. Discrete Math.* **7** (1994), no. 2, 172–183.
[523] P. Mani-Levitska and János Pach, Decomposition problems for multiple coverings with unit balls, Unpublished manuscript, 1987.
[524] Adam Marcus and Gábor Tardos, Intersection reverse sequences and geometric applications, *J. Combin. Theory Ser. A* **113** (2006), no. 4, 675–691.
[525] Jiří Matoušek, Construction of ε-nets, *Discrete Comput. Geom.* **5** (1990), no. 5, 427–448.
[526] _____, Computing the center of planar point sets, *Discrete and computational geometry (New Brunswick, NJ, 1989/1990)*, DIMACS Ser. Discrete Math. Theoret. Comput. Sci., vol. 6, Amer. Math. Soc., Providence, RI, 1991, pp. 221–230.
[527] _____, Lower bounds on the length of monotone paths in arrangements, *Discrete Comput. Geom.* **6** (1991), no. 2, 129–134.
[528] _____, Randomized optimal algorithm for slope selection, *Inform. Process. Lett.* **39** (1991), no. 4, 183–187.
[529] _____, Efficient partition trees, *Discrete Comput. Geom.* **8** (1992), no. 3, 315–334.
[530] _____, Reporting points in halfspaces, *Comput. Geom.* **2** (1992), no. 3, 169–186.
[531] _____, Epsilon-nets and computational geometry, *New Trends in Discrete and Computational Geometry* (J. Pach, ed.), Algorithms Combin., vol. 10, Springer, Berlin, 1993, pp. 69–89.
[532] _____, Linear optimization queries, *J. Algorithms* **14** (1993), no. 3, 432–448.
[533] _____, On vertical ray shooting in arrangements, *Comput. Geom.* **2** (1993), no. 5, 279–285.
[534] _____, Range searching with efficient hierarchical cuttings, *Discrete Comput. Geom.* **10** (1993), no. 2, 157–182.
[535] _____, Geometric range searching, *ACM Comput. Surv.* **26** (1994), no. 4, 422–461.
[536] _____, *Lectures on Discrete Geometry*, Graduate Texts in Mathematics, vol. 212, Springer, New York, 2002.
[537] Jiří Matoušek, János Pach, Micha Sharir, Shmuel Sifrony, and Emo Welzl, Fat triangles determine linearly many holes, *SIAM J. Comput.* **23** (1994), no. 1, 154–169.
[538] Jiří Matoušek and Otfried Schwarzkopf, On ray shooting in convex polytopes, *Discrete Comput. Geom.* **10** (1993), no. 2, 215–232.
[539] _____, A deterministic algorithm for the three-dimensional diameter problem, *Comput. Geom.* **6** (1996), no. 4, 253–262.
[540] Jiří Matoušek, Micha Sharir, Shakhar Smorodinsky, and Uli Wagner, k-sets in four dimensions, *Discrete Comput. Geom.* **35** (2006), no. 2, 177–191.
[541] Jiří Matoušek and Pavel Valtr, The complexity of the lower envelope of segments with h endpoints, *Intuitive Geometry (Budapest, 1995)*, Bolyai Soc. Math. Stud., vol. 6, János Bolyai Math. Soc., Budapest, 1997, pp. 407–411.
[542] Michael McKenna and Joseph O'Rourke, Arrangements of lines in 3-space: a data structure with applications, *Proc. Fourth Annual Symposium on Computational Geometry* (Urbana, IL), ACM, 1988, pp. 371–380.
[543] P. McMullen and G. C. Shephard, *Convex Polytopes and the Upper Bound Conjecture*, London Math. Soc. Lecture Note Series, vol. 3, Cambridge Univ. Press, London, 1971, Prepared in collaboration with J. E. Reeve and A. A. Ball.
[544] Joseph Mecke, Random tesselations generated by hyperplanes, *Stochastic geometry, geometric statistics, stereology (Oberwolfach, 1983)*, Teubner-Texte Math., vol. 65, Teubner, Leipzig, 1984, pp. 104–109.
[545] Nimrod Megiddo, Applying parallel computation algorithms in the design of serial algorithms, *J. Assoc. Comput. Mach.* **30** (1983), no. 4, 852–865.
[546] _____, Partitioning with two lines in the plane, *J. Algorithms* **6** (1985), no. 3, 430–433.
[547] E. Melchior, Über Vielseite der projektiven Ebene, *Deutsche Math.* **5** (1941), 461–475.
[548] P. G. Mezey, Molecular surfaces, *Reviews in Computational Chemistry* (Kenneth B. Lipkowitz and Donald B. Boyd, eds.), vol. 1, Wiley, Hoboken, NJ, 1990, pp. 265–294.
[549] W. H. Mills, Some Davenport-Schinzel sequences, *Proc. Third Manitoba Conference on Numerical Mathematics* (Winnipeg, Man.), Utilitas Math., 1974, pp. 307–313.
[550] J. Milnor, *Morse Theory*, Princeton Univ. Press, Princeton, NJ, 1963.
[551] _____, On the Betti numbers of real varieties, *Proc. Amer. Math. Soc.* **15** (1964), 275–280.

[552] Bhubaneswar Mishra, Computational real algebraic geometry, in Goodman and O'Rourke [**381**], pp. 743–764.
[553] Joseph S. B. Mitchell, Geometric shortest paths and network optimization, in Sack and Urrutia [**657**], pp. 633–701.
[554] Shai Mohaban and Micha Sharir, Ray shooting amidst spheres in three dimensions and related problems, *SIAM J. Comput.* **26** (1997), no. 3, 654–674.
[555] Leo Moser, On the different distances determined by n points, *Amer. Math. Monthly* **59** (1952), 85–91.
[556] Leo Moser and William O.J. Moser, Solution to problem 10, *Canad. Math. Bull.* **4** (1961), 187–189.
[557] Theodore S. Motzkin, The lines and planes connecting the points of a finite set, *Trans. Amer. Math. Soc.* **70** (1951), 451–464.
[558] _____, Nonmixed connecting lines, abstract 67t 605, *Notices of the Amer. Math. Soc.* **14** (1967), 837.
[559] Padmini Mukkamala and Mario Szegedy, Geometric representation of cubic graphs with four directions, To appear.
[560] R. C. Mullin and R. G. Stanton, A map-theoretic approach to Davenport-Schinzel sequences, *Pacific J. Math.* **40** (1972), 167–172.
[561] Ketan Mulmuley, A fast planar partition algorithm. I, *J. Symbolic Comput.* **10** (1990), no. 3–4, 253–280.
[562] _____, A fast planar partition algorithm. II, *J. Assoc. Comput. Mach.* **38** (1991), no. 1, 74–103.
[563] _____, On levels in arrangements and Voronoï diagrams, *Discrete Comput. Geom.* **6** (1991), no. 4, 307–338.
[564] _____, *Computational Geometry: An Introduction Through Randomized Algorithms*, Prentice Hall, Englewood Cliffs, NJ, 1993.
[565] Ketan Mulmuley and Sandeep Sen, Dynamic point location in arrangements of hyperplanes, *Discrete Comput. Geom.* **8** (1992), no. 3, 335–360.
[566] S. Murata, H. Kurokawa, and S. Kokaji, Self-assembling machine, *Proc. 1994 IEEE International Conference on Robotics and Automation* (San Diego, CA, USA), vol. 1, 1994, pp. 441–448.
[567] Nir Naor and Micha Sharir, Computing the center of a point set in three dimensions, *Proc. 2nd Canadian Conf. on Computational Geometry*, 1990, pp. 10–13.
[568] An Nguyen, Leonidas J. Guibas, and Mark Yim, Controlled module density helps reconfiguration planning, *Algorithmic and Computational Robotics: New Directions (Hanover, NH, 2000)*, A. K. Peters, Natick, MA, 2001, pp. 23–35.
[569] Gabriel Nivasch, An improved, simple construction of many halving edges, *Proc. 2006 AMS/IMS/SIAM Joint Summer Research Conference on Discrete and Computational Geometry: 20 Years Later* (J. Goodman, J. Pach, and R. Pollack, eds.), *Contemp. Math.*, vol. 453, Amer. Math. Soc., Providence, RI, 2008, pp. 299–305.
[570] _____, Improved bounds and new techniques for Davenport–Schinzel sequences and their generalizations, *SODA 2009*, submitted.
[571] Gabriel Nivasch and Micha Sharir, Eppstein's bound on intersecting triangles revisited, *J. Combin. Theory Ser. A*, to appear.
[572] Peter Orlik, *Introduction to Arrangements*, CBMS Regional Conference Series in Mathematics, vol. 72, Amer. Math. Soc., Proidence, RI, 1989.
[573] _____, Arrangements in topology, *Discrete and computational geometry (New Brunswick, NJ, 1989/1990)*, DIMACS Ser. Discrete Math. Theoret. Comput. Sci., vol. 6, Amer. Math. Soc., Providence, RI, 1991, pp. 263–272.
[574] Peter Orlik and Louis Solomon, Combinatorics and topology of complements of hyperplanes, *Invent. Math.* **56** (1980), no. 2, 167–189.
[575] Peter Orlik and Hiroaki Terao, *Arrangements of Hyperplanes*, Springer, Berlin, 1992.
[576] Joseph O'Rourke, Computational geometry column, *ACM SIGACT News* **20** (1989), no. 2, 10–11.
[577] _____, Visibility, in Goodman and O'Rourke [**381**], pp. 643–663.
[578] Mark H. Overmars and Chee-Keng Yap, New upper bounds in Klee's measure problem, *SIAM J. Comput.* **20** (1991), no. 6, 1034–1045.

[579] János Pach, Decomposition of multiple packing and covering, *2. Kolloquium über Diskrete Geometrie* (Salzburg), 1980, pp. 169–178.
[580] _____, Covering the plane with convex polygons, *Discrete Comput. Geom.* **1** (1986), no. 1, 73–81.
[581] _____, Notes on geometric graph theory, *Discrete and computational geometry (New Brunswick, NJ, 1989/1990)*, DIMACS Ser. Discrete Math. Theoret. Comput. Sci., vol. 6, Amer. Math. Soc., Providence, RI, 1991, pp. 273–285.
[582] _____, Geometric graphs and hypergraphs, *Graph Theory Notes N. Y.* **31** (1996), 39–43, New York Graph Theory Day, 31 (1996).
[583] _____, Crossroads in Flatland, *Combinatorics, computation & logic '99 (Auckland)*, Aust. Comput. Sci. Commun., vol. 21, Springer, Singapore, 1999, pp. 73–80.
[584] _____, Finite point configurations, in Goodman and O'Rourke [381], pp. 3–24.
[585] János Pach and Pankaj K. Agarwal, *Combinatorial geometry*, Wiley-Interscience Series in Discrete Mathematics and Optimization, John Wiley & Sons Inc., New York, 1995, A Wiley-Interscience Publication.
[586] János Pach and Dömötör Pálvölgyi, Bounded-degree graphs can have arbitrarily large slope numbers, *Electron. J. Combin.* **13** (2006), no. 1, Note 1, 4 pages.
[587] János Pach and Rom Pinchasi, Bichromatic lines with few points, *J. Combin. Theory Ser. A* **90** (2000), no. 2, 326–335.
[588] János Pach, Rom Pinchasi, and Micha Sharir, On the number of directions determined by a three-dimensional points set, *J. Combin. Theory Ser. A* **108** (2004), no. 1, 1–16.
[589] _____, Solution of Scott's problem on the number of directions determined by a point set in 3-space, *Discrete Comput. Geom.* **38** (2007), no. 2, 399–441.
[590] János Pach, Rom Pinchasi, Micha Sharir, and Géza Tóth, Topological graphs with no large grids, *Graphs Combin.* **21** (2005), no. 3, 355–364.
[591] János Pach, Richard Pollack, and Emo Welzl, Weaving patterns of lines and line segments in space, *Algorithmica* **9** (1993), no. 6, 561–571, Selections from SIGAL International Symposium on Algorithms (Tokyo, 1990).
[592] János Pach, Radoš Radoičić, and Géza Tóth, Relaxing planarity for topological graphs, *Discrete and computational geometry*, Lecture Notes in Comput. Sci., vol. 2866, Springer, Berlin, 2003, pp. 221–232.
[593] János Pach, Ido Safruti, and Micha Sharir, The union of congruent cubes in three dimensions, *Discrete Comput. Geom.* **30** (2003), no. 1, 133–160.
[594] János Pach, F. Shahrokhi, and Mario Szegedy, Applications of the crossing number, *Algorithmica* **16** (1996), no. 1, 111–117.
[595] János Pach and Micha Sharir, The upper envelope of piecewise linear functions and the boundary of a region enclosed by convex plates: combinatorial analysis, *Discrete Comput. Geom.* **4** (1989), no. 4, 291–309.
[596] _____, On vertical visibility in arrangements of segments and the queue size in the Bentley-Ottmann line sweeping algorithm, *SIAM J. Comput.* **20** (1991), 460–470.
[597] _____, Repeated angles in the plane and related problems, *J. Combin. Theory Ser. A* **59** (1992), no. 1, 12–22.
[598] _____, On the number of incidences between points and curves, *Combin. Probab. Comput.* **7** (1998), no. 1, 121–127.
[599] _____, On the boundary of the union of planar convex sets, *Discrete Comput. Geom.* **21** (1999), no. 3, 321–328.
[600] _____, Geometric incidences, *Towards a Theory of Geometric Graphs* (J. Pach, ed.), Contemp. Math., vol. 342, Amer. Math. Soc., Providence, RI, 2004, pp. 185–223.
[601] _____, Incidences, *Graph theory, combinatorics and algorithms*, Springer, New York, 2005, pp. 267–292.
[602] János Pach, Joel Spencer, and Géza Tóth, New bounds on crossing numbers, *Discrete Comput. Geom.* **24** (2000), no. 4, 623–644.
[603] János Pach, William L. Steiger, and Endre Szemerédi, An upper bound on the number of planar k-sets, *Discrete Comput. Geom.* **7** (1992), no. 2, 109–123.
[604] János Pach and Gábor Tardos, Isosceles triangles determined by a planar point set, *Graphs Combin.* **18** (2002), no. 4, 769–779.
[605] _____, On the boundary complexity of the union of fat triangles, *SIAM J. Comput.* **31** (2002), no. 6, 1745–1760.

[606] _____, Untangling a polygon, *Discrete Comput. Geom.* **28** (2002), no. 4, 585–592.
[607] _____, Forbidden paths and cycles in ordered graphs and matrices, *Israel J. Math.* **155** (2006), 359–380.
[608] János Pach, Gábor Tardos, and Géza Tóth, Indecomposable coverings, *Discrete geometry, combinatorics and graph theory*, Lecture Notes in Comput. Sci., vol. 4381, Springer, Berlin, 2007, pp. 135–148.
[609] János Pach and J. Törőcsik, Some geometric applications of Dilworth's theorem, *Discrete Comput. Geom.* **12** (1994), no. 1, 1–7.
[610] János Pach and Géza Tóth, Graphs drawn with few crossings per edge, *Combinatorica* **17** (1997), no. 3, 427–439.
[611] _____, Which crossing number is it anyway?, *J. Combin. Theory Ser. B* **80** (2000), no. 2, 225–246.
[612] _____, Conflict-free colorings, *Discrete and computational geometry*, Algorithms Combin., vol. 25, Springer, Berlin, 2003, pp. 665–671.
[613] _____, Degenerate crossing numbers, *SCG '06: Proceedings of the Twenty-second Annual Symposium on Computational Geometry* (Sedona, Arizona, USA), ACM, 2006, pp. 255–258.
[614] János Pach and Rephael Wenger, Embedding planar graphs at fixed vertex locations, *Graphs Combin.* **17** (2001), no. 4, 717–728.
[615] Amit Pamecha, Imme Ebert-Uphoff, and Gregory Chirikjian, Useful metrics for modular robot motion planning, *IEEE Trans. Robotics Automation* **13** (1997), no. 4, 531–545.
[616] Christos H. Papadimitriou, P. Raghavan, M. Sudan, and H. Tamaki, Motion planning on a graph, *Proc. 35th Annual Symposium on Foundations of Computer Science* (Santa Fe, NM, USA), IEEE, 1994, pp. 511–520.
[617] Christos H. Papadimitriou and Mihalis Yannakakis, Optimization, approximation, and complexity classes, *J. Comput. System Sci.* **43** (1991), no. 3, 425–440.
[618] Jeff Paris and Leo Harrington, A mathematical incompleteness in Peano arithmetic, *Handbook of Mathematical Logic* (Jon Barwise, ed.), North-Holland, Amsterdam, the Netherlands, 1977, pp. 1133–1142.
[619] Michael S. Paterson and F. Frances Yao, Efficient binary space partitions for hidden-surface removal and solid modeling, *Discrete Comput. Geom.* **5** (1990), no. 5, 485–503.
[620] G. W. Peck, On "k-sets" in the plane, *Discrete Math.* **56** (1985), no. 1, 73–74.
[621] Marco Pellegrini, Lower bounds on stabbing lines in 3-space, *Comput. Geom.* **3** (1993), no. 1, 53–58.
[622] _____, Ray shooting and lines in space, in Goodman and O'Rourke [**381**], pp. 839–856.
[623] Marco Pellegrini and Peter W. Shor, Finding stabbing lines in 3-space, *Discrete Comput. Geom.* **8** (1992), no. 2, 191–208.
[624] Michael J. Pelsmajer, Marcus Schaefer, and Daniel Štefankovič, Removing even crossings, *J. Combin. Theory Ser. B* **97** (2007), no. 4, 489–500.
[625] _____, Odd crossing number is not crossing number, *Graph drawing*, Lecture Notes in Comput. Sci., vol. 3843, Springer, Berlin, 2006, pp. 386–396. Also *Discrete Comput. Geom.*, in press.
[626] Gilles Pesant, Factorizations of covers, Master's thesis, School of Computer Science, McGill University, Montreal, 1989.
[627] C. R. Peterkin, Some results on Davenport-Schinzel sequences, *Proc. Third Manitoba Conference on Numerical Mathematics* (Winnipeg, Man.), Utilitas Math., 1974, pp. 337–344.
[628] I. G. Petrovskiĭ and O. A. Oleĭnik, On the topology of real algebraic surfaces, *Izv. Akad. Nauk SSSR. Ser. Mat.* **13** (1949), 389–402.
[629] Seth Pettie, Splay trees, Davenport–Schinzel sequences, and the deque conjecture, *SODA '08: Proc. Nineteenth Annual ACM-SIAM Symposium on Discrete Algorithms* (San Francisco, California), SIAM, 2008, pp. 1115–1124.
[630] _____, Origins of nonlinearity in Davenport–Schinzel sequences, *SODA 2009*, submitted.
[631] Rom Pinchasi, The minimum number of distinct areas of triangles determined by a set of n points in the plane, *SIAM J. Discrete Math.*, **22** (2008), no. 2, 828–831.
[632] Harry Plantinga and Charles R. Dyer, An algorithm for constructing the aspect graph, *27th Annual Symposium on Foundations of Computer Science (Toronto, Ontario, Canada)*, IEEE Comput. Soc. Press, 1986, pp. 123–131.
[633] Julius Plücker, On a new geometry of space, *Phil. Trans. Royal Soc. London* **155** (1865), 725–791.

[634] Richard Pollack and Marie-Françoise Roy, On the number of cells defined by a set of polynomials, *C. R. Acad. Sci. Paris Sér. I Math.* **316** (1993), no. 6, 573–577.

[635] Richard Pollack, Micha Sharir, and Shmuel Sifrony, Separating two simple polygons by a sequence of translations, *Discrete Comput. Geom.* **3** (1988), no. 2, 123–136.

[636] Helmut Pottmann and Johannes Wallner, *Computational Line Geometry*, Springer, Berlin/Heidelberg, 2001.

[637] Franco P. Preparata and Michael Ian Shamos, *Computational Geometry, An Introduction*, Texts and Monographs in Computer Science, Springer, New York, 1985.

[638] Franco P. Preparata and Roberto Tamassia, Efficient point location in a convex spatial cell-complex, *SIAM J. Comput.* **21** (1992), no. 2, 267–280.

[639] Richard Rado, Note on combinatorial analysis, *Proc. London Math. Soc.* **48** (1943), 122–160.

[640] Edgar A. Ramos, Construction of 1-d lower envelopes and applications, *SCG '97: Proceedings of the Thirteenth Annual Symposium on Computational Geometry* (Nice, France), ACM, 1997, pp. 57–66.

[641] _____, Intersection of unit-balls and diameter of a point set in R^3, *Comput. Geom.* **8** (1997), no. 2, 57–65.

[642] _____, An optimal deterministic algorithm for computing the diameter of a three-dimensional point set, *Discrete Comput. Geom.* **26** (2001), no. 2, 233–244.

[643] A. S. Rao and K. Y. Goldberg, Placing registration marks, *IEEE Trans. Industrial Electronics* **41** (1994), no. 1, 51–59.

[644] Daniel Ratner and Manfred K. Warmuth, The $(n^2 - 1)$-puzzle and related relocation problems, *J. Symbolic Comput.* **10** (1990), no. 2, 111–137.

[645] John H. Reif, Chapter 11: Complexity of the generalized mover's problem, *Planning, Geometry, and Complexity of Robot Motion* (Jacob T. Schwartz, Micha Sharir, and John E. Hopcroft, eds.), Intellect Ltd., Bristol, UK, 1987, pp. 267–281.

[646] Dušan Repovš, Arkady Skopenkov, and Fulvia Spaggiari, An infinite sequence of non-realizable weavings, *Discrete Appl. Math.* **150** (2005), no. 1-3, 256–260.

[647] P. J. de Rezende and D. T. Lee, Point set pattern matching in d-dimensions, *Algorithmica* **13** (1995), no. 4, 387–404.

[648] Frederic M. Richards, Areas, volumes, packing, and protein structure, *Annual Review of Biophysics and Bioengineering* **6** (1977), 151–176.

[649] R. Bruce Richter and Carsten Thomassen, Relations between crossing numbers of complete and complete bipartite graphs, *Amer. Math. Monthly* **104** (1997), no. 2, 131–137.

[650] Jürgen Richter-Gebert and Günter M. Ziegler, Oriented matroids, in Goodman and O'Rourke [**381**], pp. 129–152.

[651] Samuel Roberts, On the figures formed by the intercepts of a system of straight lines in a plane, and on analogous relations in space of three dimensions, *Proc London Math. Soc.* **s1-19** (1887), no. 1, 405–422.

[652] Hartley Rogers, Jr., *Theory of Recursive Functions and Effective Computability*, Second Edition, MIT Press, Cambridge, MA, 1987.

[653] D. P. Roselle, An algorithmic approach to Davenport-Schinzel sequences, *Utilitas Math.* **6** (1974), 91–93.

[654] D. P. Roselle and R. G. Stanton, Results on Davenport-Schinzel squences, *Proc. Louisiana Conf. on Combinatorics, Graph Theory and Computing)*, Louisiana State Univ., Baton Rouge, La., 1970, pp. 249–267.

[655] _____, Some properties of Davenport-Schinzel sequences, *Acta Arith.* **17** (1970/1971), 355–362.

[656] Daniela Rus and Marsette Vona, Crystalline robots: Self-reconfiguration with compressible unit modules, *Autonomous Robots* **10** (2001), no. 1, 107–124.

[657] Jörg-Rüdiger Sack and Jorge Urrutia (eds.), *Handbook of Computational Geometry*, Elsevier, Amsterdam, 2000.

[658] Marcus Schaefer, Eric Sedgwick, and Daniel Štefankovič, Recognizing string graphs in NP, *J. Comput. System Sci.* **67** (2003), no. 2, 365–380.

[659] Rolf Schneider, Tessellations generated by hyperplanes, *Discrete Comput. Geom.* **2** (1987), no. 3, 223–232.

[660] Jacob T. Schwartz and Micha Sharir, On the "piano movers" problem. II. General techniques for computing topological properties of real algebraic manifolds, *Adv. Appl. Math.* **4** (1983), no. 3, 298–351.

[661] _____, Motion planning and related geometric algorithms in robotics, *Proc. 1986 Intern. Congress Math.* (Berkeley, CA, USA), Amer. Math. Soc., Providence, RI, 1987, pp. 1594–1611.
[662] _____, Algorithmic motion planning in robotics, *Handbook of Theoretical Computer Science*, vol. A, Elsevier, Amsterdam, 1990, pp. 391–430.
[663] _____, On the two-dimensional Davenport-Schinzel problem, *J. Symbolic Comput.* **10** (1990), no. 3–4, 371–393.
[664] Otfried Schwarzkopf and Micha Sharir, Vertical decomposition of a single cell in a three-dimensional arrangement of surfaces, *Discrete Comput. Geom.* **18** (1997), no. 3, 269–288.
[665] P. R. Scott, On the sets of directions determined by n points, *Amer. Math. Monthly* **77** (1970), 502–505.
[666] Raimund Seidel, Constructing higher-dimensional convex hulls at logarithmic cost per face, *STOC '86: Proceedings of the Eighteenth Annual ACM Symposium on Theory of Computing* (Berkeley, CA, USA), ACM, 1986, pp. 404–413.
[667] _____, Exact upper bounds for the number of faces in d-dimensional Voronoi diagrams, *Applied geometry and discrete mathematics*, DIMACS Ser. Discrete Math. Theoret. Comput. Sci., vol. 4, Amer. Math. Soc., Providence, RI, 1991, pp. 517–529.
[668] _____, Small-dimensional linear programming and convex hulls made easy, *Discrete Comput. Geom.* **6** (1991), no. 5, 423–434.
[669] _____, Backwards analysis of randomized geometric algorithms, *New Trends in Discrete and Computational Geometry* (J. Pach, ed.), Algorithms Combin., vol. 10, Springer, Berlin, 1993, pp. 37–67.
[670] _____, Convex hull computations, in Goodman and O'Rourke [**381**], pp. 643–663.
[671] _____, The nature and meaning of perturbations in geometric computing, *Discrete Comput. Geom.* **19** (1998), no. 1, 1–17.
[672] Ophir Setter, Dan Halperin, and Micha Sharir, Exact and efficient construction of general Voronoi diagrams in the plane via divide and conquer of envelopes in space, in preparation.
[673] Micha Sharir, Almost linear upper bounds on the length of general Davenport-Schinzel sequences, *Combinatorica* **7** (1987), no. 1, 131–143.
[674] _____, Improved lower bounds on the length of Davenport-Schinzel sequences, *Combinatorica* **8** (1988), no. 1, 117–124.
[675] _____, The shortest watchtower and related problems for polyhedral terrains, *Inform. Process. Lett.* **29** (1988), no. 5, 265–270.
[676] _____, On k-sets in arrangements of curves and surfaces, *Discrete Comput. Geom.* **6** (1991), no. 6, 593–613.
[677] _____, Arrangements of surfaces in higher dimensions: Envelopes, single cells, and other recent developments, *Proc. 5th Canadian Conf. on Computational Geometry*, 1993, pp. 181–186.
[678] _____, Almost tight upper bounds for lower envelopes in higher dimensions, *Discrete Comput. Geom.* **12** (1994), no. 3, 327–345.
[679] _____, On joints in arrangements of lines in space and related problems, *J. Combin. Theory Ser. A* **67** (1994), no. 1, 89–99.
[680] _____, Algorithmic motion planning, in Goodman and O'Rourke [**381**], pp. 1037–1064.
[681] Micha Sharir and Pankaj K. Agarwal, *Davenport-Schinzel Sequences and their Geometric Applications*, Cambridge Univ. Press, 1995.
[682] Micha Sharir and Mark H. Overmars, A simple output-sensitive algorithm for hidden surface removal, *ACM Trans. Graphics* **11** (1992), no. 1, 1–11.
[683] Micha Sharir and Hayim Shaul, Ray shooting and stone throwing with near-linear storage, *Comput. Geom.* **30** (2005), no. 3, 239–252.
[684] Micha Sharir and Shakhar Smorodinsky, Extremal configurations and levels in pseudoline arrangements, *Algorithms and data structures*, Lecture Notes in Comput. Sci., vol. 2748, Springer, Berlin, 2003, pp. 127–139.
[685] Micha Sharir, Shakhar Smorodinsky, and Gábor Tardos, An improved bound for k-sets in three dimensions, *Discrete Comput. Geom.* **26** (2001), no. 2, 195–204.
[686] Micha Sharir and Sivan Toledo, Extremal polygon containment problems, *Comput. Geom.* **4** (1994), no. 2, 99–118.
[687] Micha Sharir and Emo Welzl, Point-line incidences in space, *Combin. Probab. Comput.* **13** (2004), no. 2, 203–220.

[688] Peter W. Shor, 1990, personal communication.
[689] _____, Geometric realization of superlinear Davenport–Schinzel sequences: I. Line segments, Unpublished manuscript, 1990.
[690] _____, Stretchability of pseudolines is NP-hard, *Applied geometry and discrete mathematics, DIMACS Ser. Discrete Math. Theoret. Comput. Sci.*, vol. 4, Amer. Math. Soc., Providence, RI, 1991, pp. 531–554.
[691] Shakhar Smorodinsky, On the chromatic number of geometric hypergraphs, *SIAM J. Discrete Math.* **21** (2007), no. 3, 676–687.
[692] Shakhar Smorodinsky, Joseph S. B. Mitchell, and Micha Sharir, Sharp bounds on geometric permutations of pairwise disjoint balls in R^d, *Discrete Comput. Geom.* **23** (2000), no. 2, 247–259.
[693] Alexandra Solan, Cutting cylces of rods in space, *SCG '98: Proceedings of the Fourteenth Annual Symposium on Computational Geometry* (Minneapolis, Minnesota, United States), ACM, 1998, pp. 135–142.
[694] József Solymosi and Gábor Tardos, On the number of k-rich transformations, *SCG '07: Proceedings of the Twenty-third Annual Symposium on Computational Geometry* (Gyeongju, South Korea), ACM, 2007, pp. 227–231.
[695] József Solymosi and Csaba D. Tóth, Distinct distances in the plane, *Discrete Comput. Geom.* **25** (2001), no. 4, 629–634.
[696] József Solymosi and Van Vu, Distinct distances in high dimensional homogeneous sets, *Towards a Theory of Geometric Graphs* (J. Pach, ed.), *Contemp. Math.*, vol. 342, Amer. Math. Soc., Providence, RI, 2004, pp. 259–268.
[697] _____, Near optimal bounds for the Erdös distinct distances problem in high dimensions, *Combinatorica* **28** (2008), no. 1, 113–125.
[698] Duncan M. Y. Sommerville, *Analytical Geometry of Three Dimensions*, Cambridge Univ. Press, 1934.
[699] Joel Spencer, Endre Szemerédi, and William Trotter, Jr., Unit distances in the Euclidean plane, *Graph theory and combinatorics (Cambridge, 1983)*, Academic Press, London, 1984, pp. 293–303.
[700] R. G. Stanton and P. H. Dirksen, Davenport-Schinzel sequences, *Ars Combinatoria* **1** (1976), no. 1, 43–51.
[701] R. G. Stanton and D. P. Roselle, A result on Davenport-Schinzel sequences, *Combinatorial theory and its applications, III (Proc. Colloq., Balatonfüred, 1969)*, North-Holland, Amsterdam, 1970, pp. 1023–1027.
[702] Jakob Steiner, Einige Gesetze über die Theilung der Ebene und des Raumes, *J. Reine Angewandte Math.* **1** (1826), 349–364.
[703] Ernst Steinitz, Polyeder und Raumeinteilungen, *Enzyklopädie der Mathematischen Wissenschaften*, vol. 3 (Geometrie) part 3AB12, Teubner, 1922, pp. 1–139.
[704] Jorge Stolfi, *Oriented Projective Geometry. A Framework for Geometric Computations*, Academic Press, Boston, MA, 1991.
[705] E. G. Straus, Some extremal problems in combinatorial geometry, *Combinatorial Mathematics (Proc. Internat. Conf. Combinatorial Theory, Australian Nat. Univ., Canberra, 1977)*, Lecture Notes Math., vol. 686, Springer, Berlin, 1978, pp. 308–312.
[706] Ivan E. Sutherland, Robert F. Sproull, and Robert A. Schumacker, A characterization of ten hidden-surface algorithms, *ACM Comput. Surv.* **6** (1974), no. 1, 1–55.
[707] Ondrej Sýkora and Imrich Vrťo, On VLSI layouts of the star graph and related networks, *Integration, the VLSI Journal* **17** (1994), no. 1, 83–93.
[708] James Joseph Sylvester, Mathematical question 11851, *Educational Times* **46** (1893), 156.
[709] László A. Székely, Crossing numbers and hard Erdős problems in discrete geometry, *Combin. Probab. Comput.* **6** (1997), no. 3, 353–358.
[710] Endre Szemerédi, On a problem of Davenport and Schinzel, *Acta Arith.* **25** (1973/74), 213–224.
[711] Endre Szemerédi and William T. Trotter, Jr., A combinatorial distinction between the Euclidean and projective planes, *European J. Combin.* **4** (1983), no. 4, 385–394.
[712] _____, Extremal problems in discrete geometry, *Combinatorica* **3** (1983), no. 3–4, 381–392.
[713] Boaz Tagansky, *The Complexity of Substructures in Arrangements of Surfaces*, Ph.D. thesis, Tel-Aviv University, Tel Aviv, 1996.

[714] _____, A new technique for analyzing substructures in arrangements of piecewise linear surfaces, *Discrete Comput. Geom.* **16** (1996), no. 4, 455–479.

[715] Hisao Tamaki and Takeshi Tokuyama, How to cut pseudoparabolas into segments, *Discrete Comput. Geom.* **19** (1998), no. 2, 265–290.

[716] _____, A characterization of planar graphs by pseudo-line arrangements, *Algorithmica* **35** (2003), no. 3, 269–285.

[717] Arie Tamir, Improved complexity bounds for center location problems on networks by using dynamic data structures, *SIAM J. Discrete Math.* **1** (1988), no. 3, 377–396.

[718] Gábor Tardos, On distinct sums and distinct distances, *Adv. Math.* **180** (2003), no. 1, 275–289.

[719] Gábor Tardos and Géza Tóth, Multiple coverings of the plane with triangles, *Discrete Comput. Geom.* **38** (2007), no. 2, 443–450.

[720] René Thom, Sur l'homologie des variétés algébriques réelles, *Differential and Combinatorial Topology (A Symposium in Honor of Marston Morse)*, Princeton Univ. Press, Princeton, NJ, 1965, pp. 255–265.

[721] Géza Tóth, Note on geometric graphs, *J. Combin. Theory Ser. A* **89** (2000), no. 1, 126–132.

[722] _____, Point sets with many k-sets, *Discrete Comput. Geom* **26** (2001), 187–194.

[723] _____, Note on the pair-crossing number and the odd-crossing number, *Discrete Comput. Geom.*, accepted.

[724] Géza Tóth and Pavel Valtr, Geometric graphs with few disjoint edges, *Discrete Comput. Geom.* **22** (1999), no. 4, 633–642.

[725] László Fejes Tóth, *Regular Figures*, Pergamon Press, New York, 1964.

[726] Paul Turán, A note of welcome, *J. Graph Theory* (1977), 7–9.

[727] W. T. Tutte, Toward a theory of crossing numbers, *J. Combin. Theory* **8** (1970), 45–53.

[728] Peter Ungar, $2N$ noncollinear points determine at least $2N$ directions, *J. Combin. Theory Ser. A* **33** (1982), no. 3, 343–347.

[729] Jorge Urrutia, Art gallery and illumination problems, in Sack and Urrutia [**657**], pp. 973–1027.

[730] Pravin M. Vaidya, Geometry helps in matching, *SIAM J. Comput.* **18** (1989), no. 6, 1201–1225.

[731] Pavel Valtr, On geometric graphs with no k pairwise parallel edges, *Discrete Comput. Geom.* **19** (1998), no. 3, 461–469.

[732] _____, Generalizations of Davenport–Schinzel sequences, *Contemporary Trends in Discrete Mathematics (DIMACS Series in Discrete Mathematics and Theoretical Computer Science, Vol. 49)* (R. L. Graham, J. Kratochvíl, J. Nesteril, and F. S. Roberts, eds.), pp. 349–389, Amer. Math. Soc., Providence, RI, 1999.

[733] _____, On the pair-crossing number, *Combinatorial and computational geometry*, Math. Sci. Res. Inst. Publ., vol. 52, Cambridge Univ. Press, Cambridge, 2005, pp. 569–575.

[734] Kasturi R. Varadarajan, A divide-and-conquer algorithm for min-cost perfect matching in the plane, *Proc. 39th Annual Symposium on Foundations of Computer Science*, IEEE Comput. Soc. Press, 1998, pp. 320–331.

[735] Amitabh Varshney, Frederick P. Brooks, Jr., and William V. Wright, Computing smooth molecular surfaces, *IEEE Computer Graphics Applications* **14** (1994), 19–25.

[736] G. A. Wade and J.-H. Chu, Drawability of complete graphs using a minimal slope set, *Computer J.* **37** (1994), no. 2, 139–142.

[737] K. Wagner, Bemerkungen zum Vierfarbenproblem, *Jahresbericht Deutsche Mahtematiker-Vereinigung* **43** (1936), 26–32.

[738] W. D. Wallis, *Introduction to Combinatorial Designs*, Second Edition, Discrete Mathematics and its Applications, Chapman & Hall/CRC, Boca Raton, FL, 2007.

[739] Paul van Wamelen, The maximum number of unit distances among n points in dimension four, *Beiträge Algebra Geom.* **40** (1999), no. 2, 475–477.

[740] Hugh E. Warren, Lower bounds for approximation by nonlinear manifolds, *Trans. Amer. Math. Soc.* **133** (1968), 167–178.

[741] Kym S. Watson, Sylvester's problem for spreads of curves, *Canad. J. Math.* **32** (1980), no. 1, 219–239.

[742] K. Weiler, Edge-based data structures for solid modeling in curved-surface environments, *IEEE Computer Graphics and Applications* **5** (1985), no. 1, 21–40.

[743] Emo Welzl, More on k-sets of finite sets in the plane, *Discrete Comput. Geom.* **1** (1986), no. 1, 95–100.
[744] Rephael Wenger, Upper bounds on geometric permutations for convex sets, *Discrete Comput. Geom.* **5** (1990), no. 1, 27–33.
[745] ———, Helly-type theorems and geometric transversals, in Goodman and O'Rourke [**381**], pp. 73–96.
[746] John E. Wetzel, On the division of the plane by lines, *Amer. Math. Monthly* **85** (1978), no. 8, 647–656.
[747] Ady Wiernik and Micha Sharir, Planar realizations of nonlinear Davenport-Schinzel sequences by segments, *Discrete Comput. Geom.* **3** (1988), no. 1, 15–47.
[748] Dan E. Willard, Polygon retrieval, *SIAM J. Comput.* **11** (1982), no. 1, 149–165.
[749] V. C. Williams, Mathematical Notes: A proof of Sylvester's theorem on collinear points, *Amer. Math. Monthly* **75** (1968), no. 9, 980–982.
[750] Ernst Witt, Ein kombinatorischer Satz der Elementargeometrie, *Math. Nachr.* **6** (1952), 261–262.
[751] D. R. Woodall, Thrackles and deadlock, *Combinatorial Mathematics and its Applications (Proc. Conf., Oxford, 1969)*, Academic Press, London, 1971, pp. 335–347.
[752] Andrew C. Yao and Frances F. Yao, A general approach to d-dimensional geometric queries, *STOC '85: Proceedings of the Seventeenth Annual ACM Symposium on Theory of Computing* (Providence, RI, USA), ACM, 1985, pp. 163–168.
[753] Mark Yim, Ying Zhang, John Lamping, and Eric Mao, Distributed control for 3d metamorphosis, *Autonomous Robots* **10** (2001), no. 1, 41–56.
[754] E. Yoshida, S. Murata, A. Kamimura, K. Tomita, H. Kurokawa, and S. Kokaji, A motion planning method for a self-reconfigurable modular robot, *Proc. 2001 IEEE/RSJ International Conference on Intelligent Robots and Systems* (Maui, HI, USA), vol. 1, 2001, pp. 590–597.
[755] Thomas Zaslavsky, Facing up to arrangements: face-count formulas for partitions of space by hyperplanes, *Mem. Amer. Math. Soc.* **1** (1975), no. 1, 154, vii+102.
[756] ———, A combinatorial analysis of topological dissections, *Adv. Math.* **25** (1977), no. 3, 267–285.
[757] Yunhong Zhou and Subhash Suri, Geometric permutations of balls with bounded size disparity, *Comput. Geom.* **26** (2003), no. 1, 3–20.
[758] Binhai Zhu, Computing the shortest watchtower of a polyhedral terrain in $O(n \log n)$ time, *Comput. Geom.* **8** (1997), no. 4, 181–193.
[759] Günter M. Ziegler, *Lectures on Polytopes*, Graduate Texts in Mathematics, vol. 152, Springer, New York, 1995.
[760] Rade T. Živaljević and Siniša T. Vrećica, The colored Tverberg's problem and complexes of injective functions, *J. Combin. Theory Ser. A* **61** (1992), no. 2, 309–318.

Index

Ackermann's function, 81
  inverse, *see* inverse Ackermann function
Adamec, Radek, 81
adversary (to online algorithm), 182
affine copy (of point set), 143–144
Agarwal, Pankaj K., 15, 26, 27, 37–39, 48, 51, 58–63, 68, 74, 80, 81, 85, 88, 111, 112, 115, 117, 126, 142, 163, 166, 168
Ajtai, Miklós, 35, 106, 124
Ajwani, Deepak, 180
Alevizos Panagiotis, 90
allowable sequence, 4
Alon, Noga, 111, 125, 188
Alt, Helmut, 36
Amato, Nancy M., 96
angle
  determined by point set, 144
angular resolution (of graph), 126–127
animal (set of grid cubes), 196
annulus, smallest width of, 67
antipodality, 35
Apfelbaum, Roel, 142, 145
APX-hardness, 186, 187
Arkin, Esther M., 91
Aronov, Boris, 27, 28, 31, 32, 37–39, 41, 46–48, 59, 60, 63, 111, 112, 114, 126, 161, 166, 168
arrangement, 13–71, 73, 99, 145, 150, 165
  and Davenport–Schinzel sequences, 89–98
  applications, 13, 49, 56, 63–70
  combinatorial complexity of, 16, 89
  complexity of cell in, 14, 27–29, 90–92, 99
  complexity of many cells in, 37–39, 97, 99, 101, 103, 108, 109
  computing, 56–58, 89
  computing substructures in, 58–63
  decomposition of, 49–54, 156, 157
  definition, 13, 16, 89
  history, 15
  in complex space, 15
  joint in, *see* lines in space, joint
  lattice, *see* lattice arrangement
  level in, *see* level
  of algebraic surface patches, 15, 19, 28, 55, 56, 61, 168
  of arcs, 37, 58, 61, 63, 74, 90, 92, 96, 97
  of circles, 15, 18, 38, 39, 63, 66, 97, 101
  of graphs of polynomials, 55
  of hyperplanes, 14, 15, 18, 31, 37, 39, 54, 56, 57, 60–64, 96, 98
  of lines, 8, 15, 18, 34, 37, 56, 62, 74, 90, 96–98, 104, 108, 147, 170
  of parabolas, 37
  of planes, 18, 37, 62
  of polytope boundaries, 19
  of pseudo-circles, 101, 111
  of pseudo-parabolas, 37
  of pseudo-planes, 37
  of pseudo-segments, 101
  of pseudolines, 2, 36, 37
  of quadrics, 148
  of rays, 90
  of segments, 37, 38, 55, 57, 63, 74, 90, 91, 96, 97, 110
  of semi-pfaffian sets, 15
  of simplices, 28, 31, 56
  of spheres, 14, 18, 53
  of triangles, 28, 37, 53, 61, 62
  representation of, 54–56
  zone in, *see* zone
art gallery problem, 166
aspect graph, 165–166
  orthographic, 165
  perspective, 165
assembly, 69, 147
Atallah, Mikhail J., 73
Avis, David, 57, 61
Avital, Shmuel, 125

Badent, Melanie, 128
Balaban, Ivan J., 57
Balogh, József, 36
Bar-Noy, Amotz, 182
Bárány, Imre, 36, 37
Barát, János, 127
Basu, Saugata, 19, 28, 29, 32, 69
Bereg, Sergey, 189

de Berg, Mark T., 59, 61–63, 164, 166
Bern, Marshall, 96
Bezout's theorem, 17
Bien, Lilach, 146
binary space partition, 160
bisection width (of graph), 128, 129
block design, 3
Blokhuis, Aart, 4
Blundon, William J., 173
Boissonnat, Jean-Daniel, 41, 48, 90
"book proof", 2, 3
Boolean formula, 15, 16, 55
Böröczky, Károly, Sr., 190
Brass, Peter, 136
Bremner, David, 61
brick factory problem, 120
Brönnimann, Hervé, 66
Buck's theorem, 18
Buck, R. Creighton, 18

CAD, 147
Cairns, Grant, 121
Călinescu, Gruia, 186, 187
Canham, Raymond J., 38
Canny, John F., 55, 68
cell complex, 50
cell-tuple data structure, 56, 60
cellular phone network, 178, 179
center point, 67, 68
center-transversal, 68
Černý, Jakub, 125
Chakerian, Gulbank Don, 5
Chan, Timothy M., 37, 61, 62
Chazelle, Bernard, 37, 51, 54, 57, 58, 61, 64, 66, 93, 160, 163
Chebychev system, 74
Cheilaris, Panagiotis, 182
Chen, Ke, 182
Chen, Xiaomin, 180
Cheong, Otfried, 169
Chew, L. Paul, 40
Chung, Fan R. K., 101
Chvátal, Václav, 35, 106, 124
Clarkson, Kenneth L., 38, 54, 57, 60, 92, 104
Clarkson–Shor theorem, 33
Clarkson–Shor technique, 23, 33, 57, 98, 180
coin
  in graph, 185–187
  sliding in the plane, 187–194
Cole, Richard, 62, 66, 68, 165
Collins decomposition, see cylindrical algebraic decomposition
Collins, George E., 50
coloring problem, see geometric coloring problem
coloring space, 135, 137, 141, 143, 144

coloring the plane, 138, 139, 146
coloring, biased (of point set), 5
coloring, conflict-free, see conflict-free coloring
combination lemma, 95
computer graphics, 127–128, 147, 149, 158
computer vision, 147
configuration space, 13, 43, 66
conflict-free coloring, 178–180
  online, 181–182
congruent copy (of point set)
  in higher dimensions, 139–141
  in the plane, 137–139
congruent simplices (determined by point set), 100, 117–118, 140
contact surface, 13
converter (in graph), 121, 122
convex chain, 35, 36
convex hull
  computing, 60
Conway's thrackle conjecture, 120
Conway, John H., 3
cover-decomposable family, 175–178
covering, k-fold, 173–175
crossing in graph, see graph, crossing in
Crossing Lemma, 35, 100, 106, 116, 124, 125
  generalization, 116
  proof, 106, 129
crossing number, see graph, crossing number of
cutting, 53–54, 104–106, 112, 162
cycle, elementary (of lines in space), 163
cylindrical algebraic decomposition, 50, 55

Da Silva, Ilda P. F., 10
Danzer's conjecture (transversals of balls), 170
Davenport, Harold, 73, 80, 173
Davenport–Schinzel sequence, 16, 20, 73–98, 170
  and arrangements, 89–98
  chain in, 82
  definition, 73
  generalized, 81
  geometric realization, 76, 78, 89
  lower bounds, 86–89
  sharp upper bounds, 81–85
  simple upper bounds, 79–81
degeneracy, handling, 17
degenerate set of surfaces, 17
Delaunay graph, 178–180
  for axis-parallel rectangles, 180
(δ, β)-covered object, 46
Demaine, Erik D., 195
Demaine, Martin L., 195
dense point set, 36
depth (of sequence), 80

depth order (of lines in space), 149, 158–163
deque conjecture (for splay trees), 81
descendent hyperedge, 176–178
description complexity (of surface), 15, 16
Dey, Tamal K., 35–37
Di Giacomo, Emilio, 128
diameter (of point set), 66
Dilworth's theorem, 125
Dirac's conjecture, 5
  weak, 5
Dirac–Motzkin conjecture, 2, 3
direction
  determined by point set, 2–5
distance function, 40, 41, 47, 59
distance selection, 66
Distinct Distances problem, 100, 101,
    115–116, 134, 137, 138
  in higher dimensions, 140
double wedge, 7, 8, 10
  avoiding, 7, 8
duality transform, 7, 9, 64, 92, 112, 115,
    145
Dumir, Vishwa C., 173
Dumitrescu, Adrian, 116, 145, 146, 186,
    187, 189, 196
Dwyer, Rex A., 41

Edelsbrunner, Herbert, 15, 29, 31, 36–40,
    45, 51, 56–58, 61, 62, 70, 93, 104, 112,
    160, 163
Efrat, Alon, 45, 46, 48, 61, 66
El Din, Mohab Safey, 41
Elbassioni, Khaled, 180
Elekes, György, 102
Eppstein, David, 37, 96
ε-net, 54, 156
Erdős's theorem (connecting lines), 3
Erdős, Paul, 5, 34, 36, 100, 103, 125, 126,
    134, 135, 139, 145, 173
Euler's polyhedral formula, 11, 126
Even, Guy, 178–180
Everett, Hazel, 41, 61
expander graph, 65, 66
extremal combinatorics, 133–146
extremal polygon placement, 66
Ezra, Esther, 46–48

face-incidence lattice, see incidence graph
Facello, Michael A., 70
fat object, 45–49, 169
Fejes Tóth, László, 173, 187–194
  life of, 194
  work in Discrete Geometry, 194
Feldman, Sharona, 162
Felsner, Stefan, 5, 36
Fiat, Amos, 182
Fifteen Puzzle, 183–185
  generalization, 184, 185
finite projective plane, 3

Finschi–Fukuda counterexample, 10, 11
fixed-area triangle (determined by point
    set), 100, 116–117, 145, 146
fixed-perimeter triangle (determined by
    point set), 100, 116–117, 146
fixturing, 69
Formann, Michael, 126
frequency allocation, 178–180
Friedman, Joel, 54, 57
Fu, Ping, 70
Fukuda, Komei, 10, 18, 57
functional inverse, 81
Füredi, Zoltan, 37

Gallai, Tibor, 1
Gallai–Sylvester theorem, 3
Gallai–Witt theorem, 143
general position, 17
generalized exponentional, see
    Ackermann's function
"genetics" of symmetric objects, 194
geometric coloring problem, 173–182
geometric matching, 67
geometric optimization, 65–68, 115
geometric pattern
  definition, 133
  finding, 134, 137, 139, 141, 143–145
  maximum number of occurrences,
    133–146
  monochromatic, 135, 137, 139, 141, 143,
    144, 146
geometric permutation, see permutation,
    geometric
geometric transversal, 64–65, 149, 167–170
Goaoc, Xavier, 169
Goddard, Wayne, 126
Goodman, Jacob E., 4, 34
Goodrich, Michael T., 58, 96
Govindarajan, Sathish, 180
Graham, Ronald L., 135, 139
graph
  angular resolution of, see angular
    resolution
  bisection width of, see bisection width
  crossing in, 100, 106, 119
  crossing number of, 100, 106–109, 111,
    119–131
    applications, 127–128
    definition, 119
  cubic, 127
  degenerate crossing number of, 124, 125
  drawing of, 119, 121
  drawn in the plane, 44, 106
  geometric, 35, 106, 125–126, 128
    leftmost edge of, 125
  multiple crossing in, 124
  odd-crossing number of, 122, 123

pairwise crossing number of, 122, 123, 128
planar, 106, 119, 121, 128
  definition, 119
rectilinear crossing number of, 122
simple, 106, 107
simple drawing of, 124
slope number of, see slope number
straight-line drawing of, see graph, geometric
topological, 106
trifurcation in, see trifurcation
Graph Reconfiguration problem, 186, 187
Grassmann–Plücker relation, 14
Grassmannian hypersurface, see Plücker hypersurface
Grassmannian manifold, 150, 170
Greenstein, Erez, 169
grid (of segments in the plane), 160
Grünbaum, Branko, 5, 15
Guibas, Leonidas J., 27, 37, 38, 45, 51, 52, 56, 58, 61, 62, 64, 90, 92–95, 104, 112, 160, 163, 187

Hadwiger, Hugo, 135
Hadwiger–Nelson problem, 135, 138
Hadwiger-type theorem, 149, 167
Hagerup, Torben, 58, 126
Hales–Jewett theorem, 143
half-edge data structure, 54
Halperin, Dan, 21, 22, 28, 32, 46, 52, 70, 91, 166
halving segment, 35
ham sandwich cut, 68
Hanani, Haim, 125
Hanani–Tutte theorem, 44, 123
Hans-Gill, Rajinder J., 173
Har-Peled, Sariel, 160, 180
Haralambides, James, 126
Hart, Sergiu, 73, 80, 81, 86
Haussler, David, 39
Heintz, Joos, 19
Helly-type theorem, 149, 167, 169
Heppes, Aladár, 173, 187
Hershberger, John E., 38, 45, 78
hidden surface removal, 147, 158, 159, 165
Hill, Dale, 4
history dag, 58, 60, 93, 94
Holmsen, Andreas, 169
homogeneous coordinates, 151, 152
homogeneous point set, 101
homogeneous space, 148
homothetic copy (of point set), 143–144
Hopcroft's problem, 99, 115
Hopf–Pannwitz–Erdős theorem, 125
Hurtado, Ferran, 36
Huttenlocher, Daniel P., 80
hypergraph

coloring of, 175–178
descendent hyperedge in, see descendent hyperedge
dual of, 175
dual realization of, 176–178
geometric, 126
monochromatic edge in, 175, 176
planar realization of, 176, 177
sibling hyperedge in, see sibling hyperedge
two-colorable, 175, 176
two-edge-colorable, 175, 176

incidence, 39, 99–118
and crossing numbers, 106–109
applications, 114–118
computing, 62–63, 100, 115
counting, 100, 115
definition, 99
history, 100
in higher dimensions, 99, 101, 112–114
lower bounds, 102–104
partition technique, 104–106, 108, 112
upper bounds, 104–106
with circles, 99, 101, 103, 105, 108, 110, 111, 115
with graphs of polynomials, 103
with lines, 99, 102–104, 111, 115
with parabolas, 102, 111
with pseudo-circles, 110
with unit circles, 99, 103, 105, 107
incidence graph, 55, 56, 60, 105, 106, 153
inclination (of line), 113, 114
infinitesimal
  in perturbation, 17
intersection point
  simple, 1
inverse Ackermann function, 20, 73, 81
  definition, 81
inversion (of permutation), 66
inversion transform, 103
isosceles triangle (determined by point set), 100, 116–117
isotopy, 167–170
isotopy class (of lines in space), 149, 150, 153, 156

Jadhav, Shreesh, 68
Jamison, Robert E., 4
joint of lines in space, see lines in space, joint
Jordan arc, 37, 44, 89, 90, 92, 94–96, 100, 106
  definition, 89
Jordan curve, 24, 43, 90, 91, 96, 98, 110, 112
  definition, 89
Jung, Hermann, 58

k-border (of cell), 30, 31
k-fold covering, *see* covering, k-fold
k-fold packing, *see* packing, k-fold
k-level, *see* level
k-set, 33, 98
  of dense point set, *see* dense point set
  of random point set, 36
Kapelushnik, Lior, 62
Kaplan, Haim, 169, 182
κ-curved object, 46, 48
κ-round object, 48, 49
Katchalski, Meir, 169, 188
Katz, Matthew J., 46, 66, 169
Katz, Nets Hawk, 116, 134
Kaufmann, Michael, 126
Kedem, Klara, 40, 44, 80, 91
Kelly, Leroy M., 2
Klawe, Maria M., 36
Klazar, Martin, 81, 85
Klein hypersurface, *see* Plücker
  hypersurface
Kleinberg, Jon M., 80
Kleitman, Daniel J., 124, 126
Klugerman, Michael, 126
Koltun, Vladlen, 26, 27, 40, 41, 48, 51, 52,
  114, 161
Kornhauser, Daniel, 184
Kővári–Sós–Turán theorem, 104
van Kreveld, Marc, 61, 63, 166
Kuratowski's theorem, 119

Las Vegas algorithm, 92
Las Vergnas, Michel, 18
Last, Hagit, 111
lattice, 55
lattice arrangement, 173, 174
lattice covering, k-fold, 173
lattice packing, k-fold, 173
lattice theory, 18
Lazard, Daniel, 41
Lazard, Sylvain, 41
Lee, Der-Tsai, 64
Leighton, Frank Thomson, 35, 106, 124, 126
lens, 103, 110, 111, 114
  definition, 110
Lenz' construction, 139–142
level, 32–37, 97–98
  computing, 61–62, 98
Levy, Meital, 182
Lewis, Ted, 169
Liang, Jie, 70
line
  at infinity, 10
  ordinary, 1–3, 10
line transversal, 149, 167–170
linear complex, 152
linearization, 24, 25, 27, 29, 40, 53, 54, 64
lines in space, 147–171
  cutting, 149, 158, 160, 170
  depth order of, *see* depth order
  isotopy class of, *see* isotopy class
  joint, 113, 149, 161–163, 170
  orientation class of, *see* orientation class
  Plücker coordinates of, *see* Plücker
    coordinates
  relative orientation, *see* orientation,
    relative
  representing, 149
  separating, 158
  towering property of, *see* towering
    property
  upper envelope, *see* upper envelope, of
    lines in space
  weaving pattern, 161
Liotta, Guseppe, 128
Lipton–Tarjan separator theorem, 128
Livné, Ron, 44
Lo, Chi-Yuan, 68
Local Lemma, *see* Lovász' Local Lemma
local ratio algorithm, 187
locally finite (arrangement, etc.), 173, 176
Lotker, Zvi, 178–180
Lovász' Local Lemma, 173, 174
Lovász, László, 34, 36, 37, 121
lower envelope, 20–27, 34, 40, 42, 67,
  73–79, 98, 166
  breakpoint of, 74
  combinatorial complexity of, 21
  computing, 58–60, 78–79
  definition, 20, 74
  of arcs, 21, 89
  of conic sections, 89
  of degree-4 polynomials, 89
  of hypersurfaces, 24, 25
  of multivariate functions, 74
  of partially defined functions, 77–78
  of piecewise-linear functions, 80
  of pseudo-planes, 24
  of pseudo-spheres, 24
  of segments, 45, 77–79, 85, 88, 89
  of spheres, 26
  of surface patches, 20, 21, 24
lower-envelope sequence, 75, 76, 78
Loyd, Sam, 183–185
  life of, 183

Möbius inversion formula, 18
Malitz, Seth, 126
Mani-Levitska, Peter, 174
many cells in arrangement, *see*
  arrangement, complexity of many cells
  in
Marcus, Adam, 111
marked cell
  computing, 62–63, 66, 89, 100, 115
Markov's inequality, 182

matching (in hypergraph), 110
matching, geometric, *see* geometric matching
Matoušek, Jiří, 31, 36, 37, 39, 45, 52, 57, 61, 62, 64, 68, 92, 127, 163, 182
maximization diagram, 21, 75
mean curvature, 48
Megiddo, Nimrod, 65, 68
Melchior's inequality, 11
MEMS (micro electronic mechanical systems), 69
metamorphic system, 194
   reconfiguration of, 195
metric, 40, 41, 47
   Euclidean, 40, 41
milestone (in motion planning), 69
Miller, Gary, 184
Milnor–Oleĭnik–Petrovskiĭ–Thom theorem, 19
minimization diagram, 21, 22, 40, 42, 51, 65, 166
   boundary vertex of, 22
   breakpoint of, 75
   computing, 58, 78
   definition, 21, 75
   inner vertex of, 22
   overlay of, 26, 27
Minkowski sum, 43, 47–49, 53
Mitchell, Joseph S. B., 91, 168
model of computation, 56, 78, 79, 89, 92
molecule
   model of, 14, 69
   pocket of, 70
   tunnel of, 70
   void of, 70
monochromatic copy of pattern, *see* geometric pattern, monochromatic
monotone matrix, 65, 89
Montgomery, Peter L., 135, 139
morphing (computer graphics), 128
Morse decomposition (of cell), 29
Moser, Leo, 101, 135
Moser, William O. J., 2, 135
Mossel, Elchanan, 182
motion planning, 14, 28, 43, 66, 68, 147, 184, 185
   for metamorphic system, 194
Motzkin, Theodore S., 3, 5
Mukhopadhyay, Asish, 68
Mulmuley, Ketan, 57, 61
multiple covering, *see* covering, k-fold
multiple packing, *see* packing, k-fold

Na, Hyeon-Suk, 169
Naor, Nir, 68, 91
Nevo, Eran, 37, 111
Newborn, Monroe M., 35, 106, 124
Nikolayevsky, Yuri, 121

Nivasch, Gabriel, 36, 37, 74, 80, 81, 85, 88
non-target move (in coin puzzle), 185, 188, 190
Noy, Marc, 36
NP-completeness, 123, 184, 186
NP-hardness, 166, 167, 184, 186

obstacle (in graph puzzle), 184, 185
Olonetsky, Svetlana, 182
$(1/r)$-cutting, *see* cutting
1-skeleton, 54, 55, 58
optimization, geometric, *see* geometric optimization
orientation class (of lines in space), 147, 149, 153–156
   computing, 154
   definition, 152
orientation, relative (of two lines), 147, 148
   definition, 151
oriented matroid, 15, 71
O'Rourke, Joseph, 29, 56
Overmars, Mark H., 63, 166

Pach, János, 4, 10, 21, 35, 37, 44–47, 107, 111, 117, 121, 123, 125–129, 131, 174, 176, 179, 180, 182, 186, 187, 189, 196
packing, k-fold, 173–174
Painter's algorithm, 159
pairwise inflatable balls, 169
Pálvölgyi, Dömötör, 127
Papadimitriou, Christos H., 184
Papakostas, Achilleas, 126
parallel algorithm, 58
   CRCW model, 58
   CREW model, 58
parametric searching, 65, 66, 164
partition tree, 164
Paterson, Michael S., 36
path compression scheme, 86
path planning, 69
pattern recognition, 133–146
Pellegrini, Marco, 27, 31, 168
Pelsmajer, Michael J., 122
perfect matching, 187
Perles, Micha A., 125
permutation, geometric, 149, 167–170
perturbation, 17, 150
Pesant, Gilles, 176
Petitjean, Sylvain, 169
Piano Movers' problem, 184
Pinchasi, Rom, 4, 10, 37, 111, 145
Pippenger, Nicholas, 36
plane cover (of point), 113
Plassmann, Paul, 96
Plücker coefficients, 151
Plücker coordinates, 148, 149, 151, 152, 157, 162
   definition, 151

Plücker hypersurface, 152, 153, 155, 156, 164
point at infinity, 2
point location, 56–58, 60, 63–66, 94, 145
Pollack, Richard, 4, 19, 34, 45, 69, 90, 126, 160
polygon
    regular, 2, 4, 10
polygonal path, 36
polyhedral formula, *see* Euler's polyhedral formula
polyhedral terrain, 163, 165, 166
    guarding, 167
popular face (of cell), 29, 32
power diagram, 42
Preparata, Franco P., 90
Preparata–Tamassia point-location algorithm, 60
projective space, 150, 152
projective space, oriented, 148, 152
pseudo-circle, 110
pseudo-disk, 42, 45
pseudo-parabola, 37
pseudo-segment, 109–112, 115
pseudo-trapezoid, 60, 92
pseudoline, 2, 4
    sweeping by, 56
PSPACE-completeness, 68
Puiseux series, 17
Pulleyblank, William R., 188
Purdy, George, 5, 134, 142

quad-edge data structure, 54, 55

Rabin, Michael, 5
radial ratio (of set of balls), 169, 170
Raghavan, Prabhakar, 184
Ramos, Edgar A., 67, 96
Ramsey-set, 141
Ramsey-type problem, 126, 135, 137–139, 141, 143–146
range searching, 64, 115, 149, 164
rational affine space, 136
Ratner, Daniel, 184
ray shooting, 56, 64, 115, 149, 155, 163–167, 170
    in the plane, 163
ray tracing, 147
Ray, Saurabh, 180
Regev, Oded, 36
regularity lemma, *see* Szemerédi's regularity lemma
regulus, 150, 153, 162
    ruling in, 153
relative orientation, *see* orientation, relative
Repeated Distances problem, 100, 101, 103, 106, 133, 134, 137, 138
    in 3-space, 140
repeated pattern, 133–146

roadmap, 55, 68, 69
    probabilistic, 69
Robert, Jean-Marc, 61
Roberts, Samuel, 18
robot system, 13, 43, 68
Rogers, C. Ambrose, 173
Ron, Dana, 178–180
Rotschild, Bruce L., 135, 139
Roy, Marie-Françoise, 19, 69
Rubin, Natan, 169

Safruti, Ido, 47
Saito, Shigemasa, 18
Salowe, Jeffrey S., 66
sandwich region, 26, 27, 48, 59, 65, 66, 76, 167, 168
Schaefer, Marcus, 122
Schiffenbauer, Robert, 166
Schinzel, Andrzej, 73, 80
Schulman, Leonard J., 126
Schwartz, Jacob T., 68
Schwarzkopf, Otfried, 26, 51, 58, 59, 61–63
Scott, Paul R., 4
segment
    avoiding, 6, 8
segment center, 66
Seidel, Raimund, 17, 29, 31, 40, 45, 56, 61, 160
semi-pfaffian set, 15
semialgebraic set, 16
Sen, Sandeep, 57
separator theorem, *see* Lipton–Tarjan separator theorem
Seress, Ákos, 4
Set Cover problem, 186, 187
Sharir, Micha, 4, 15, 21, 22, 24, 26–29, 31, 32, 37–41, 44–48, 51, 52, 58–63, 66, 68, 73, 74, 80, 81, 85, 88, 90, 92–95, 104, 111–117, 126, 142, 145, 146, 160–166, 168, 169, 182
Shaul, Hayim, 164
Shelton, Christian R., 70
Shor, Peter W., 57, 60, 74, 80, 81, 85, 88, 168
sibling hyperedge, 176–178
Sifrony, Shmuel, 27, 45, 61, 90, 92, 94, 95
sign sequence, 19, 55, 154
similar copy (of point set), 141–143
similar simplices (determined by point set), 117–118, 142, 143
Simmons, Gustavus J., 34, 36
simple set of hyperplanes, 17
single cell
    computing, 60–61, 89, 92–96
sliding model (of coins), 187
slope number (of graph), 126–127
slope selection, 66

Smorodinsky, Shakhar, 37, 111, 168, 178–180, 182
Smyth, Clifford, 36
Snoeyink, Jack, 38, 45, 61, 63, 93, 160
Solan, Alexandra, 160
Solernó, Pablo, 19
solid modeling, 147
solvent accessible model (of molecule), 14
Solymosi, József, 101, 116, 141, 143
Spencer, Joel, 101, 128, 131, 134, 135, 139
sphere packing, 173–182
spherically convex set, 169
Spirakis, Paul, 184
splay tree, 81
squares, pushing, 194–196
stabbing triangles in space, 151
Štefankovič, Daniel, 122
Steiger, William L., 35, 36, 66, 68
Steiner, Jakob, 15, 18
Stolfi, Jorge, 163
straight chain (in metamorphic system), 196
stratification, 16, 59, 68
Straus, Ernst G., 34, 36, 135, 139
Sudan, Madhu, 184
surface area
  computing, 63
surface patch, 15, 16
  arrangement of, see arrangement, of algebraic surface patches
  domain boundary of, 18
  domain of, 18
  lower envelope of, see lower envelope, of surface patches
Suri, Subhash, 169
sweep-line algorithm, 56, 61, 89, 95
Sylvester, James Joseph, 1–3
  life of, 1
Sylvester–Gallai problem, 1
Sylvester–Gallai theorem, 1, 2, 11
  colored versions, 10
Symvonis, Antonios, 126
Szegedy, Mario, 121, 180
Székely, László A., 100, 106, 108, 116
Szemerédi's regularity lemma, 133
Szemerédi–Trotter theorem, 100, 116
Szemerédi, Endre, 35, 36, 39, 66, 80, 100, 101, 106, 124, 134

Tagansky, Boaz, 28, 40, 41, 47, 48, 52, 63
Tamaki, Hisao, 36, 37, 110, 184
Tamura, Akihisa, 18
Tardos, Gábor, 37, 45, 111, 116, 117, 129, 134, 143, 176, 180
target move (in coin puzzle), 185, 188
terrain, see polyhedral terrain
terrain analysis, 147
thrackle, 120–122
  generalized, 121
  straight-line, 125
Tokuyama, Takeshi, 18, 36, 37, 110
topological sweep, 56, 57
Törőcsik, Jenő, 125
Tóth, Csaba D., 101, 116, 145, 146
Tóth, Géza, 36, 107, 122, 123, 125, 126, 128, 131, 176, 179
towering property (of lines in space), 149, 156–158
translate (of point set), 135–137
transversal number (of hypergraph), 110
transversal space, 167
transversal, geometric, see geometric transversal
transversal, line, see line transversal
triangle
  equivalence relation for, 144–146
triangulation, 32, 49–50, 53, 62
  bottom-vertex, 49, 50, 53
trifurcation (in graph), 121
trisector (of lines in space), 41
Trotter, William T., Jr., 39, 100, 101, 134
Turán, Paul, 119, 120
Turán-type problem, 135

Ungar's theorem, 4–7
  generalized, 6, 7
union of objects, 42–49
  computing, 63
  regular vertex of, 46
union-find, 60
Upper Bound Theorem, 21, 25, 27, 42, 148, 152
upper envelope, 20, 73, 75
  of lines in space, 149, 154
    complexity, 156

Vaidya, Pravin M., 67
Valtr, Pavel, 36, 81, 92, 126
Van der Waals model (of molecule), 14, 70
Varadarajan, Kasturi R., 67, 169
Verrill, Helena A., 195
vertical decomposition, 50–55, 58, 60–62, 65, 67, 92, 93
visibility, 165–167, 170
visibility map, 165
VLSI, 120
volume
  computing, 63
Voronoi diagram, 40–42, 47, 52, 53
  additive-weight, 42
  computing, 59, 60
  dynamic, 41
Vu, Van, 101, 141

Wagner, Uli, 37, 182
Warmuth, Manfred K., 184
warping (computer graphics), 128

Warren, Hugh E., 19
watchtower, 167
weaving pattern of lines, *see* lines in space, weaving pattern
Welzl, Emo, 36, 38, 40, 45, 58, 62, 68, 104, 113, 126, 163, 182
Wenger, Rephael, 37, 128
Wenk, Carola, 27
White, Neil, 153
Whitney stratification, 68
width (of point set), 67
Wiernik, Ady, 86, 88
Wiernik–Sharir construction, 21, 45, 86, 90
winged-edge data structure, 54
Woeginger, Gerhard J., 126
Wood, David R., 127

Yao, F. Frances, 96, 187
Yap, Chee-Keng, 61–63, 68
Yvinec, Mariette, 41, 48

z-buffer, 159
Zarankiewicz's conjecture, 119
Zaslavsky, Thomas, 18
zenith point, 154, 155
Zhou, Yunhong, 169
zone, 27, 29–32, 56, 58, 63, 96–97
    complexity of, 29, 96
    definition, 29, 96
zone theorem, 29, 31
    extension of, 31
zonotope, 15

Titles in This Series

152 **János Pach and Micha Sharir,** Combinatorial geometry and its algorithmic applications: The Alcála lectures, 2009

151 **Ernst Binz and Sonja Pods,** The geometry of Heisenberg groups: With applications in signal theory, optics, quantization, and field quantization, 2008

150 **Bangming Deng, Jie Du, Brian Parshall, and Jianpan Wang,** Finite dimensional algebras and quantum groups, 2008

149 **Gerald B. Folland,** Quantum field theory: A tourist guide for mathematicians, 2008

148 **Patrick Dehornoy with Ivan Dynnikov, Dale Rolfsen, and Bert Wiest,** Ordering braids, 2008

147 **David J. Benson and Stephen D. Smith,** Classifying spaces of sporadic groups, 2008

146 **Murray Marshall,** Positive polynomials and sums of squares, 2008

145 **Tuna Altinel, Alexandre V. Borovik, and Gregory Cherlin,** Simple groups of finite Morley rank, 2008

144 **Bennett Chow, Sun-Chin Chu, David Glickenstein, Christine Guenther, James Isenberg, Tom Ivey, Dan Knopf, Peng Lu, Feng Luo, and Lei Ni,** The Ricci flow: Techniques and applications, Part II: Analytic aspects, 2008

143 **Alexander Molev,** Yangians and classical Lie algebras, 2007

142 **Joseph A. Wolf,** Harmonic analysis on commutative spaces, 2007

141 **Vladimir Maz'ya and Gunther Schmidt,** Approximate approximations, 2007

140 **Elisabetta Barletta, Sorin Dragomir, and Krishan L. Duggal,** Foliations in Cauchy-Riemann geometry, 2007

139 **Michael Tsfasman, Serge Vlăduţ, and Dmitry Nogin,** Algebraic geometric codes: Basic notions, 2007

138 **Kehe Zhu,** Operator theory in function spaces, 2007

137 **Mikhail G. Katz,** Systolic geometry and topology, 2007

136 **Jean-Michel Coron,** Control and nonlinearity, 2007

135 **Bennett Chow, Sun-Chin Chu, David Glickenstein, Christine Guenther, James Isenberg, Tom Ivey, Dan Knopf, Peng Lu, Feng Luo, and Lei Ni,** The Ricci flow: Techniques and applications, Part I: Geometric aspects, 2007

134 **Dana P. Williams,** Crossed products of C^*-algebras, 2007

133 **Andrew Knightly and Charles Li,** Traces of Hecke operators, 2006

132 **J. P. May and J. Sigurdsson,** Parametrized homotopy theory, 2006

131 **Jin Feng and Thomas G. Kurtz,** Large deviations for stochastic processes, 2006

130 **Qing Han and Jia-Xing Hong,** Isometric embedding of Riemannian manifolds in Euclidean spaces, 2006

129 **William M. Singer,** Steenrod squares in spectral sequences, 2006

128 **Athanassios S. Fokas, Alexander R. Its, Andrei A. Kapaev, and Victor Yu. Novokshenov,** Painlevé transcendents, 2006

127 **Nikolai Chernov and Roberto Markarian,** Chaotic billiards, 2006

126 **Sen-Zhong Huang,** Gradient inequalities, 2006

125 **Joseph A. Cima, Alec L. Matheson, and William T. Ross,** The Cauchy Transform, 2006

124 **Ido Efrat, Editor,** Valuations, orderings, and Milnor K-Theory, 2006

123 **Barbara Fantechi, Lothar Göttsche, Luc Illusie, Steven L. Kleiman, Nitin Nitsure, and Angelo Vistoli,** Fundamental algebraic geometry: Grothendieck's FGA explained, 2005

122 **Antonio Giambruno and Mikhail Zaicev, Editors,** Polynomial identities and asymptotic methods, 2005

121 **Anton Zettl,** Sturm-Liouville theory, 2005

TITLES IN THIS SERIES

120 **Barry Simon,** Trace ideals and their applications, 2005
119 **Tian Ma and Shouhong Wang,** Geometric theory of incompressible flows with applications to fluid dynamics, 2005
118 **Alexandru Buium,** Arithmetic differential equations, 2005
117 **Volodymyr Nekrashevych,** Self-similar groups, 2005
116 **Alexander Koldobsky,** Fourier analysis in convex geometry, 2005
115 **Carlos Julio Moreno,** Advanced analytic number theory: L-functions, 2005
114 **Gregory F. Lawler,** Conformally invariant processes in the plane, 2005
113 **William G. Dwyer, Philip S. Hirschhorn, Daniel M. Kan, and Jeffrey H. Smith,** Homotopy limit functors on model categories and homotopical categories, 2004
112 **Michael Aschbacher and Stephen D. Smith,** The classification of quasithin groups II. Main theorems: The classification of simple QTKE-groups, 2004
111 **Michael Aschbacher and Stephen D. Smith,** The classification of quasithin groups I. Structure of strongly quasithin K-groups, 2004
110 **Bennett Chow and Dan Knopf,** The Ricci flow: An introduction, 2004
109 **Goro Shimura,** Arithmetic and analytic theories of quadratic forms and Clifford groups, 2004
108 **Michael Farber,** Topology of closed one-forms, 2004
107 **Jens Carsten Jantzen,** Representations of algebraic groups, 2003
106 **Hiroyuki Yoshida,** Absolute CM-periods, 2003
105 **Charalambos D. Aliprantis and Owen Burkinshaw,** Locally solid Riesz spaces with applications to economics, second edition, 2003
104 **Graham Everest, Alf van der Poorten, Igor Shparlinski, and Thomas Ward,** Recurrence sequences, 2003
103 **Octav Cornea, Gregory Lupton, John Oprea, and Daniel Tanré,** Lusternik-Schnirelmann category, 2003
102 **Linda Rass and John Radcliffe,** Spatial deterministic epidemics, 2003
101 **Eli Glasner,** Ergodic theory via joinings, 2003
100 **Peter Duren and Alexander Schuster,** Bergman spaces, 2004
99 **Philip S. Hirschhorn,** Model categories and their localizations, 2003
98 **Victor Guillemin, Viktor Ginzburg, and Yael Karshon,** Moment maps, cobordisms, and Hamiltonian group actions, 2002
97 **V. A. Vassiliev,** Applied Picard-Lefschetz theory, 2002
96 **Martin Markl, Steve Shnider, and Jim Stasheff,** Operads in algebra, topology and physics, 2002
95 **Seiichi Kamada,** Braid and knot theory in dimension four, 2002
94 **Mara D. Neusel and Larry Smith,** Invariant theory of finite groups, 2002
93 **Nikolai K. Nikolski,** Operators, functions, and systems: An easy reading. Volume 2: Model operators and systems, 2002
92 **Nikolai K. Nikolski,** Operators, functions, and systems: An easy reading. Volume 1: Hardy, Hankel, and Toeplitz, 2002
91 **Richard Montgomery,** A tour of subriemannian geometries, their geodesics and applications, 2002
90 **Christian Gérard and Izabella Łaba,** Multiparticle quantum scattering in constant magnetic fields, 2002

For a complete list of titles in this series, visit the
AMS Bookstore at www.ams.org/bookstore/.